Engineering Economics of Life Cycle Cost Analysis

Engineering Economics of Life Cycle Cost Analysis

John Vail Farr and Isaac Faber

CRC Press
Taylor & Francis Group
Boca Raton London New York

CRC Press is an imprint of the
Taylor & Francis Group, an **informa** business

CRC Press
Taylor & Francis Group
6000 Broken Sound Parkway NW, Suite 300
Boca Raton, FL 33487-2742

First issued in paperback 2021

ISBN-13: 978-0-367-78074-6 (pbk)
ISBN-13: 978-1-138-60678-4 (hbk)

Library of Congress Cataloging-in-Publication Data

Names: Farr, John V., author. | Faber, Isaac, author.
Title: Engineering economics of life cycle cost analysis / by John Vail Farr and Isaac Faber.
Description: Boca Raton, FL : CRC Press/Taylor & Francis Group, 2018. | Includes index.
Identifiers: LCCN 2018029920| ISBN 9781138606784 (hardback : acid-free paper) | ISBN 9780429466304 (ebook)
Subjects: LCSH: Life cycle costing. | Engineering economy. | Systems engineering--Cost effectiveness.
Classification: LCC TA177.7 .F368 2018 | DDC 620.0068/1--dc23
LC record available at https://lccn.loc.gov/2018029920

Visit the Taylor & Francis Web site at
http://www.taylorandfrancis.com

and the CRC Press Web site at
http://www.crcpress.com

Contents

Preface

Engineering has changed dramatically in the last century. The advent of modern computing systems and instantaneous communication, the elimination of low and middle management who tracked and managed information, and the implementation of extremely efficient supply chains has dramatically affected the roles and responsibilities of engineers at all levels. With today's globalization, technology, and complexity, few engineers sit at a desk and design at the component level. We are mainly integrators of systems. This 21st-century paradigm has given engineers the opportunity to use their technical training and problem-solving skills as points of departure into spiral and rapid prototyping, creation of disruptive products and services, and complex and interdisciplinary product development in non-traditional engineering fields such as sales, law, medicine, and corporate leadership. The future will require systems that are more complex, providing for increased opportunities for engineers to be business leaders first and engineers second. The marketplaces of the information age will require employees at all levels to be both business and technically savvy. They must be able to develop accurate cost estimates based on defensible cost analysis. Even at the component level, engineers must conduct meaningful trade space studies to make informed decisions. Engineers continue to be poised for tremendous career opportunities, but the landscape will require different job skills. It is against this backdrop that we wrote this book.

At the undergraduate level, most engineers are, at best, introduced to engineering economics as their only exposure to the costing and estimating of projects. Many universities introduce economic planning material in some type of capstone class. Few require an entire course on the subject. At the graduate level, most programs are domain-centric and may not deliver any exposure to cost estimation and management, especially at the systems level. Engineers must pursue an MBA or learn the appropriate skills via on-the-job training to become competent in business affairs. Unfortunately, the courses traditionally taught in an MBA program—mainly strategy, finance, and accounting—do not provide the skills needed to be a program/product manager, to bring products to market on time and within budget, to understand the true costs of the product, and to conduct trade-off analysis. Thus, most engineers are ill prepared to enter and excel in the job market and work on interdisciplinary projects where cost analysis, estimation, and management skills are not only valued but required. Our formal training has failed us; we must learn the requisite costing, accounting, and management skills via on-the-job training or return to school to pursue a business or related degree.

There are some outstanding and very mature engineering economics textbooks in the marketplace. The methods, processes, and tools (MPTs) presented in these textbooks are as relevant now as they were 20 years ago. Most of the leading textbooks do an outstanding job of presenting the mechanics of engineering economics. However, the work environment has dramatically changed; it now requires more than simply understanding the mathematics of engineering economy. First and foremost, engineers often enter the market as leaders of integrated product teams (IPTs). Often early in their career they are responsible for on-time and on-budget products that are produced by interdisciplinary teams across multiple time zones. They are responsible, in part or directly, for profit and loss at all levels or products/services development. Technology has increased productivity and the span of control. When combined with complexity, risk is also increased. Few engineers have the luxury of becoming component-level experts and assuming more responsibility as their level of experience grows. Thus, entry-level and junior engineers need more than the math of engineering economics; they need to understand the business of being an engineer. Traditional engineering economics should be one of the tools at their disposal. They must understand not only how to make capital investment decisions throughout a product life cycle but also the economics of operating a business, contracting, complex analysis techniques, and, more importantly, making risk trade-offs.

This book is a textbook adaptation, evolution, and updating of the *Systems Life Cycle Costing: Economic Analysis, Estimation, and Management* published in 2011. Though marketed as a reference book, it was adopted by many universities as a text for graduate classes in engineering economics. Unlike the mature knowledge encompassed by the traditional engineering disciplines, the techniques and tools for costing complex systems are rapidly evolving and being driven mainly by the commercial sector. Modern engineers need MPTs beyond classic engineering economics, must understand life cycle cost analysis (LCCA), and must apply these to complex systems including infrastructure, cars, planes, information technology systems, and so on. First and foremost, we are interested in the cost analysis, estimation, and management of systems. Though components are important, the more challenging and relevant problems are systems. Second, engineers must worry about life cycle or total ownership costs (TOC) and not simply development costs. All too often, we only cost our piece of the life cycle and do not make sound TOC acquisition decisions using quantifiable and defendable analysis. Lastly, an engineer must be able to conduct economic analysis, estimate the cost of the system, lead and manage people and resources, and understand the big picture as part of the business enterprise. Also, the MPTs and techniques are often not presented in the open literature because of the competitive advantage afforded any company that can accurately estimate the life cycle costs (LCC) of a product. Thus, much of the MPTs were gleaned from government sources, especially the Department of Defense (DoD) and NASA. This was not our desire; however, the DoD and NASA are in many ways the intellectual thought leaders on costing and estimating of complex systems because of the sheer size and complexity of their projects and programs.

When Dr. Farr wrote the original LCCA reference, he had hoped to improve on the engineering economy material taught in most engineering programs. He originally hoped that book could also be used in graduate programs in engineering management, mechanical engineering, systems engineering, and the like. As we collected feedback from students and other customers, we became convinced that this was the right set of MPTs and should replace engineering economy at both the undergraduate and graduate levels. In our software-centric workplace, Isaac Faber is able to bring some modern relevance to the traditional engineering economic topics based on his software and data science background. First and foremost, tools like Excel and @Risk have made teaching real-world cost problems possible. Second, most engineers work in an interdisciplinary world and must understand the TOC of their products. Entry-level engineers need a broader perspective to be successful than the mathematics of the time value of money which is taught in traditional engineering economics courses. Life cycle costing is, at best, an immature, industry-driven discipline. Unfortunately, textbooks are often long on theory and short on meaningful examples, real data, and so on.

Our former colleagues at West Point along with the students and faculty at Stevens Institute of Technology, University of Central Florida, and Clarkson University contributed to much of the material presented herein. Having made every effort to correctly reference the material, we suspect there are phrases and other elements of the book that are not properly referenced. If you encounter a phrase, figure, or anything else that is not correctly referenced, please send us the correct information. Also, please send us any ideas, problems, mistakes, new material, or anything that you find relevant. We are maintaining the website www.systemscosting.com as a way to contact us and provide feedback. Through the website, we can provide you solution manuals, Excel and @Risk solutions to the problems in the textbook, and so on.

John V. Farr
Professor Emeritus
United States Military Academy at West Point

Isaac F. Faber
Assistant Professor
US Army

Acknowledgments

We would like to thank some of the many people who contributed to the publishing of this textbook. First and foremost, we would like to thank our former colleagues at the United States Military Academy at West Point, particularly James, Kenny, Tim, Jeff, and Travis, and many of Dr. Farr's colleagues at Stevens Institute of Technology, specifically Dinesh, Brian, Rob, Mike, Kate, and Don, who inspired him to write the original life cycle costing reference and this textbook adaptation. Jacqueline Harris from West Point deserves a special thank-you, as she developed many of the problems used in this book. Dr. Farr's exposure to systems engineering at both West Point and Stevens helped shape his motivation to move beyond traditional engineering economics and write the original reference book. In addition, Dr. Farr's current colleagues at the University of Central Florida, especially Tim, and Clarkson University continue to inspire us to deliver high-quality, relevant, and connected material to our students. We also owe a debt of gratitude to the selfless servants in the US Army whom we worked with for the majority of our professional careers. Their high standards for professionalism sustained us when writing this book and performing all the jobs of an academician became tedious. We would be remiss if we did not thank the many students who encouraged us to write the book and contributed to much of the material used in the text. Their projects formed the basis for many of the class problems, examples, and case studies. There is no way to recognize each of them individually, but we wish to thank them all collectively.

Most importantly, we would like to thank our wives, Michele and Kerry. Dr. Farr owes a debt of gratitude to his two sons, Michael and David, for their patience during the many nights he worked late on both editions of this book and for understanding as he traveled around the world to teach the course for which it was developed. Isaac also owes a debt to his four children, who sacrificed many hours of time with their father to make this manuscript possible.

About the authors

Dr. John V. Farr is currently an adjunct faculty member at the University of Central Florida and in the School of Business at Clarkson University; a principal subject-matter expert for Applied Research Associates, conducting cost and risk analysis; and a professor emeritus of engineering management at the United States Military Academy at West Point. He was a professor and the founding director of the Center for Nation Reconstruction and Capacity Development from 2010 until 2017, and from 2007 to 2010 he was a professor of systems engineering and engineering management and associate dean for academics in the School of Systems and Enterprises at Stevens Institute of Technology. He was the founding director of the Department of Systems Engineering and Engineering Management at Stevens from 2000 to 2007. Before coming to Stevens in 2000, he was a professor of engineering management at the United States Military Academy at West Point, where he was the first permanent civilian professor in engineering and director of their Engineering Management Program. Dr. Farr is a past president and fellow of American Society for Engineering Management (ASEM) and a fellow of the American Society of Civil Engineers (ASCE). He is a former editor of the *Journal of Management in Engineering* and the founder of the *Engineering Management Practice Periodical*. He has authored or edited over 200 technical publications, including 5 books, 7 book chapters, and over 90 refereed publications mainly on cost analysis, infrastructure, engineering education, engineering management, and systems engineering. Dr. Farr earned his undergraduate degree from Mississippi State University and his masters and PhD in civil engineering from Purdue and the University of Michigan, respectively. Dr. Farr is also a member of Chi Epsilon, a founding member of Epsilon Mu Eta, and a member of Phi Kappa Phi honor societies. He is a member of the International Council of Systems Engineering, ASCE, and ASEM. He is also a registered civil engineer in Florida and Mississippi and a certified project management professional. Dr. Farr has served on numerous defense national and academic advisory boards including membership on the Army Science Board from 2002 through 2010 and on the Air Force Studies Board of the National Academies from 2006 through 2012. He has served as a consultant to numerous companies and government agencies and has worked in Afghanistan, East Africa, Vietnam, and the Marshall Islands on a wide variety of economic and capacity development and assessment projects. He taught at the University of Technical Education, Ho Chi Minh City, Vietnam, in 2013 as a Fulbright Specialist.

Isaac J. Faber is an active duty officer in the United States Army and is attending Stanford University through the Army Operations Research Systems Analysis advanced civil schooling program. His research focuses on advanced techniques for cybersecurity, early warning of complex attacks, and the costing of complex computer systems. While at Stanford, he was involved in the startup of a company focused on managing data-mining solutions. He previously served as the lead data scientist at the Army's new cyberwarfare unit. In that role, he guided the development and architecture of an operational cyberanalytics platform. He also served a 3-year tour at the United States Military Academy at West Point, where he was promoted to the rank of assistant professor. At West Point, he served in the Department of Systems Engineering and taught courses in decision analysis and engineering economy. He also maintained an active sponsored research program in logistics and cyberwarfare. Though still an active duty Army officer, he has found time to produce four peer-reviewed journal articles, nine peer-reviewed conference articles, and six working papers. He is a Ranger qualified combat veteran of Iraq with a Bronze Star and Combat Infantry badge. He has served as the Institute of Industrial and Systems Engineering chair of the Engineering Economy Division. He holds a BS degree in computer information systems from Arizona State University and a MS degree in industrial engineering from the University of Washington.

List of abbreviations

AC	actual cost
ACWP	actual cost of work performed
ADM	arrow diagramming method
AEW	annual equivalent worth
APR	annual percentage rate
BAC	budget at completion
BCWP	budgeted cost of work performed
BCWS	budgeted cost of work scheduled
CAGR	calculated compound annual growth ratio
CASE	computer-aided software engineering
CD	certificate of deposit
CDF	cumulative distribution function
CER	cost estimating relationships
CFD	cash flow diagram
CLOC	commented lines of code
COCOMO	constructive cost model
COCOMO II	constructive cost model version II
COCOTS	constructive commercial off-the-shelf model
COSYSMO	constructive systems engineering cost model
COTS	commercial off-the-shelf
CPG	continuous process generator
CPI	consumer price index or cost performance index
CPM	critical path method
CPP	conventional payback period
CSI	cost schedule index
CUT	code and unit test
CV	conventional carrier or cost variance
DD	detailed design
DoD	department of defense
DPG	discrete process generator
DPP	discounted payback period
EAC	estimate at completion
EAF	effort adjustment factor

ELOC	effective lines of code
ETC	estimate to complete
EV	earned value
FFP	firm-fixed price
FP	function points
FV	future value
GOTS	government off-the-shelf
IPT	integrated product teams
IRR	internal rate of return
ITM	inverse transformation method
KDSI	thousands of delivered source instructions
KLOC	thousands of lines of code
LCC	life cycle costs
LCCA	life cycle cost analysis
LDC	large defense contractor
LOGPACS	logistic packages
M&S	modeling and simulation
MACRS	modified accelerated cost recovery system
MARR	minimum attractive or acceptable rate of return
MODA	multi-objective decision analysis
MOTS	modifiable or military off-the-shelf
MPTs	methods, processes, and tools
NCLOC	non-commented lines of code
NFV	net future value
NFW	new future worth
NPV	net present value
NPW	net present worth
O&M	operation and maintenance
O&S	operations and support/sustainment
PCE	parametric cost estimating/estimate
PDF	probability distribution function
PDM	precedence diagramming method
PERT	program evaluation review technique
PM	program/project management
PMI	project management institute
PS	percent spent
PSM	professional staff months
PV	planned value
R&D	research and development
RDT&E	research, development, testing, and evaluation
ROI	return on investment
SBC	simulation-based costing
SE/PM	systems engineering/project management
SLOC	source lines of code

SoS	system of systems
SV	systems view or schedule variance
TCF	technical complexity factor
TDEV	time to develop
TOC	total ownership costs
TRL	technical readiness level
TVM	time value of money
UCP	use case points
UFP	unadjusted function point
VAC	variance at completion
WACC	weighted average cost of capital
WBS	work breakdown structure
YTM	yield to maturity

chapter one

Overview of systems life cycle costing

1.1 *Overview of engineering economic analysis*

With the globalization of the manufacturing industry and much of the services profession, the efficiencies derived from advances in technology (and the subsequent decrease in middle management positions), and the shifting of our economy to be service based, the roles of the technical organization and of engineers have dramatically changed. In the 21st century, technical organizations must be concerned with

- Maintaining an agile, high-quality, and profitable business base of products or services in an unstable and global economy,
- Hiring, managing, and retaining a highly qualified and trained staff of engineers, scientists, technicians, and support personnel in a rapidly changing technological environment, and
- Demonstrating a high level of innovation, entrepreneurship, and capability maturity, usually with an ever-increasing amount of government oversight and regulation.

A company's basic objective is to grow its earnings quickly and sustainably by providing goods and services that meet stakeholder requirements. In kind, government and non-profit agencies must have strong cost aversions due to resource restraints and the nature of public service. Engineers support this overarching objective by developing products and services that attain their targeted contribution to corporate earnings and/or agency cost savings. Thus, engineers must be able to do more than simply produce economical and creative designs; they must be able to

- Understand and track both the technical and profitability aspects of a project,
- Communicate with the financial people, such as accountants and auditors, and to senior management,
- Read and understand financial statements to assess the health of a potential partner as well as one's own corporation or agency,

- Understand the financial and legal liabilities of their designs and contracts,
- Have a life cycle perspective including long-term obligations such as lease versus buy, warranties, equipment replacement, true costs of employees, and so on, and
- Understand the execution of an engineering business and that operational and strategic decisions can affect profitability and business operations.

Much has been written about innovation and entrepreneurship for engineers and scientists. The National Science Foundation and many others have written about producing engineers with a better understanding of cultures, communication skills, interdisciplinary learning, and business and management. Traditional engineering economy is only a small component of the formal financial education needed by entry-level engineers.

As conveyed with the systemigram in Figure 1.1, the world that a modern engineer must operate in is complex and interrelated. Much has changed in the last 20 years, and unfortunately engineering education has been slow to evolve. In practice, we have become more interdisciplinary, and new engineering disciplines (systems, software, biomedical, robotics, mechatronics, etc.) have emerged. Journeymen engineers no longer focus solely on technical issues within traditional engineering fields and grow into the non-design financial aspects of business as they progress into the

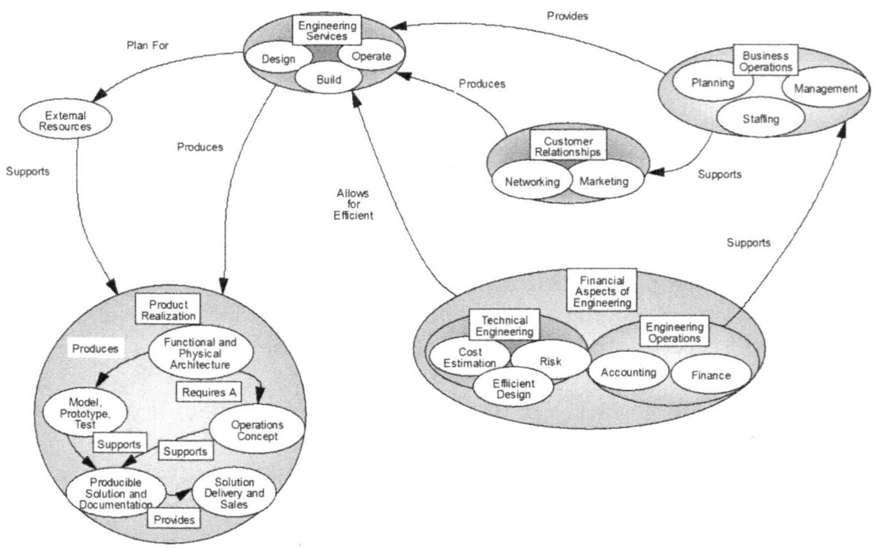

Figure 1.1 Systemigram showing the financial skills needed by an engineer

ranks of management. Given the number of small design firms, the interdisciplinary nature of engineering, the flattening of organizations because of technology, and computer-based design requiring engineers to be involved throughout the project life cycle, engineers must perform most of the functions shown in Figure 1.1 earlier if not immediately in their career.

In Figure 1.1, we divided the financial skills into four main functions: (1) technical engineering, or those skills needed for successful product realization, (2) managing customer relationships, including expectations and growing the customer base, (3) engineering operations, or those skills needed to successfully operate within a modern business, and (4) technical and engineering operations skills. Each of these areas requires differing skill sets and abilities related to costs and economics of an engineering project. For example, an engineer must understand the financial motivations for selecting a project for production. This information will influence planning, design, and ultimately project outcomes. In general, we contend that an engineer needs technical skills, a financial understanding, and the ability to communicate about both areas.

Figure 1.2 presents just some of the competencies needed for good cost analysis, estimating, and management. These skills are needed to execute business operations, build and sustain customer relationships, run a technical business, and deliver a high-quality, cost-effective product. Though probably not totally complete, the figure does demonstrate the diverse skills needed to conduct accurate estimates and manage a project.

These competencies are required for most engineers to be successful, even early in their careers. In particular, the relationships between the skills, such as how financial motivations affect technical work, are critical and lend to engineers' ability to provide value to their organizations.

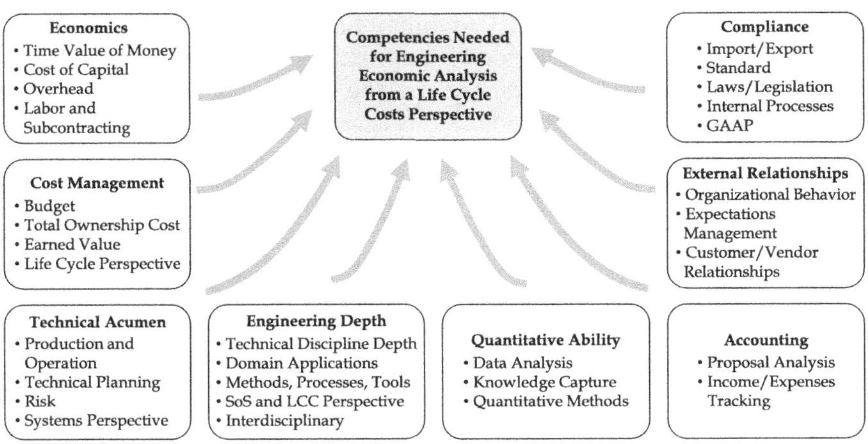

Figure 1.2 Competency model for costing

1.1.1 Business operations

Every engineering job is different, and every company or organization has different expectations. However, in most entry-level jobs, from your first day you will be held accountable for a budget and for your time, and you will be involved in customer relations and expectations. Within a very short time, you will be involved in

- Market and customer analysis,
- Budgeting and forecasting,
- Beating the competition,
- High-tech marketing and sales,
- Planning and managing for profitability, and
- Partnering with other businesses.

These skills are as important to your success as developing cost-effective and efficient products that meet customers' needs.

Obviously, the smaller the company the more an entry-level engineer must be focused on business operations and customer relationships. Large defense, information technology, and manufacturing companies often hire engineers for mainly technical work. However, as engineers grow into management positions, understanding the policies and finances of the organization is key to their success. Indeed, the selection of candidates for higher-level management is based on engineers' understanding of the business process and how it relates to technical work.

Companies are typically broken up into sales and marketing; research and development; operations, including all aspects of support, such as quality, logistics, human resources, and legal; and major product lines. Depending on the company, they can have divisions centered on products, divisions or regions, and corporate. Engineers can be employed in any area and at any level, with each requiring different knowledge bases and skills. In today's complex engineering world, the ability to understand cost and economic considerations across various divisions of a firm can add tremendous value.

1.1.2 Customer relationships

Engineers are often the front line between an organization and its customers. They collect requirements, iterate on designs, make technical presentations, write proposals, and, for smaller companies, they can be the face of the organization. Beyond listening and communicating, they must work with stakeholders to ensure their expectations are executable within the financial constraints of the project. Customer relations is focused on making the customer feel both valued and welcomed, which requires excellent listening and communication skills. Being able to communicate the language of finances is an important element of not only managing but also cultivating customer relations. Engineers who possess only technical knowledge can often unnecessarily intimidate customers, leading to a loss in confidence and potentially a loss in revenues.

Customer relations involves working with stakeholders not only on engineering services but also on ensuring that business operations such as billing, contracts, and other service obligations are met. Again, this shows that engineers must be involved in business operations even if their primary focus is technical work. Often customers make decisions primarily based on cost factors, so engineers who understand this motivation can be more effective in tailoring their technical efforts.

1.1.3 *Engineering services*

Engineers deliver a wide variety of products and services. They design, build, and/or operate a wide variety of products and services through low-cost, best value, or sole source bidding. They use existing scientific principles to create outcomes that meet stakeholder needs. One of the core needs of most stakeholders is low-cost design that meets stated requirements.

Numerous design processes can be found in the literature, and all consist of identifying the problem and stakeholder requirements, brainstorming ideas, generating solutions, building a model or prototype, developing selection criteria and choosing the best alternative, and collecting feedback from the customer.

Design-build is a method of project delivery in which a single firm (the designer-builder) is contracted to provide a finished product. Operations lumped with design-build for product realization leads to estimates for total life cycle costs (LCCs) or total ownership costs (TOC) and have become popular because one contractor assumes all the risk. The LCC implications of providing for all types of project delivery (i.e., any combination of design, build, and/or operation) are important and complex, requiring engineers to be part of a team of lawyers, accountants, bankers, and others.

1.1.4 *Product realization*

Product realization is most often defined as the work that the organization does to develop, manufacture, and deliver the finished goods or services that effectively and efficiently meet stakeholder requirements. Figure 1.3 shows the various factors that can contribute to an on-time and on-budget project or one with schedule and cost overruns.

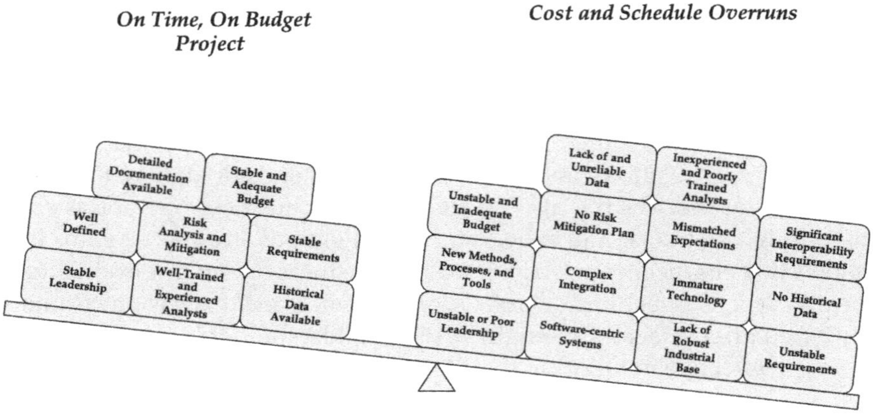

Figure 1.3 Challenges cost estimators typically face (modified from Government Accounting Office, 2009)

1.2 Engineering challenges for the 21st century

Engineers are trained to innovate, design, and produce products. Our formal education system focuses primarily on the skills needed for product realization.

According to the Institute of Electrical and Electronics Engineers (2011), about one-third of all masters of business administration candidates are engineers. Many will argue that there is too much content in undergraduate programs and that engineering degrees must follow the path of other professional programs, such as law, architecture, and medicine. However, we see the exact opposite trend. Many states are mandating that the number of credit hours be capped at 120 in order to minimize costs.

Engineering has traditionally been defined as "the creative application of scientific principles to design or develop structures, machines, apparatus, or manufacturing processes, or works utilizing them singly or in combination; or to construct or operate the same with full cognizance of their design; or to forecast their behavior under specific operating conditions; all as respects an intended function, economics of operation or safety to life and property" (Engineers' Council for Professional Development, 1947). The National Academy of Engineering (2008) states, "No profession unleashes the spirit of innovation like engineering. From research to real-world applications, engineers constantly discover how to improve our lives by creating bold new solutions that connect science to life in unexpected, forward-thinking ways. Few professions turn so many ideas into so many realities. Few have such a direct and positive effect on people's everyday lives. We are counting on engineers and their

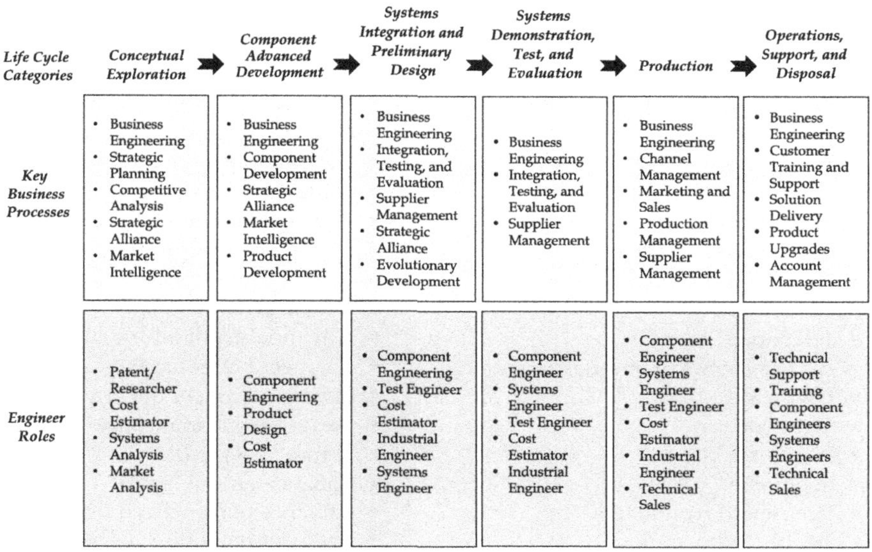

Figure 1.4 Generic life cycle model with business process and engineering

imaginations to help us meet the needs of the 21st century." Another definition often quoted about engineering comes from the Quality Assurance Agency (2015) in the United Kingdom and states that the context of "engineering is concerned with developing, providing and maintaining infrastructure, products, processes and services for society. Engineering addresses the complete life cycle of a product, process or service, from conception, through design and manufacture, to decommissioning and disposal, within the constraints imposed by economic, legal, social, cultural and environmental consideration." Figure 1.4 shows one generic life cycle model and the role of engineers throughout the life cycle.

The core function of the engineering profession is to develop solutions. Solutions are what systems engineers produce to solve a specific problem or to help close a performance gap. From a user perspective, these gaps are either known (e.g., the need to bring fresh water to a community) or unknown (e.g., the need to listen to on-demand, tailored music, anytime and almost anywhere via a smart device). The National Academy of Engineering determined the top 20 engineering achievements of the 21st century, listed in Table 1.1, which offer solutions to make life easier, to connect people, and to explore new frontiers. As we move forward, the National Academy of Engineering has also defined the next set of engineering challenges to be solved. In developing solutions to these challenges, the engineering organization and the engineer must make decisions about how to use resources and which path is the best to solve the challenge.

Table 1.1 Engineering achievements and challenges

Top 20 Engineering Achievements	Grand Challenges
• Electrification • Automobile • Airplane • Water Supply and Distribution • Electronics • Radio and Television • Agricultural Mechanization • Computers • Telephone • Air Conditioning and Refrigeration • Highways • Spacecraft • Internet • Imaging • Household Appliances • Health Technologies • Petroleum and Petrochemical Technologies • Laser and Fiber Optics • Nuclear Technologies • High-Performance Materials	• Make solar energy affordable • Provide energy from fusion • Develop carbon sequestration methods • Manage the nitrogen cycle • Provide access to clean water • Restore and improve urban infrastructure • Advance health informatics • Engineer better medicines • Reverse-engineer the brain • Prevent nuclear terror • Secure cyberspace • Enhance virtual reality • Advance personalized learning • Engineer the tools for scientific discovery

To help connect the need or requirement to the solution, the engineer must be able to provide credible answers to a set of questions:

- How strategic is this solution? How important is it for us to solve this challenge?
- How well are we providing a solution? How does this solution solve the challenge?
- How well can we deliver?
- Will we make money?
- Can we win?

Answering these questions ensures that the solution will be profitable and able to make a difference. The focus of this text is on the fourth question, Will we make money?, which in many ways drives the other questions.

In today's global business environment, engineers integrate hardware, software, people, processes, and interfaces to produce economically viable and innovative products and services while ensuring that all pieces of the enterprise work together. The engineer's real challenge is to make all the pieces work together to solve the real problems of the enterprise using cost-effective solutions.

The notion of cost-effective solutions has been gaining increasing scrutiny in engineering projects due to traditional industries that have seen competition erode historical profit margins and government agencies whose budgets have been constrained or have come under scrutiny. A recent example of this is the successful landing of a series of rovers on Mars. The four vehicles employed in interplanetary space exploration have been heralded almost as much for their cost-effectiveness as for the technical engineering involved in designing them. Indeed, on NASA's website (see NASA, n.d.), the first of the four rovers is described as "a demonstration of the technology necessary to deliver a lander and free-ranging robotic rover to the surface in a cost-effective and efficient manner." The implication of this statement is that the rover would be thought of as a failure if its design was not cost-effective and efficient. It is becoming ever more important that engineers understand the cost implications of their designs and decisions in order to continue to fulfill their roles.

Note that in all the definitions of engineering discussed above, there exists an economic element. This element has grown in importance over time; notice that cost is not predominant in the first definition by the Engineers' Council for Professional Development but is clearly elaborated in more recent definitions, such as the one from the Quality Assurance Agency. Financial and business issues will continue to do more than directly affect technical work done each day; more than ever, they will also affect the careers of engineers. Also, technical employees who contribute to the resolution of business issues are considered more valuable and have faster career progressions. Many engineers transition from technical to business jobs as their career develops, including starting a business or leading firms in managerial roles.

As the process of globalization continues, leading to improved standards of living and increased demand for goods and services provided by engineers, it is critical that technical employees understand the economic considerations of design at every stage of their careers. However, there is a disconnect with the current engineering education curriculum and this important consideration of cost. Many traditional engineering programs, such as mechanical, chemical, civil, and electrical, only require the most basic courses in finance, often in the form of an introductory course in engineering economics or a module as part of a senior design class. Many graduates are surprised at how important and pervasive cost and economics are throughout an engineer's career in any domain and how much emphasis is placed on these considerations in the "real world." As shown in Figure 1.5, engineering at all levels involves financial aspects. This book seeks to provide the engineer with a guidebook for understanding how to incorporate the challenges of economic considerations at each of these levels.

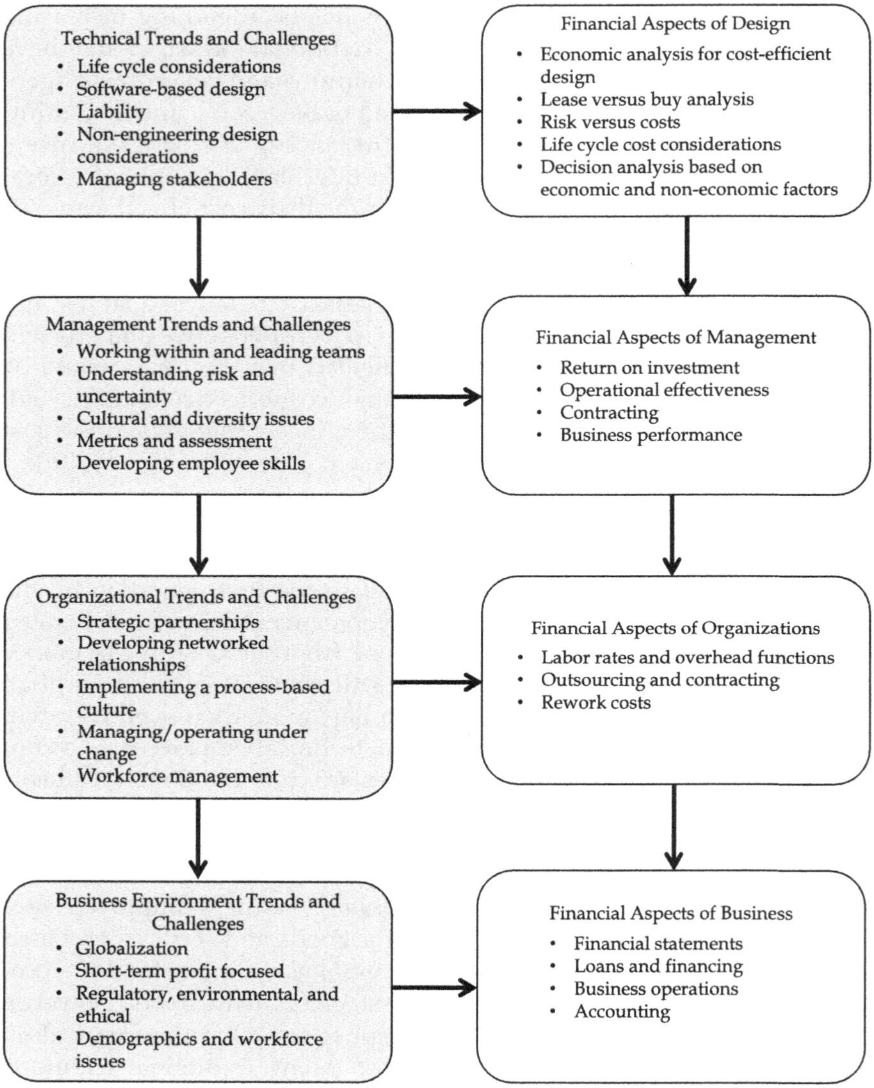

Figure 1.5 Trends and challenges for engineers and the associated financial aspects

Each of these distinct levels (see Figure 1.5) brings its own challenges and unique approaches to addressing them, ranging from traditional engineering economics to advanced cost analysis. While this text is not intended to be comprehensive, it will provide engineers with a set of tools to handle cost and economic issues related to each respective stage in their career.

Given the increasing complexity of the current environment, engineers must provide innovative products and services. In general, engineers perform a wide variety of services, including (modified from Pavarini, 2007)

- Design and delivery of technology-based products, including research and development,
- Product integration (hardware, software, human factors, etc.),
- Corporate information technology (infrastructure and/or applications),
- Sales engineering (technical sales and pre-sales support),
- Customer technical support (post-sales support),
- Technical consulting,
- Marketing and/or product management of technology-based products,
- Sales,
- Education and training,
- Logistics or support operations,
- Business or management consulting,
- Financial analysis and consulting,
- Leading technology-based companies/divisions, and hopefully,
- Founding high-tech businesses.

Cost estimating and management operates across all these business areas and environmental considerations. More than ever, engineers must consider the financial aspects of the services they provide. As the success and failure of engineering projects is closely tied to economic factors, it is critical for engineers to understand how their decisions affect outcomes. The following material presents methods, techniques, and lessons learned and can serve as a reference book or textbook for any technical career field.

1.3 Introduction to systems life cycle costing

In today's global business environment, engineers, information technology professionals and practitioners, and other related product development professionals integrate hardware, software, people, and interfaces to produce economically viable and innovative applications while ensuring that all pieces of the enterprise work together. No product or service is immune from cost, performance, schedule, quality, and risks and tradeoffs. Yet engineers spend most of their formal education focused on performance and most of their professional careers worrying about resources and schedules. Too often we become fixated on the technical performance required to meet the customer's expectations without worrying about the downstream costs that contribute to the total LCC of a system.

Unfortunately, in many cases the LCC or TOC are ignored because either the total costs would make the project untenable (especially for large government projects) or the increased acquisition costs needed to reduce the LCC would make the project unacceptable.

We have an extensive array of economic techniques and tools at our disposal to predict and monitor costs and schedules, yet overruns are commonplace, especially for software-enabled systems. We do not understand either the technical or non-technical aspects of LCC and the associated risks. The technical domain of the engineer continues to grow in complexity through the requirements of more complex products. This growth also increases the considerations of cost and finances on their products. Figure 1.6 shows some of the external and internal factors that we must tackle in conducting cost analysis and then must address when managing the program in the most effective manner.

Engineering has changed dramatically in our technology-driven global economy. Beyond technical excellence, understanding the economics or business aspects of modern engineering is key to success. In a profession where network-centric systems of systems are commonplace, engineers must be involved in all aspects of new product development. Accounting, project management, leadership, and marketing are just some of the necessary skills we need to succeed in the workforce. It is against this backdrop that this book was written to serve as a collection of methods, processes, and tools (MPTs) and terminology needed to understand the business aspects of modern engineering.

We divided this book into three broad groupings: analysis, estimation, and management. Under analysis, we present techniques such

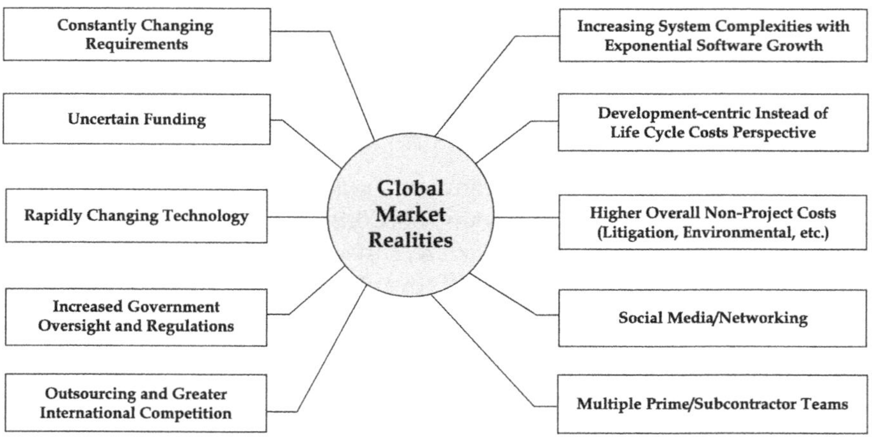

Figure 1.6 Some of the factors that can affect the cost of a system (modified from Stevens Institute of Technology, 2008)

as engineering economics and simulation-based costing focusing on total life cycle understanding and perspective. This analysis section is titled "The Mathematics of Engineering Economy" and includes Chapters 2 to 6; it presents techniques for detailed analysis that are relevant for modern complex systems. Under estimation, which we titled "Estimation of Complex Systems," Chapters 7 to 9 present a discussion of complex systems, software, and estimating techniques. Lastly, we grouped management techniques, which we titled "Cost Management"; Chapters 10 to 13 discuss commercial off-the-shelf systems, the cost of quality, the role of project management in LCC management, and the use of metrics and ratios in cost management.

1.3.1 Systems life cycle costing

Life cycle costs are all the anticipated costs associated with a project or program throughout its life. LCC is the sum total of the direct, indirect, recurring, non-recurring, and other related costs incurred, or estimated to be incurred, in the design, research and development, investment, operations, maintenance, retirement, and any other support of a product over its life cycle (i.e., its anticipated useful lifespan). All relevant costs should be included regardless of funding source, business unit, management control, and so on. LCC is important for systems because the acquisition is a small part in relation to the true cost or TOC associated with owning and operating the systems.

There are four generally accepted methods for determining LCC:

- Engineering cost methods: direct estimation at the component level leading to a detailed engineering build of the system,
- Cost accounting: modern cost management systems to track and allocate expenses,
- Analogy: an estimate using historical results from similar products or components, and
- Parametric: based on mathematical relationships between costs and some product- and process-related parameters.

Any combination of these four can be used to develop TOCs.

Analogy and parametric cost estimates (PCEs) are considered top-down estimating techniques by which the costs are estimated by looking at the project based on the customer's requirements. Engineering costs (or detailed engineering builds) and accounting techniques are bottom-up methods and refer to estimating and tracking costs by breaking the project down into elements using work breakdown structure and physical and functional architectures.

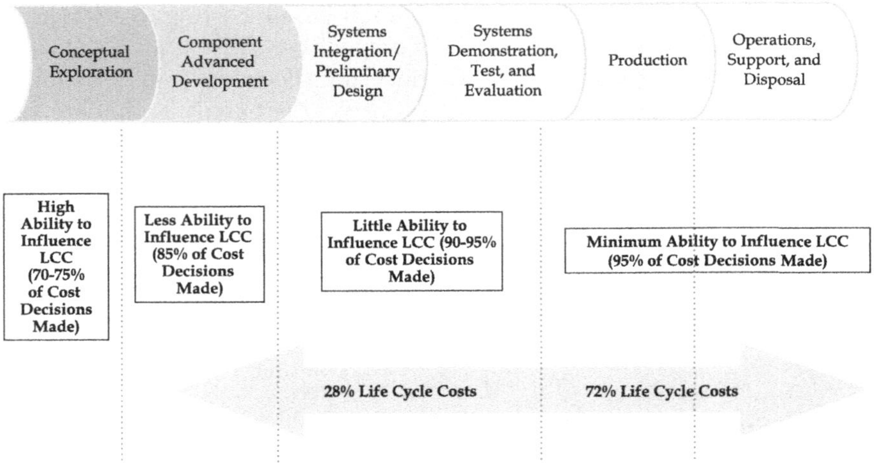

Figure 1.7 Costs incurred and committed during our systems life cycle acquisition process (modified from Andrews, 2003)

Figure 1.7 presents a standard new product development process. This figure shows how costs are committed and incurred as a function of phases in our LCC model. Modern products can have life cycles that range from several years to decades long. One of the main challenges that engineers encounter is that early in the product development phases cost considerations are often ignored or downplayed in favor of performance factors. However, it is during these early phases that decisions have the largest impact on the LCC of a project. This tendency can lead to downstream cost overruns and potentially project failure. It is an engineer's responsibility to ensure that design factors presented early in the life cycle before costs are realized include financial considerations.

In its simplest form, LCC is composed of initial and future expenses. However, clearly defining initial costs can be a challenge. For example, take a typical new product development process as shown in Figure 1.7; how do we allocate investments to the industrial base for major projects? What about spare parts? Should they be categorized as production or operations and support? Very quickly you become mired in determining how and when to allocate expenses. Figure 1.7 also shows the standard LCC terminology and model we adopted for this text and how costs are committed and incurred as a function of phases in our LCC model. We will discuss this figure throughout the text because of its importance in illustrating how programs and funds are committed early in a program and what and when various techniques should be used.

The MPTs and terminology presented are important in that they can be used to

- Estimate the TOC to the various stakeholders;
- Reduce and capture TOC through using LCC trade-offs in the systems engineering and product development processes;
- Control cost through using LCC contractual provisions in procurements; and
- Assist in day-to-day acquisition management actions by providing timely, consistent, and relevant cost information.

1.4 Cost analysis, estimation, and management

1.4.1 Cost analysis

All engineers make trade-offs in the four domains shown in Figure 1.8. Good engineers follow a disciplined and structured approach when developing a product or service. Costing hardware, software, and integration requires an understanding of many MPTs and terminology that few engineers have received formal training in. Once technical characteristics have been ascertained from the requirements, selecting the right MPTs is critical to accurately determining costs early in the development cycle and estimating realistic LCC.

Because of the complexity and the rapid pace of technology evolution, the costing of complex systems has become a tremendous challenge. We understand how to cost hardware and to a lesser extent software; however, we are still developing tools and processes for costing the integration of complex systems. As we scale to larger and more complex systems, systems of systems, and enterprises, our ability to determine costs becomes less relevant and reliable.

Figure 1.8 Domains in which engineers make trade-offs for any project

1.4.2 Cost estimation

Cost estimation techniques can be divided into three categories: PCEs, analogies, and detailed engineering builds. Figure 1.9 shows their applicability throughout the life cycle as well as how this course is organized. We will discuss PCEs, analogies, and engineering builds in the conduct of life cycle analysis. Engineering economy and simulation-based costing are critical tools needed to conduct any type of meaningful LCC. As stated, we do not discuss accounting, but like the other tools, this is an important technique for ascertaining costs. However, capturing expenses in a formal manner is certainly the best way to ascertain costs. Obviously, developing true costing amounts and using good cost management requires good accounting practices and the tracking of expenses using activity-based costing techniques.

1.4.2.1 Parametric cost estimation

PCEs are usually based on mathematical equations or models. Simple mathematical relationships, such as linear and non-linear regression, are mainly used. Often, they are based on historical data from similar projects. Tools such as Microsoft Excel can easily be used to fit these relationships. The biggest challenge is determining the relationships between the dependent and independent variables and their range of usefulness.

1.4.2.2 Analogy cost estimating methodology

Analogy estimates are performed on the basis of comparison and extrapolation using like items or efforts. In many instances, this can be

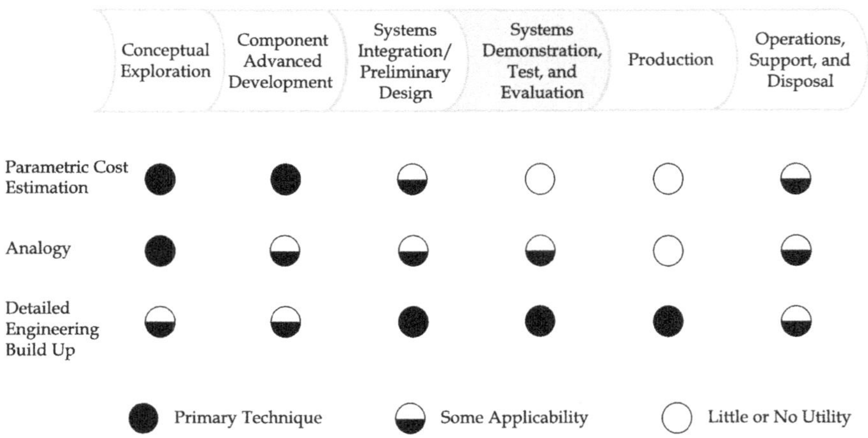

Figure 1.9 Cost estimation techniques throughout the life cycle (Courtesy of NASA, 2015)

accomplished using simple relationships or equations representative of detailed engineering builds of past projects. Obviously, this is the preferred means to conduct a cost estimate based on past programs that are technically representative of the program to be estimated. Cost data is then subjectively adjusted upward or downward depending on whether the subject system is felt to be more or less complex than the analogous program (NASA, 2015).

1.4.2.3 Engineering build-up methodology

Sometimes referred to as "bottom-up" estimating, the engineering build-up methodology rolls up individual estimates for each element, item, and component into the overall cost estimate. This can be accomplished at the work breakdown structure element level or at the component level. In this costing methodology, the cost of a work breakdown structure element is computed by estimating at the lowest level of detail and computing quantities and effort levels to determine the total system cost. Obviously, this is the most accurate means to develop a cost estimate. The challenge is that early in the systems development a bottom-up approach cannot be used because the systems haven't been fully designed. Ideally, you would like to take bottom-up estimates and scale them based on experience.

The techniques used in estimating software (we chose to treat software separately from systems) are much more mature than systems. At best, the tools commonly used are estimates and analogies and have little mathematical basis. Whether purely a service-centric or a physical system, most products now have a significant software element. The methodology for estimating software has been around for over 30 years and can be classed as a PCE tool. However, because of new languages, hardware and software integration challenges, computer-aided software tools, and so on, techniques and algorithms must be continually updated. Software estimating is still dominated by experience supplemented with quantitative techniques.

1.4.3 Cost management

Engineering cost management can be defined as the process of identifying, allocating, and tracking the resources needed to meet the stakeholder's requirements. An integrated, process-centered, documented process backed with quantifiable data provides real and tangible benefits to all stakeholders. Engineering cost management can best be described as an integrated, process-centered, measurable, and disciplined approach to LCC and management to make the trade-offs between cost, performance, schedule, and risk. Good cost management practices, supported by sound analysis, can lead to (modified from NASA, 2015)

- Complete, unambiguous, and documented functional requirements in order to meet LCC goals;
- Bounded and clearly defined product functional expectations and acceptance criteria, understood and agreed to by all stakeholders;
- More accurate, credible, and defensible scope, cost, and schedule estimates with realistic assessments of risk;
- More complete and timely risk identification, leading to more effective risk mitigation;
- A basis for properly quantifying, evaluating, and controlling the acceptance and timing of changes to requirements (i.e., precluding "scope creep");
- Final products that deliver better reliability, adaptability, usability, performance, maintainability, supportability, and functionality—in short, higher quality and value;
- Insight into near-, mid-, and long-term technology, design, infrastructure, and operational investment needs as they relate to different effects on the phases and trade-offs within the life cycle;
- Earlier and more consistent visibility of problems (fewer surprises);
- An understanding of the costs for each step in the development process;
- More efficient project management; and
- Organizational credibility and reputation.

Engineers play a critical role in corporate or business planning. Engineers are involved in cost management from top-level corporate planning to costing components and subsystems. All require the same basic understanding of the time value of money, risk, and life cycle perspective.

Engineering cost management is employed as a means of balancing a project's scope and expectations of risk, quality, and technical performance to ensure that the most cost-effective solution is delivered; it consists of three steps:

1. Define the requirements, level of quality desired, and the budget.
2. Ensure that the domains of performance, cost, schedule, and risk/quality are aligned with the budget.
3. Monitor and manage the balance of these four domains throughout the life of the project by using sound engineering techniques.

The ability to use analysis techniques such as those discussed in this chapter allow an engineer to conduct defensible analysis that can not only provide true ownership costs but can also allow for conducting meaningful tradeoff analysis.

1.5 Summary

Because of technology, globalization, and complexity most engineers have become product development professionals integrating hardware, software, people, and interfaces to meet the stakeholder's requirements. Few products or services are immune from cost, performance, schedule, quality, and risk and trade-off considerations, and all must address environmental, sociological, and political concerns. Unfortunately, engineers spend much of their formal education focused on technical performance and most of their professional careers focused on communication, resources and schedules. Too often we become fixated on the technical performance required to meet the customer's requirements without worrying about the downstream costs. In essence we ignore the problem because funding the operations and maintenance, disposal, etc., becomes someone else's problem.

One of the unintended consequences of globalization has been the outsourcing of core competencies. In essence, many engineers and companies have become marketers, packagers, and integrators of technology. More so than for companies with a strong core technology as their product, engineers in the global economy must understand costs as a business driver.

In order to be successful, we must understand the role of finances in every aspect of our daily job. This needs to be taught at the undergraduate level. As they enter the job market, engineers must understand at least some of the financial aspects of engineering business operations and considerations that can affect product realization.

In our global service-sector economy, engineers now serve as key systems integrators, enablers, and managers of people, technology, and processes to produce economically viable and innovative products and services. Engineers, and not just technical experts, must play a key role in all phases of new product development. Most importantly, we must understand how to perform good cost analysis, estimation, and management as well as their limitations.

Too often, performance (features and functionality) is used as the entire measure of system effectiveness. As shown in Figure 1.10, there must be consideration given to LCC. The broader point is that often in designing systems, we focus most of our attention on the functions to be provided, the operational requirements, but only when we take LCC into account do we have operational effectiveness.

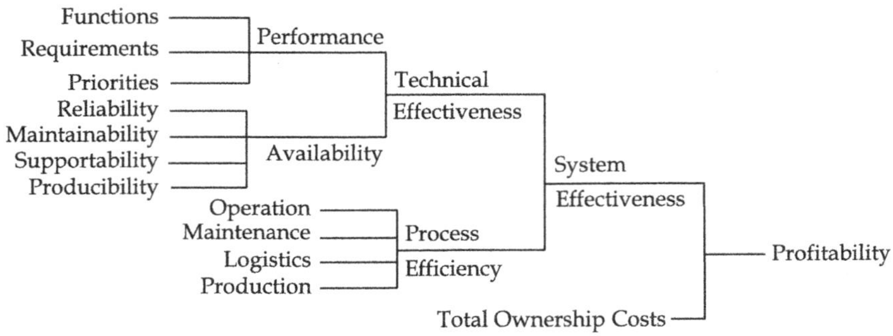

Figure 1.10 Components of operational effectiveness (modified from Stevens Institute of Technology, 2008)

QUESTIONS

1-1. Your subcontractor company has teamed with a large defense contractor (LDC) and has been awarded the new Super Fighter—the largest procurement contract in defense history. List three areas for each of the following stakeholders that should be your primary focus when monitoring, billing, and paying for the Super Fighter contract:

- Program manager for the LDC,
- Program manager for your subcontractor company,
- LDC corporate headquarters,
- Defense sponsoring agency, and
- Legislative body.

1-2. When we buy cars, homes, major appliances, and so on, we are mainly focused on the upfront costs (mainly purchase price) and seldom assess the LCCs of a major investment. Unfortunately, our decisions are often driven solely by performance. From your own buying experience, write down your thought process for buying a new car and weigh the major components of your decision (upfront costs, trade-in value, gas mileage, looks, accessories, etc.). List and assign weights (must add up to 100%) to each component of upfront and recurring costs.

1-3. One of the key challenges for buyers is that we fixate on development costs with little or no regard to downstream LCC costs. Briefly explain why you think this occurs. Is this more of a problem for large government programs than for private projects?

1-4. Firm-fixed price (FFP) contracts provide a pre-established price, which places more risk and responsibility for costs and resulting profit or loss on the contractor and provides more incentive for efficient and economical performance (modified from Government Accounting Office, 2008). Our everyday life is governed by FFP contracts (home construction, car maintenance, etc.), yet few large contracts are FFP. What cultural obstacles must be overcome to institutionalize FFP contracts for government?

1-5. What are the grand challenges of the next 10 years for engineers? What are the common characteristics of these problems?

1-6. What is the difference between interdisciplinary and multidisciplinary problems?

1-7. Let's assume that the role of engineering-centric business is to provide customers with a better price or better value proposition than competitors while operating the business to attain targeted profits. What are the exceptions? How would you restate this for government?

1-8. For almost any industry engineers conduct trade-off studies for their designs that address the critical elements of risk, schedule, cost, and performance. For example, NASA will not compromise risk to meet schedule deadlines. Can you give some industry specific examples where risk, schedule, cost, and performance may or may not be compromised?

1-9. The Boeing Company Dreamliner (787) is a long-range jet airliner that was developed by Boeing for the intended purpose of servicing commercial airline customers. The airplane is the first to employ a lightweight composite material for structural components and as such realizes a significant fuel savings for its operators. However, the delivery of the first airplane was initially slated for 2008 but due to several project delays the first in-service flight did not take place until 2011. The airplane also suffered from numerous technical problems in the years following its initial service. All these technical problems and delays led to dramatic cost overruns from initial estimates. Research public sources, and describe the cost problems of the Dreamliner. Discuss the systems that could have been put in place to prevent these problems.

References

Andrews, Richard, 2003, "An Overview of Acquisition Logistics," Fort Belvoir, VA: Defense Acquisition University, Course Notes, 2007.

Engineers' Council for Professional Development, 1947, "Canons of Ethics for Engineers."

Government Accounting Office, March 2008, "2008 Defense Acquisitions, Assessments of Selected Weapon Programs," GAO-08-467SP.

Government Accounting Office, March 2009, "Cost Estimating and Assessment Guide Best Practices for Developing and Managing Capital Program Costs," GAO-09-3SP.

Institute of Electrical and Electronics Engineers, February 2011, "Engineers and Business School—A Match Made in Heaven," accessed 1 April 2013 at http://spectrum.ieee.org/at-work/education/engineers-and-business-schoola-match-made-in-heaven.

NASA, 2015, "Cost Estimating Handbook," accessed 1 August 2018 at https://www.nasa.gov/pdf/263676main_2008-NASA-Cost-Handbook-FINAL_v6.pdf.

NASA, n.d., "Mars Pathfinder Overview," accessed 10 April 2018 at https://www.nasa.gov/mission_pages/pathfinder/overview.

National Academy of Engineering, 2008, "Changing the Conversation: Messages for Improving Public Understanding of Engineering," National Academies Press, Washington, DC.

Pavarini, Carl, 2007, "Entrepreneurship and Business for Engineers and Scientists," Stevens Institute of Technology, Class TG 401.

Quality Assurance Agency, 2015, "QAA Subject Benchmark Statement for Engineering," accessed 1 August 2018 at http://www.qaa.ac.uk/docs/qaa/subject-benchmark-statements/sbs-engineering-15-masters.pdf?sfvrsn=fb91f681_16.

Stevens Institute of Technology, 2008, "SYS 625 Fundamentals of Systems Engineering Class Notes."

section one

The mathematics of engineering economy

chapter two

The science of engineering economics: understanding the time value of money

2.1 *Introduction and the time value of money*

The need for an in-depth understanding of economic analysis is twofold. First, in our flat, global, and complex world, engineers work on multidisciplinary projects that encompass "cradle to grave" dimensions, including finance, design, construction, operation, maintenance, and retirement. To compete in international markets, all industries face and must respond to increased competition in the form of lower-cost operations and competition from non-traditional sources. Second, economic analysis is important from the traditional perspectives of life cycle costing, analysis of alternatives, business operations, and so on. Economic analysis, in many cases, is as important as technical requirements. Many parameters can influence an economic analysis (Figure 2.1). In all these cases, an in-depth understanding of the economic factors affecting a project must be understood in order to remain competitive in today's ever-changing markets.

Engineering economics uses relatively simple mathematical techniques to make decisions about capital projects by comparing various alternatives. Engineering economy techniques allow for comparisons by accounting for the time value of money (TVM).

Spreadsheets have dramatically changed how we conduct the economic analysis of alternatives. What once involved manipulation of equations and tables can now be modeled in a spreadsheet using only a few basic commands. The use of spreadsheets is ideal because

- Most problems involve repetitive calculations that can be expressed using simple formulas as a function of time. Note that Excel has built-in functions for most engineering economy equations.
- Sensitivity analysis is key to conducting good analysis, and in a properly designed spreadsheet the parameters can be changed and plots easily developed.

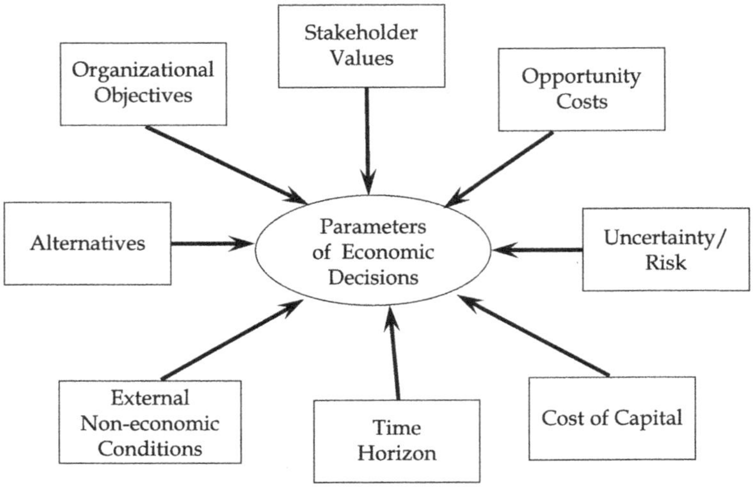

Figure 2.1 Parameters of economic decisions

- Complex models can be rapidly and easily built and are, for the most part, self-documenting.
- The user can develop professional reports and plots using the functionality in most spreadsheets.

Because of the inherent way spreadsheet models are developed, problem formulation is often more comprehensive and transparent than in traditional equation-based models.

2.2 The capital budgeting decision

2.2.1 Basic concepts in capital budgeting

The ability to plan and evaluate economically major projects is an important requirement of most government agencies and engineering firms. Most agencies have guidelines and regulations to evaluate rationally proposed projects with regard to their economic feasibility. Almost all these guidelines and regulations require that the benefits of the proposed project exceed its costs. Because corporate investment in projects has serious consequences in the financial viability of a corporation, private-sector projects are often easier to evaluate than their public-sector counterparts. Corporate policies require rational and deliberate analysis of capital budgeting decisions before projects are approved. Corporate investment is different from government investment in major capital projects in that corporate investment must also consider the source of funding to pay for the capital investment. In many instances, the ability to finance a project (in lieu of the most economical alternative) determines its feasibility. In some cases, capital projects may be financed through the use of corporate bonds or other vehicles. Invariably, when the project nears construction, financing will depend on borrowed money. The method of financing must be considered in the corporate capital budgeting decision since it may be the factor that determines the viability of the project.

It is important to understand that the economic evaluation of alternatives and the evaluation of alternative financing for a project are key to the viability of that project. Economic evaluation involves developing the cash flows representing the benefits and costs associated with the acquisition and/or operation of the system. The cash flow over the life cycle is often referred to as the economic cash flow. Most economic analysis of a program or project should include the financing plan, or the cash flow representing the incomes and expenses as a result of adopting a specific financing plan for funding the project.

The object of the economic evaluation is to select the most cost-effective alternative that will satisfy the stated goals and objectives. (It may be to approve or disapprove a single project.) Therefore, prior to performing the economic analysis, the alternatives must be identified. To analyze the investment under consideration, a life cycle must be established. For some agencies and types of projects, the life cycle is established by regulation. For instance, most federal and state flood control projects have a life cycle of 50 years. A co-generation plant (a plant that produces steam for a customer and sells excess energy, perhaps as electricity, to another organization) may have a life cycle of 20 years. The life cycle should bear some relationship to the life of the product.

To determine whether or not a project is feasible, its rate of return should be compared to a minimum attractive rate of return (MARR) or the required rate of return for the project. The MARR should be greater than the rate of return that one may obtain with roughly the same risk in another venture. A project should be considered feasible if its rate of return exceeds the MARR. Obviously, MARR has different implications for public- and private-sector projects.

In economic evaluations, project alternatives are analyzed with respect to their cash flow profiles over n periods in the life cycle. This type of analysis is usually shown in the spreadsheet format and will be referred to as cash flow schedules. The interest period is traditionally, but need not be, in years.

Once money is invested in a project, those funds are no longer available for alternative investment. The term "opportunity cost" refers to the return that could have been realized by investing in the next best alternative, if defined, or another opportunity that becomes known after the decision is made. In general, the MARR reflects the opportunity cost of capital, the market interest rates for borrowing and lending, and the risks associated with the investment in the capital project. For public agencies, policy or regulation may specify the MARR.

2.2.2 Benefit and cost development

For commercial products, the benefits or revenues of a capital project are relatively easy to determine. The basic revenue is what you can sell the product for, the rental income it can produce, the depreciation or other asset available for tax credit, the rental cost to avoid by relocating staff, and so on. For the co-generation plant, it is the steam revenue from the primary user, the revenue for the electricity sold to the grid, and so on. For an office building, it is the rental avoidance and the rental income provided. Both may include some tax credits.

For public capital projects, such as roads, dams, bridges, mental health, and the like, the benefits may be harder to ascertain. Certainly, the benefits of a new office building may include rent avoidance. But for most public capital projects, the benefits are more difficult to quantify. For a flood control project, the expected benefits are the lack of property loss, the lack of disruption of lives, the lack of social cost of public facilities being out of commission, and similar factors. The social benefits are more difficult to quantify.

The costs, on the other hand, are more easily quantified. Certainly, all the capital costs (labor, equipment, materials, supplies, financing, etc.) enter into the cost. In addition, the cost of operating and maintaining the facility may be calculated. Finally, the most complicated cost may be the amount and type of financing involved. The former cost will have a great impact on the cash flow analysis. Depending on the discount rate, the method of debt repayment may make the difference between a viable and non-viable project.

2.3 Time value of money

Engineering economics is simply a subset of economic analysis as applied to engineering projects that typically consist of systems and/or services. Engineers consider risk, technical aspects or performance, and cost when developing projects. The most basic concept of analysis useful in engineering economics is an understanding of the TVM.

The most basic explanation for the TVM is that money today is worth more than money tomorrow, or "a nearby penny is worth more than a future dollar." Or, simply, the money available in the now is worth more than the same amount in the future due to its earning capacity.

This is the most basic principle of financial planning and analysis. Were this not the case, all investments would make little sense because the population would be incentivized to simply hold money and allow it to increase in value. While this is a very simple explanation, breaking down the origins of this commonly accepted principle of economics is far more involved.

Most capital investment and some personal finance decisions are typically based on the TVM. To explore this concept further, we propose the following definitions:

i ≡ interest rate per payment period
N ≡ total number of interest periods
n ≡ number of a specific interest period
P ≡ present value/present worth (time zero)
F_n ≡ future value/future worth (time n)
V_n ≡ value at time n
A ≡ annuity or uniform payment occurring at uniform time
G ≡ linear gradient payment (constant growth each time period)

Note that for most problems $N = n$, and so they are often used interchangeably. Normally n is different from N when we are examining a specific time period not equal to the total number of interest periods for the problem. For example, what is the interest portion of a payment at the end of the 24th month ($n = 24$) for a 30-year mortgage ($N = 360$).

Most engineering economy texts use cash flow diagrams (CFDs) as a way to define and visualize the solution concept. Like the free body diagrams, they are an important step in turning a word problem into a form that can be analyzed readily using mathematical techniques. Cash flow diagrams are an important way to visualize information about an engineering project or financial transaction. Figure 2.2 shows the steps in solving most engineering economics problems. An example of a CFD is shown in Figure 2.3.

Note that arrows pointing in the positive y direction (up) show positive cash flow (receipts, savings, etc.), whereas arrows pointing in the negative y direction (down) denote negative cash flows (disbursements, costs, etc.). One of the key assumptions implied by this diagram is that all interest occurs at the end of a time period. Otherwise, the analysis would be too complex to perform. Also, an arrow means that cash or money changes hands. If money were deposited in a savings account, the interest paid monthly would not be represented with positive arrows on a CFD. A down arrow would be drawn when the money is withdrawn from the bank. The interest rate and compounding period should also be labeled on the CFD.

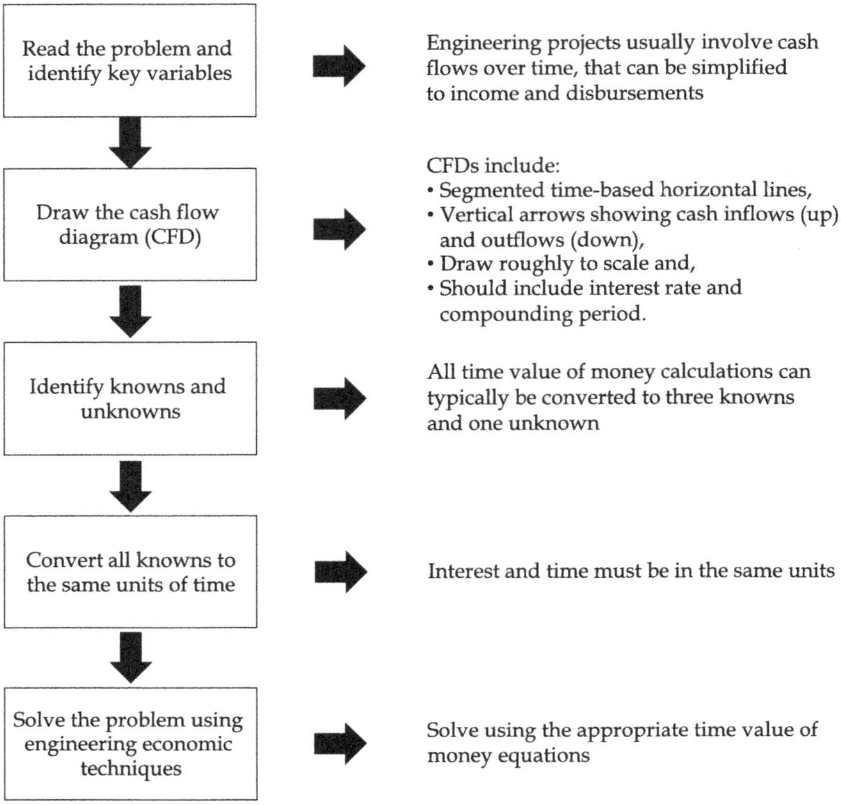

Figure 2.2 Steps in solving an engineering economics problem

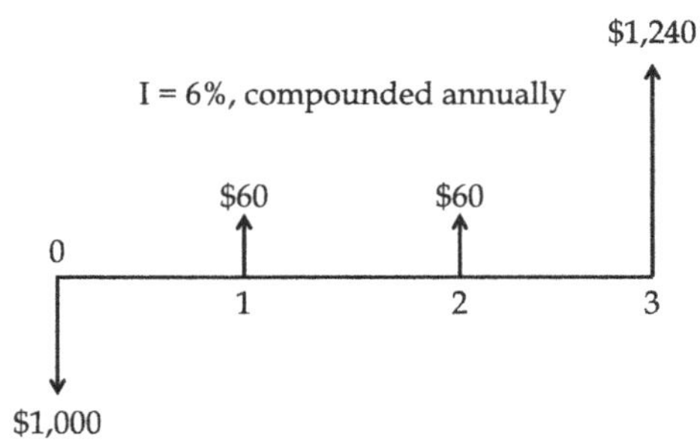

Figure 2.3 Typical CFD

2.3.1 Cash flow diagram rules

When drawing a CFD, the following vectors and information should be labeled:

- The first period of the CFD,
- The last period of the CFD,
- The appropriate negative or positive vector for each cash flow,
- The period in which the cash flows occur,
- The dollar value of each cash flow, and
- The interest rate with compounding period.

Some other basic diagraming rules include the following:

- For traceability, it is better to show two or more cash flows at a point in time than to take their sum and show just one vector.
- The diagram should be roughly to scale, but only if it doesn't detract from conveying the information.
- Almost every problem has a mirror image; for example, one person's cash inflow (i.e., receiving a loan) is another person's outflow (i.e., bank loaning the money). Make sure you draw the CFD from the right perspective.
- Time 0 is defined by the problem and can be today, the year of the investment, or at a point in the past or future.
- For CFDs, the end of period n is equal to the start of period n + 1.

2.4 Interest

To understand the TVM, one must understand cash flows, time, and the cost of money or interest. Interest is simply the cost of money. It is either the rent you pay on money borrowed from a bank or bonds issued by a corporation, or the money you receive for investing money in a bank or some other type of financial institution. Interest rates are normally stated on an annual basis. For example, when the federal discount or bank rate is stated to be 6.5%, it is generally understood to be an annual percentage rate (APR).

2.4.1 Simple interest

Simple interest is called simple because it ignores the effects of compounding. The interest charge is always based solely on the original principal, so interest on interest is not included. However, it is useful to demonstrate the concept of interest. As an example, a loan might be given to an entry-level college student by his or her parents. The parents lend the student an amount of money now, or P. The student must pay back his or her parents an amount of money in the future (interest plus principal), F, at the end of n years if the interest rate is i.[1] For simple interest

$$F = P(1 + ni) \tag{2.1}$$

For example, if you borrow \$20,000 from your parents for tuition, for a period of 4 years at 5% simple interest, you must repay them upon graduation:[2]

$$F = \$20,000(1 + 4 * .05) = \$24,000$$

Note that simple interest assumes the total loan amount is held for the entire life of the loan, even though periodic payments may be made. Thus, the borrower pays the interest on the original balance. It is important to remember that when calculating simple interest, one does not earn interest on interest earned during previous time periods (i.e., compounding) but only on the beginning account balance. This is why, from an investor perspective, we generally prefer compound interest to simple interest.

For illustrative purposes, assume that your parents give you \$10,000 today (P). Using simple interest and I = 6%, accrued annually, how much do you have in the future (F)? Table 2.1 illustrates a portion of the interest schedule. Note that the time period and interest rate must be in the same time units. In Example 2.1, we cannot mix years and months. Since the payments are monthly, we must use a monthly interest rate.

This equation can be used to determine one unknown given three known inputs. For example, using Equation 2.1, how long would it take for the initial investment to double (F = \$10,000)?

$$\$20,000 = \$10,000\ (1 + n(0.06))$$

This produces an n value of 16.67 years.

Table 2.1 Illustration of simple interest

Time Period	Interest Earned in the Time Period	Total Value
0	-	\$10,000
1	\$600	\$10,600
2	\$600	\$11,200
3	\$600	\$11,800
4	\$600	\$12,400

[1] Different books to include this text use different notations for interest. Often i is used to express interest as a fraction and I to express it as a percentage.

[2] Significant figures are numbers that imply precision of the calculation. Most engineering economics textbooks simply use dollars and cents. Obviously, using factor notation, Excel, or the mathematical formulas produces different levels of precision. Engineers should be careful when presenting results to ensure that they do not imply a level of accuracy that does not exist.

EXAMPLE 2.1

Draw the CFD for the following scenarios:

a. Receive \$1,000 today and immediately give \$200 to charity.

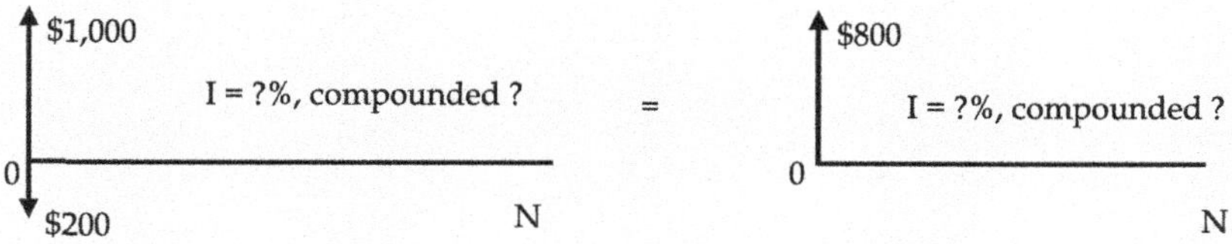

b. Invest \$2,000 per year for the next 20 years in an individual retirement account (IRA) earning 7% annually.

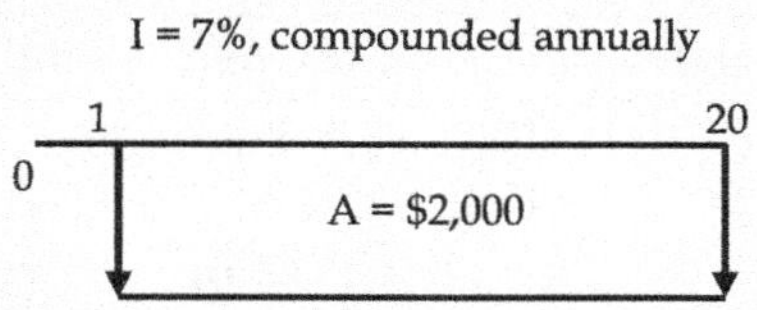

c. Pay off your \$30,000, 5-year, 6% APR (compounded monthly) car loan in 60 monthly installments of?

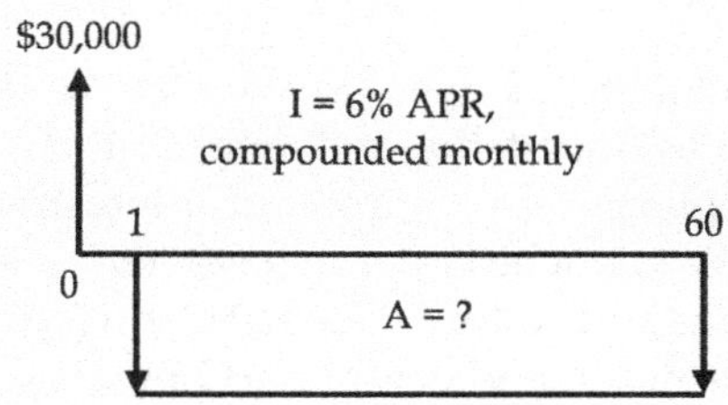

2.4.2 *Compounded interest*

Compounded interest means that the interest is paid on the capital and the accumulated interest. This is often referred to as the power of compounding. All business interest is compounded. Using current technology (calculators and spreadsheets), students of the time value of money should know one basic equation for compound interest. All other equations related to compound interest can be derived from this:

$$F = P(1+i)^n \tag{2.2}$$

Example 2.2 demonstrates the construction of the seminal compound interest equation for present and future values.

EXAMPLE 2.2

Derive by induction the future payment equation from the single payment equation.

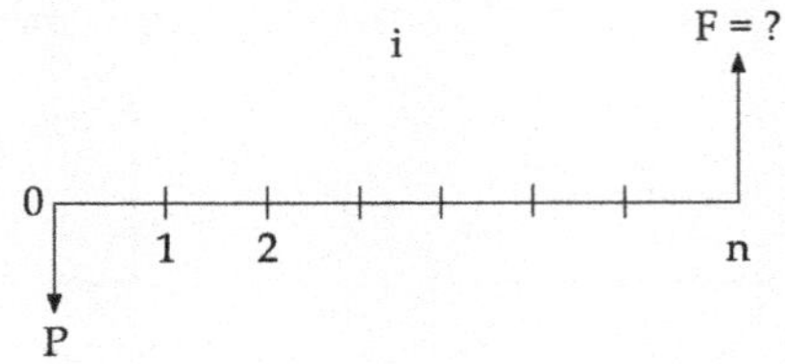

Solution

Establish a pattern for n = some finite number:

Period	Capital	Interest	Future Worth
0	P	0	P
1	P	Pi	P(1 + i)
2	P(1 + i)	P(1 + i)i	$P(1 + i)^2$
3	$P(1 + i)^2$	$P(1 + i)^2 i$	$P(1 + i)^3$

From the patterns shown above, we can easily deduce that $F = P(1 + i)^n$.

Note that Equation 2.2 dates to the start of the 18th century, when Jakob Bernoulli developed the irrational number e (i.e., approximately 2.71828) while exploring the compound interest formula. He asked this simple question: Would it be better to have 100% compounded once a year or 50% twice a year and so forth? (See O'Connor and Robertson, 2001, for a discussion of Bernoulli and compound interest.)

Let's use the equation based on compound interest for the same example. Again, assume you plan to invest $10,000 into a compound interest account earning a 6% rate accrued annually. Table 2.2 illustrates a portion of the interest schedule. Examples 2.3 and 2.4 are some simple examples that demonstrate the effects of compound interest on growing wealth.

Table 2.2 Example demonstrating the power of compounding interest

Time Period	Interest Earned in the Time Period	Total Value $F = P(1 + i)^n$
0	0	$\$10,000(1 + 0.06)^0 = \$10,000$
1	$10,000 × 6% = $600	$\$10,000(1 + 0.06)^1 = \$10,600$
2	$10,600 × 6% = $636	$\$10,000(1 + 0.06)^2 = \$11,236$
3	$11,236 × 6% = $674.16	$\$10,000(1 + 0.06)^3 = \$11,910.20$
4	$11,576.30 × 6% = $714.61	$\$10,000(1 + 0.06)^4 = \$12,624.80$

EXAMPLE 2.3

How long would it take for the initial investment to double (F = \$20,000) at an interest rate of 3%?

Using Equation 2.2,

$$\$20,000 = \$10,000(1+0.06)^n \text{ or } 2 = (1.06)^n$$

Using Excel, the function would be log(2,1.06). Also, we know that for $n = a^x$ that the $x = \log_n/\log_a$ or x = 11.9 or 4.8 years quicker than with simple interest.

EXAMPLE 2.4

Let's assume that we invest \$200,000 in an interest-earning account for a period of 20 years with an interest rate of 12%. At the end of the 20 years, the difference between a simple interest contract and a compound interest contract is stark. Simple interest produces a balance in year 20 of \$680,000, whereas compound interest produces a balance in year 20 of \$1,929,259. The yearly balances are illustrated in Figure 2.4.

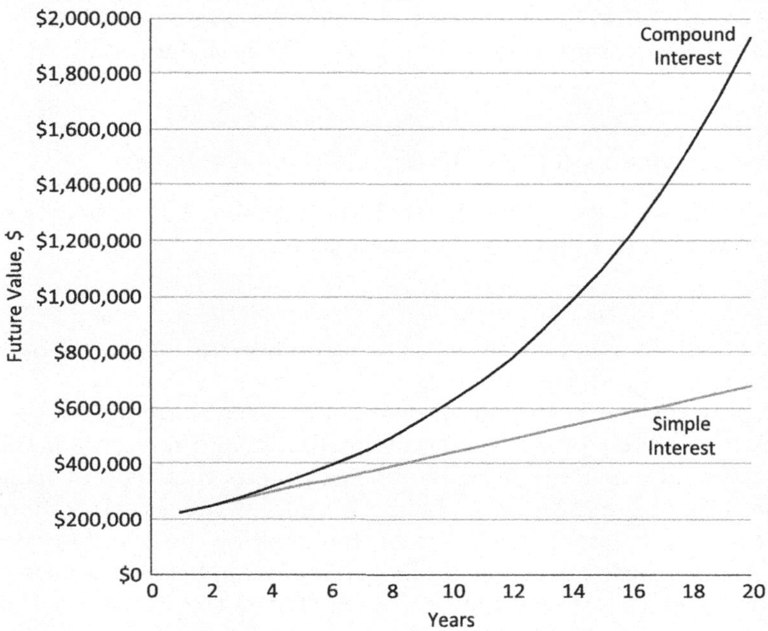

Figure 2.4 Graph showing the power of compounding interest

Table 2.3 Demonstration of the Rule of 72

Interest Rate	Actual	Rule of 72
1%	69.7	72.0
2%	35.0	36.0
5%	14.2	14.4
7%	10.2	10.3
9%	8.0	8.0
10%	7.3	7.2
15%	5.0	4.8
20%	3.8	3.6
25%	3.1	2.9
30%	2.6	2.4

2.4.3 *The "Rule of 72"*

A useful shortcut that simultaneously demonstrates the power of compounding and provides a quick way to evaluate compound interest rates is known as the "Rule of 72." In order to determine how many years it will take for an initial investment to double when deposited into a compounded interest account, divide 72 by the interest rate. For example, dividing 72 by 12% would yield a doubling period of 6 years. Table 2.3 is a comparison of the Rule of 72 and the actual doubling periods.

2.4.4 *Compound interest notation and interest tables*

Most textbooks and reference manuals use a shorthand notation to describe compound interest problems. For example:

$$(F/P, n, i) \tag{2.3}$$

Expanding this notation results in the future value given present value for n time periods with i interest rate.

Using the shorthand notation not only saves space but also allows the use of interest tables common to textbooks and reference manuals. The Appendix at the end of the book contains this type of table for some standard interest rates. Table 2.4 contains the short notation factors for an interest rate of 6%. Other than making the calculations easy to perform, these tables are no longer really applicable because most interest rates are compounded daily, which generates a fractional interest rate. The use of these tables can be a significant time saver that can be very valuable in a time-constrained environment or to do a rough order of magnitude analysis.

Table 2.4 Shorthand notation factors

Interest Rate	6.00%							
N	F/P	P/F	F/A	A/F	P/A	A/P	A/G	P/G
1	1.0600	0.9434	1.0000	1.0000	0.9434	1.0600	0.0000	0.0000
2	1.1236	0.8900	2.0600	0.4854	1.8334	0.5454	0.4854	0.8900
3	1.1910	0.8396	3.1836	0.3141	2.6730	0.3741	0.9612	2.5692
4	1.2625	0.7921	4.3746	0.2286	3.4651	0.2886	1.4272	4.9455
5	1.3382	0.7473	5.6371	0.1774	4.2124	0.2374	1.8836	7.9345
6	1.4185	0.7050	6.9753	0.1434	4.9173	0.2034	2.3304	11.4594
7	1.5036	0.6651	8.3938	0.1191	5.5824	0.1791	2.7676	15.4497
8	1.5938	0.6274	9.8975	0.1010	6.2098	0.1610	3.1952	19.8416
9	1.6895	0.5919	11.4913	0.0870	6.8017	0.1470	3.6133	24.5768
10	1.7908	0.5584	13.1808	0.0759	7.3601	0.1359	4.0220	29.6023
11	1.8983	0.5268	14.9716	0.0668	7.8869	0.1268	4.4213	34.8702
12	2.0122	0.4970	16.8699	0.0593	8.3838	0.1193	4.8113	40.3369
13	2.1329	0.4688	18.8821	0.0530	8.8527	0.1130	5.1920	45.9629
14	2.2609	0.4423	21.0151	0.0476	9.2950	0.1076	5.5635	51.7128
15	2.3966	0.4173	23.2760	0.0430	9.7122	0.1030	5.9260	57.5546
16	2.5404	0.3936	25.6725	0.0390	10.1059	0.0990	6.2794	63.4592
17	2.6928	0.3714	28.2129	0.0354	10.4773	0.0954	6.6240	69.4011
18	2.8543	0.3503	30.9057	0.0324	10.8276	0.0924	6.9597	75.3569
19	3.0256	0.3305	33.7600	0.0296	11.1581	0.0896	7.2867	81.3062
20	3.2071	0.3118	36.7856	0.0272	11.4699	0.0872	7.6051	87.2304
21	3.3996	0.2942	39.9927	0.0250	11.7641	0.0850	7.9151	93.1136
22	3.6035	0.2775	43.3923	0.0230	12.0416	0.0830	8.2166	98.9412
23	3.8197	0.2618	46.9958	0.0213	12.3034	0.0813	8.5099	104.7007
24	4.0489	0.2470	50.8156	0.0197	12.5504	0.0797	8.7951	110.3812
25	4.2919	0.2330	54.8645	0.0182	12.7834	0.0782	9.0722	115.9732
26	4.5494	0.2198	59.1564	0.0169	13.0032	0.0769	9.3414	121.4684
27	4.8223	0.2074	63.7058	0.0157	13.2105	0.0757	9.6029	126.8600
28	5.1117	0.1956	68.5281	0.0146	13.4062	0.0746	9.8568	132.1420
29	5.4184	0.1846	73.6398	0.0136	13.5907	0.0736	10.1032	137.3096
30	5.7435	0.1741	79.0582	0.0126	13.7648	0.0726	10.3422	142.3588
31	6.0881	0.1643	84.8017	0.0118	13.9291	0.0718	10.5740	147.2864
32	6.4534	0.1550	90.8898	0.0110	14.0840	0.0710	10.7988	152.0901
33	6.8406	0.1462	97.3432	0.0103	14.2302	0.0703	11.0166	156.7681
34	7.2510	0.1379	104.1838	0.0096	14.3681	0.0696	11.2276	161.3192
35	7.6861	0.1301	111.4348	0.0090	14.4982	0.0690	11.4319	165.7427
36	8.1473	0.1227	119.1209	0.0084	14.6210	0.0684	11.6298	170.0387
37	8.6361	0.1158	127.2681	0.0079	14.7368	0.0679	11.8213	174.2072
38	9.1543	0.1092	135.9042	0.0074	14.8460	0.0674	12.0065	178.2490
39	9.7035	0.1031	145.0585	0.0069	14.9491	0.0669	12.1857	182.1652
40	10.2857	0.0972	154.7620	0.0065	15.0463	0.0665	12.3590	185.9568
41	10.9029	0.0917	165.0477	0.0061	15.1380	0.0661	12.5264	189.6256
42	11.5570	0.0865	175.9505	0.0057	15.2245	0.0657	12.6883	193.1732
43	12.2505	0.0816	187.5076	0.0053	15.3062	0.0653	12.8446	196.6017
44	12.9855	0.0770	199.7580	0.0050	15.3832	0.0650	12.9956	199.9130
45	13.7646	0.0727	212.7435	0.0047	15.4558	0.0647	13.1413	203.1096
46	14.5905	0.0685	226.5081	0.0044	15.5244	0.0644	13.2819	206.1938
47	15.4659	0.0647	241.0986	0.0041	15.5890	0.0641	13.4177	209.1681
48	16.3939	0.0610	256.5645	0.0039	15.6500	0.0639	13.5485	212.0351
49	17.3775	0.0575	272.9584	0.0037	15.7076	0.0637	13.6748	214.7972
50	18.4202	0.0543	290.3359	0.0034	15.7619	0.0634	13.7964	217.4574

For instance, given present value = \$10,000, n = 4, and i = 6%, the future value notation would be the following:

$$(F/P, n, i) \text{ or } (F/\$10000, 4, 6\%)$$

Use the chart to find a factor of 1.2625, which produces FV = \$10,000 × 1.2625 = \$12,625.

Few people, other than academics, use these charts because they are more of an artifact of the way calculations were conducted before the use of spreadsheets. They serve some purpose to illustrate TVM and to make calculations much simpler; however, given that most interest rates are compounded daily, thus producing few whole number rates, they are of little value for real problems.

2.4.5 *Interest compounded other than yearly*

Most real-world problems have interest-compounding periods other than yearly. In fact, most interest is compounded daily. The interest must be in the same time units as the payments. Examples up to now have dealt with the value of money at different points in time with various payment methods using the same units of time. Now we will determine the real interest rates used in most business transactions. However, interest is usually compounded more frequently than once per year. Interest is usually expressed in terms of APR (compounded yearly). When compounding occurs more frequently than annually, the APR does not reflect the true interest accumulated in a year. Thus, an effective interest rate must be used. Depending on the frequency of compounding, the effective interest rate can differ significantly from the APR, or nominal rate.

There are several methods to convert interests to various time units. The simplest is probably the two-step method. To use the two-step method, we present these definitions:

$r \equiv$ APR
$i_{per} \equiv$ periodic interest rate
$i_{eff} \equiv$ annual effective interest rate
$m \equiv$ compounding frequency or the number of interest periods per year
$k \equiv$ number of payment periods per year
$c \equiv$ number of interest periods per payment period

The first step is to relate the annual effective interest rate to the APR:

$$i_{eff} = \left(1 + \frac{r}{m}\right)^m - 1 \tag{2.4}$$

Financial institutions can use different terms to include annual percentage yield and annual effective rate in lieu of i_{eff}. They mean the same thing, which is the rate of interest that is earned when taking into consideration the effect of compounding.

For example, assume that the APR = 6% but that the compounding occurred monthly:

$$i_{eff} = (1 + .06/12)^{12} - 1 = 6.1678\%$$

The second step is to convert the annual effective interest rate to the equivalent interest for the time period needed for the calculation:

$$i_{per} = (1 + i_{eff})^{\frac{1}{k}} - 1 \tag{2.5}$$

Suppose you needed to determine a quarterly interest rate for the 6% APR compounded monthly:

$$i_{per} = (1 + 0.061678)^{\frac{1}{4}} - 1 = 1.5.75\%$$

Some authors collapse this into one step, which produces the following formula:

$$i_{per} = \left(1 + \frac{r}{ck}\right)^{c} - 1 \tag{2.6}$$

As one would expect, the annual effective rate is larger than the APR when compounded more often than annually. Examples 2.5 and 2.6 demonstrate how to use these equations to calculate the appropriate interest rates.

EXAMPLE 2.5

You have three different mutual fund options to invest your IRA.

- Fund A: 8% APR compounded annually,
- Fund B: 7.75% APR compounded quarterly, or
- Fund C: 7% APR compounded daily.

Where would you invest your money?

Using Equation 2.2, we need to determine the largest effective annual interest rate.

Fund	r	m	i_{eff}
A	.08	1	8.00%
B	.0775	4	7.98%
C	.07	365	7.25%

Fund A is the best option.

EXAMPLE 2.6

You want to save for your employees' Christmas bonuses. The bank has quoted you a nominal interest rate of 8% APR compounded daily. You plan to make monthly deposits to the bonus fund. What interest will you use in your calculations?

Using Equation 2.4, where

$r \equiv$ nominal interest rate expressed as an APR = 0.08
$k \equiv$ number of payment periods per year = 12
$c \equiv$ number of interest periods per payment period = 365/12 = 30.42

or

$$i_{per} = \left(1+\frac{r}{ck}\right)^{c} - 1 = \left(1+\frac{.08}{365}\right)^{30.42} - 1 = 0.6688\%$$

Using the two-step method will produce the same results:

$$i_{eff} = \left(1+\frac{r}{m}\right)^{m} - 1 = \left(1+\frac{.08}{365}\right)^{365} - 1 = 0.0833 = 8.33\%$$

$$i_{per} = (1+i_{eff})^{\frac{1}{k}} - 1 = (1+0.0833)^{\frac{1}{12}} - 1 = .6688\%$$

2.5 *Single period cash flow*

The most basic form of cash flow series is a single period cash flow. Example 2.7 illustrates the concept of a single period cash flow. The single period cash flow can occur in the present time or at some point in the future. When drawing a single period CFD, we always construct it from the perspective of the firm, just like with all other CFDs. From the firm's perspective, it can be a cash outflow or a cash inflow. Any time money leaves the firm, known as a cash outflow, we illustrate this with a downward arrow. Conversely, when money arrives at the firm, known as a cash inflow, we illustrate this with an upward arrow. These concepts are shown in Figure 2.5.

2.6 *Common cash flow series*

It is imperative to grasp the concept of drawing a proper single cash flow series before moving on to some of the more complex but common cash flow series. The three common cash flow series other than a simple single period cash flow addressed in this text are uniform, linear gradient, and irregular.

EXAMPLE 2.7

Assume you receive a \$10,000 payment at year 4. The interest rate equals 6%, compounded annually. The appropriate CFD is:

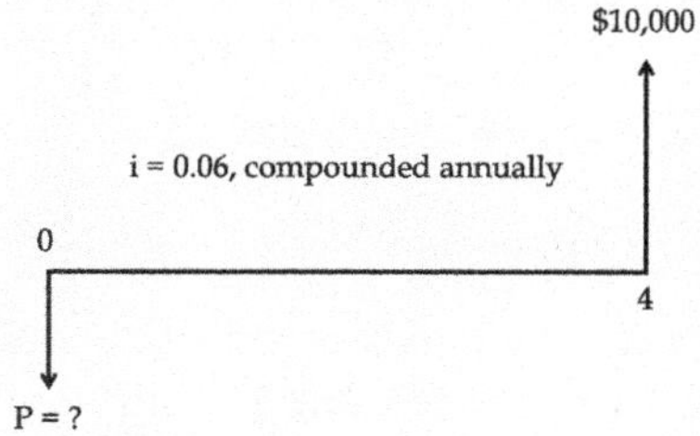

For a single period cash flow series, we simply apply the basic compounding equation: $F = P(1 + i)^n$ or the inverse, which is known as the present worth formula, where we discount a future cash flow back to the present ($t = 0$):

$$P = F(1+i)^{-n}$$

What is the present worth of the CFD with a \$10,000 payment in year 4?

$$P = \$10,000(1+.06)^{-4} = \$7,920.94$$

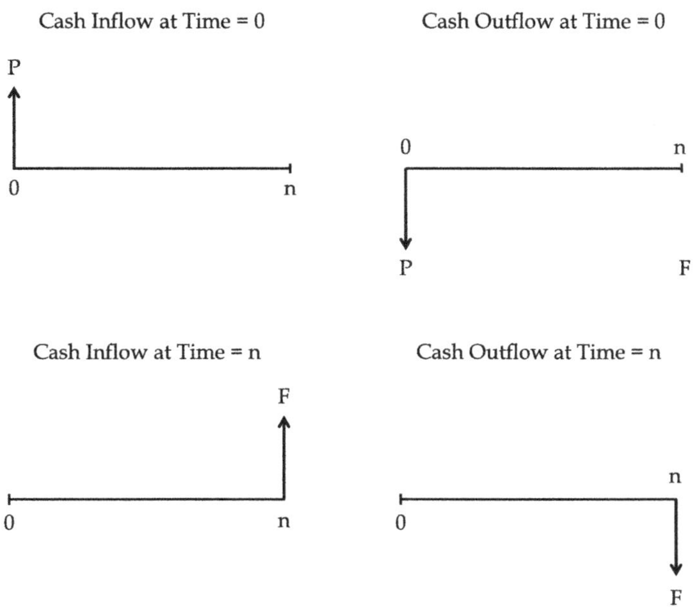

Figure 2.5 Single period CFDs

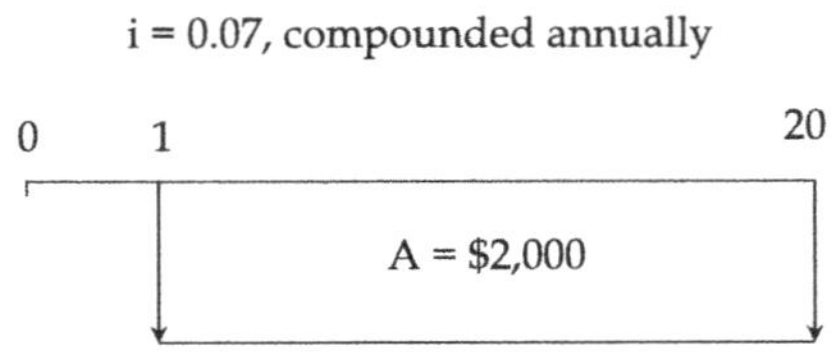

Figure 2.6 CFD for a uniform cash flow series

2.6.1 *Uniform cash flow series*

Uniform cash flow series describes a repeating cash flow amount over multiple time periods. We use the term "annuity" to describe this repeating cash flow scenario. Students often find it helpful to relate "annuity" to "annual." However, it is a bad analogy because payments can be monthly, quarterly, and so on. We use the variable A to indicate the uniform cash flow value. The most commonly misunderstood element of an annuity is that, in a standard annuity, the first payment occurs at the end of the first time period. This is illustrated in Figure 2.6.

Example 2.8 demonstrates the derivation of the compound equation for the future value of an annuity.

This inverse of the equation in Example 2.8 or

$$A = F\left[\frac{i}{(1+i)^n - 1}\right] \tag{2.7}$$

is often referred to as the sinking fund or the constant periodic amount at a constant interest rate that must be deposited to accumulate a future value.

Other annuity equations include the present worth

$$P = A\left[\frac{(1+i)^n - 1}{i(1+i)^n}\right] \tag{2.8}$$

and the uniform payment series, which is also called the capital recovery factor or the series of uniform payments that will recover an initial investment.

$$A = P\left[\frac{i(1+i)^n}{(1+i)^n - 1}\right] \tag{2.9}$$

2.6.2 *Linear gradient cash flow series*

A linear gradient cash flow series describes the situation where cash flows in a series are increased each time period by a constant value (G). As with the uniform series, all derived equations for linear gradient are based on the standard form where the corner of the triangle is located at the end of time period 1 (i.e., increases with time) as shown in Figure 2.7.

EXAMPLE 2.8

Derive the equal payment or uniform series equation, and use it in a simple example.

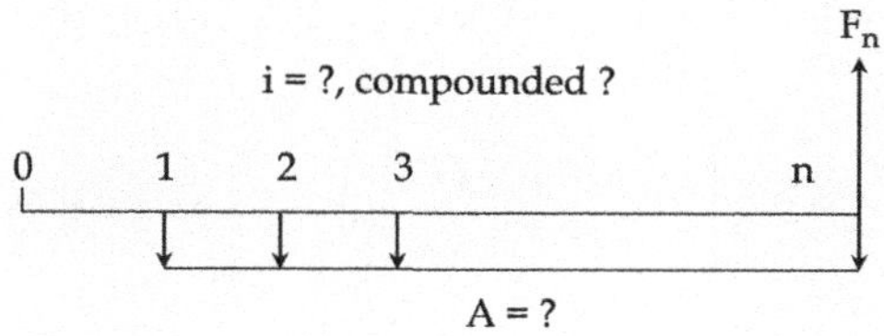

Let

$$F_n = A + A(1+i)^1 + A(1+i)^2 + A(1+i)^3 + ... + A(1+i)^{n-1}$$

or

$$F_n = A[1 + (1+i)^1 + (1+i)^2 + (1+i)^3 + ... + (1+i)^{n-1}]$$

$$(1+i)F_n = A[(1+i)^1 + (1+i)^2 + (1+i)^3 + ... + (1+i)^n]$$

Subtracting these equations

$$F_n(1+i) - F_n = A(1+i)^n - A$$

or

$$F_n = A\left[\frac{(1+i)^n - 1}{i}\right]$$

Suppose we plan to invest $2,000 per year for the next 20 years in an IRA earning 7% annually. How much will be available in 20 years?

$$F = \$2000\left[\frac{(1+.07)^{20} - 1}{.07}\right] = \$81{,}991$$

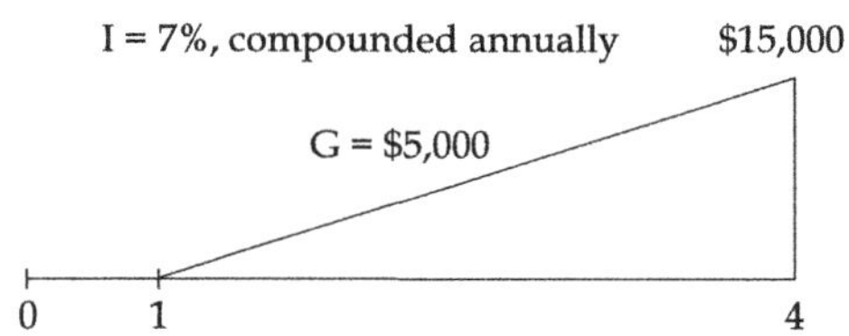

Figure 2.7 Linear gradient present worth

Where the net present worth (NPW) is calculated using

$$P = G\left[\frac{(1+i)^n - in - 1}{i^2(1+i)^n}\right] \tag{2.10}$$

For the example CFD, (G = $5,000, i = 7%, n = 4) the present worth is calculated using:

$$P = \$5{,}000\left[\frac{(1+.07)^4 - (.07)(4) - 1}{.07^2(1+.07)^4}\right] = \$23{,}973.60$$

The linear gradient future worth equation is

$$F = G\left[\frac{(1+i)^n - 1}{i^2} - \frac{n}{i}\right] \tag{2.11}$$

$$F = \$5{,}000\left[\frac{(1+.07)^4 - 1}{.07^2} - \frac{4}{.07}\right] = \$31{,}424.50$$

and the linear gradient equal payment is

$$A = G\left[\frac{(1+i)^n - in - 1}{i[(1+i)^n - 1]}\right] \tag{2.12}$$

Note that Equations 2.10 and 2.11 only work for increasing cash flows. If the cash flows are decreasing, similar to Example 2.9, we must subtract a linear increasing gradient for a uniform series.

2.6.3 *Irregular cash flow series*

The final type of cash flow series we will cover is the irregular cash flow series. Unfortunately, there are no shortcut equations to make this task easier. To solve these problems, you simply solve each individual cash flow separately and then combine the value. This is demonstrated in Example 2.10.

2.7 *Concept of economic equivalence*

Economic equivalence is probably the primary tenet of engineering economics. Table 2.5 summarizes each of the previously presented equations used to conduct analysis of cash flows. Note that the table also includes the Excel equations that can be used in lieu of the equations presented. The equations in Table 2.5 are used to make cash flows equivalent. Interest, time, and payments are used to calculate these equivalent cash flow. Note that the

EXAMPLE 2.9

The estimated savings from an alternative energy project will decrease as the equipment becomes older. Initially you will save $45,000 per year, which decreases by $5,000 per year for the next 3 years. At the end of the fourth year, new technology will be available, and you will replace this system. Using an interest rate of 6% APR compounded annually, what is the maximum amount you should pay for this technology?

Your CFD decomposed to show the gradient would be

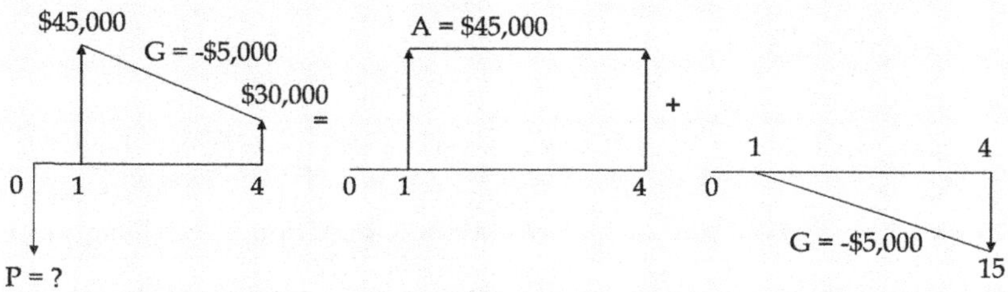

Thus, the present value using the tables in the Appendix at the end of the book would be

$$P = 45{,}000(P/A,6\%,4) - 5{,}000(P/G,6\%,4) =$$
$$45{,}000 * 3.4651 - 5{,}000 * 4.9455 = \$131{,}202$$

PV, FV, and PMT functions in Excel return negative values. Care must be taken to ensure that you have the right signs for your cash flow problems.

Consider two purchasing options for a $30,000 vehicle. The first option allows us to make 60 monthly payments at a 6% APR compounded daily and will produce a car payment of $580.19. The second option is to

EXAMPLE 2.10

Consider the following cash flows. These are typical for irregular cash flows that are key for conducting life cycle analysis.

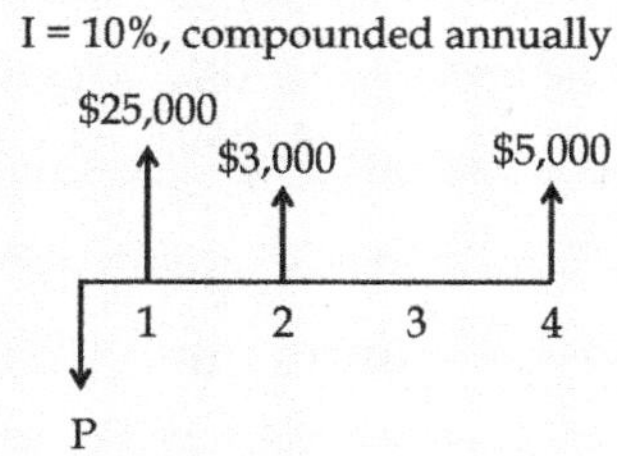

(continued)

We simply solve for the present worth of each cash flow separately and then combine.

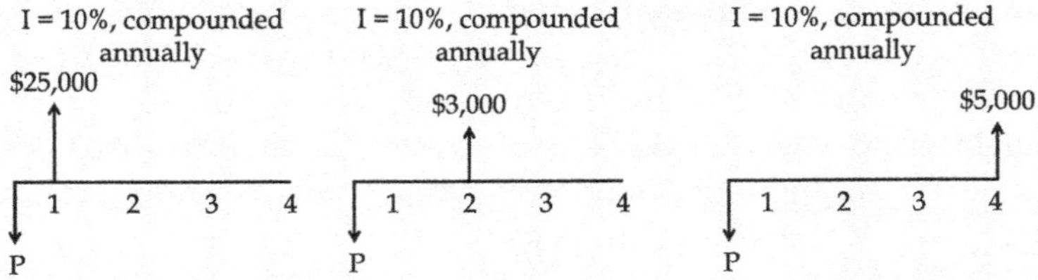

$P_1 = \$25,000\ (P/F, 10\%, 1)$ $P_2 = \$3,000(P/F, 10\%, 2)$ $P_3 = \$5,000\ (P/F, 10\%, 4)$

Thus $P = P_1 + P_2 + P_{3.}$

Table 2.5 Summary of cash flow analysis equation and Excel functions

Cash Flow Type	Formula	Factor Notation	Excel Function	Cash Flow Diagram
Single	$P = \frac{F}{(1+i)^n}$	(P/F, I, n)	=PV(i,n,,F)	
Single	$F = P(1+i)^n$	(F/P, I, n)	=FV(i,n,,P)	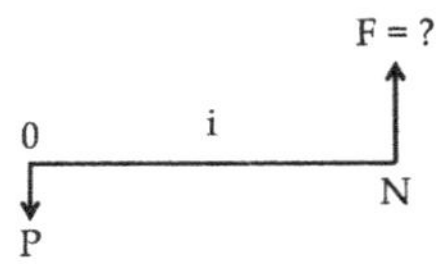
Uniform	$A = P\left[\frac{i(1+i)^n}{(1+i)^n - 1}\right]$	(A/P, I, n)	=PMT(i,n,P)	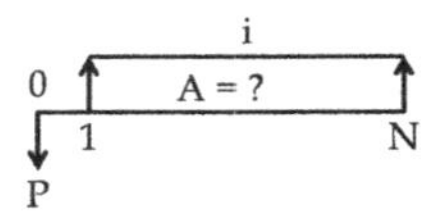
Uniform	$P = A\left[\frac{(1+i)^n - 1}{i(1+i)^n}\right]$	(P/A, I, n)	=PV(i,n,A)	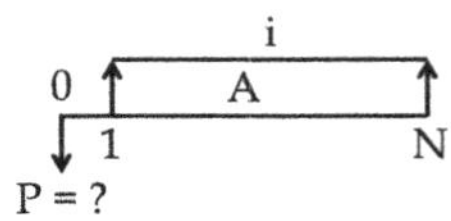
Uniform	$A = F\left[\frac{i}{(1+i)^n - 1}\right]$	(A/F, I, n)	=PMT(i,n,,F)	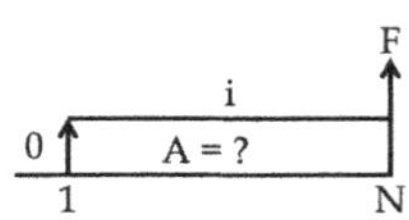
Uniform	$F = A\left[\frac{(1+i)^n - 1}{i}\right]$	(F/A, I, n)	=FV(i,n,A)	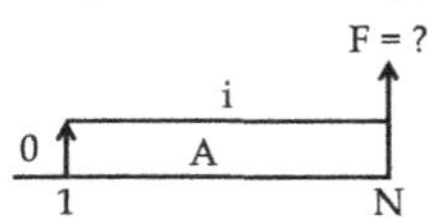
Linear Gradient	$P = G\left[\frac{(1+i)^n - in - 1}{i^2(1+i)^n}\right]$	(P/G, I, n)	NA	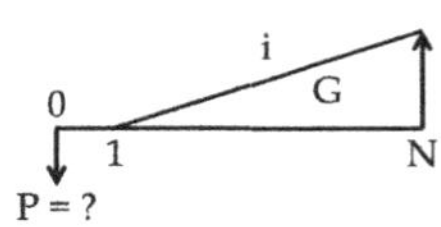

make 36 monthly payments at 4% APR compounded daily for 3 years for a car payment of $885.81. Both of these payment plans are equivalent to a lump sum amount of $30,000.

Economic equivalence is needed to compare projects. This concept allows us to capture the value of income and expenses over the life of a project and then convert them to an equivalent value at a comparable point in time.

Consider the example of two investment opportunities for $10,000 for 3 years. One pays 10% APR compounded annually and the other, less risky, option returns 6% APR compounded annually. The first option will produce $13,310, and the second will yield $11,910. Both of these are equivalent to the $10,000 in today's dollars but yield different returns in 3 years because of the higher interest rates.

2.8 Amortization schedules

"Amortization" comes from the Latin word *admortire*, meaning "to kill." When we amortize a loan, we trace the life of the loan and retire it. To understand the concept of amortization, we present these definitions:

$B_n \equiv$ remaining balance at the end of period n, with $B_0 = P$
$I_n \equiv$ interest payment in dollars in period n, where $I_n = B_{n-1}\, i_{per}$
$P_n \equiv$ principal payment in period n

Note that the annuity uniform payment is composed of both an interest and principal component:

$$A = P_n + I_n \tag{2.13}$$

All popular spreadsheet programs have an amortization function.

Obviously, the tabular method shown in the preceding example could not be used for problems with large numbers of payment periods. For example, this tabular method would be tedious if we wanted to determine the balance after the 217th payment for a 30-year mortgage. The remaining balance method can be used to solve these types of problems.

The tabular method shown in Example 2.11 would be cumbersome to use for large numbers of months or when we want to know the interest and principal for a specific point in time. For example, if I have a 30-year mortgage, what is the payout for the loan 10 years into repayments? Equation 2.14 is often called the remaining balance method. If we know the remaining balance of the load, we can easily determine I_n and P_n. This is visually shown in Figure 2.8.

$$B_n = A\left[\frac{(1+i)^{N-n}-1}{i(1+i)^{N-n}}\right] \tag{2.14}$$

EXAMPLE 2.11

Suppose you need to borrow \$100,000 to purchase a computer system. Your banker quotes you an APR of 10%, compounded daily, for these types of capital equipment loans. Develop an amortization table assuming you will pay the loan off in three equal annual payments. Consider the following CFD:

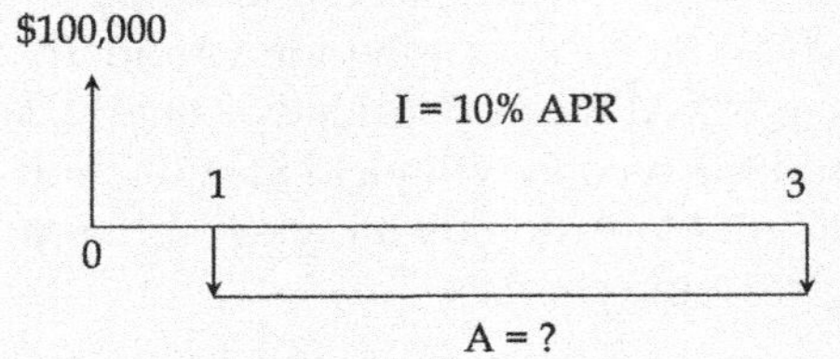

Calculate the effective annual interest rate:

$$i_{eff} = \left(1+\frac{.10}{365}\right)^{365} - 1 = 10.516\%$$

Calculate your annual payments:

$$A = 100,000\left[\frac{.10516(1+.10516)^3}{(1+.10516)^3 - 1}\right] = \$40,576.98 \text{ per year}$$

Develop the amortization table:

End of Period	Annual Payment	Interest Payment $I_n = (B_{n-1})i_{per}$	Reduction In Principal $P_n = A - I_n$	Balance
0	-	-	-	\$100,000.00
1	\$40,576.98	\$10,515.58	\$30,061.40	\$ 69,938.60
2	\$40,576.98	\$ 7,354.45	\$33,222.53	\$ 36,716.07
3	\$40,576.98	\$ 3,860.91	\$36,716.07	\$0

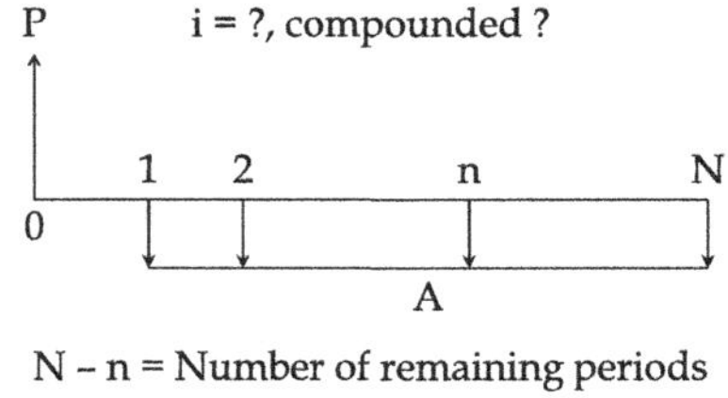

Figure 2.8 Concept of the remaining balance method

Example 2.12 demonstrates the usage of the remaining balance method.

Note that Excel can perform these calculations using the built in functions shown in Table 2.6.

EXAMPLE 2.12

Consider the following CFD, representative of borrowing $77,000 for a new beach house. At age 55, you want the tax write-off but would like to pay off the balance of the loan at your planned retirement age of 62. After looking at a calendar, you see that it is 84 months until retirement. What will be the loan payout?

a. Draw the CFD.

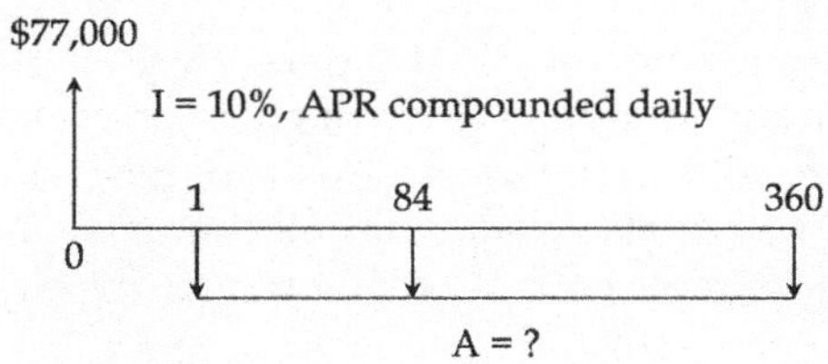

b. Calculate the effective annual interest rate.

$$i_{eff} = \left(1+\frac{.10}{365}\right)^{365} - 1 = 10.516\%$$

c. Calculate the periodic interest rate.

$$i_{per} = (1+.10516)^{1/12} - 1 = 0.8367\% = 0.008367$$

d. Calculate your annual payments.

$$A = 77,000\left[\frac{.008367(1+.008367)^{360}}{(1+.008367)^{360}-1}\right] = \$678.03 \text{ per month}$$

e. Determine the balance after the 84th payment.

$$B_n = A\left[\frac{(1+i)^{N-n}-1}{i(1+i)^{N-n}}\right] = 678.03\left[\frac{(1+.008367)^{276}-1}{.008367(1+.008367)^{276}}\right] = \$72,909.05$$

Note that we have only reduced the principal by a little over $4,000 in 7 years!

Table 2.6 Excel function for remaining loan balance and interest

Calculation	Excel Function
Monthly Payment	PMT
Interest Portion of the Payment	IPMT
Principal Portion of the Payment	PPMT
Cumulative Interest Paid over the Payment	CUMIPMT
Cumulative Principal Paid over the Payment	CUMPRINC

Thus, for our problem, CUMPRINC(rate,nper,pv,start_period,end_period,type) or CUMPRINC(.008367,360,77000,1,84,0) produces $4,090.94, producing a load balance of $72,909.06.

2.9 Summary

Understanding the science of the TVM is the first step in conducting meaningful economic analysis. The principles presented are the building blocks for more sophisticated analysis. An understanding of not only the concept of TVM but also the actual mechanics is necessary. The following two chapters and the rest of this text build on the methodology presented.

QUESTIONS

2-1. Modern engineers must be able to conduct cost analysis/estimation and manage the economic aspects of a complex interdisciplinary project. They must accurately bid projects and then manage to that cost to ensure profitability. They must be able to read financial statements and understand strategic investments in new markets. Discuss why each of these is important to both a company and to your own career management.

2-2. Relate the concept of the TVM to personal computers and automobiles. What is the big difference?

2-3. Define the TVM in simple terms. What are the two types of "power" that contribute to TVM?

PROBLEMS

2-1. You are evaluating two potential IT systems. Each system has an initial cost, hardware upgrade at 24 months, monthly maintenance charge, and a salvage value after 4 years. Cash flow diagrams for each option are shown below. Which option is the best investment?

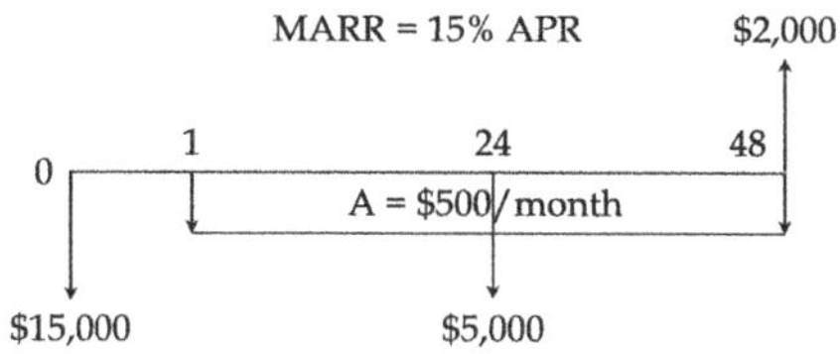

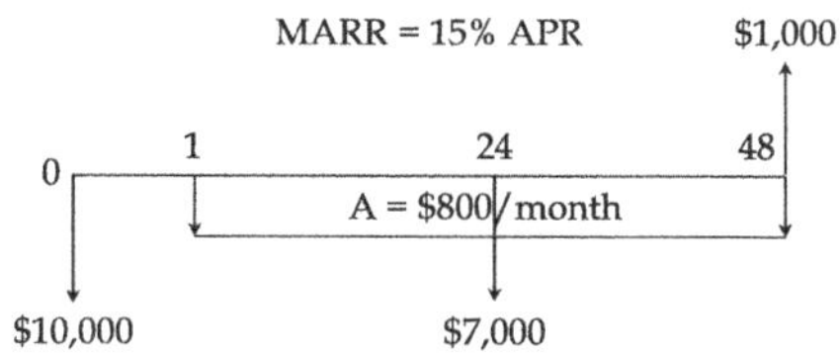

2-2. The cost estimate of a project is $3.5 million. Annual costs for maintaining and operating the facility are forecast as $250,000 per year. After 8 years, it is anticipated that the facility will be sold for $2 million. If the owner requires a 15% return on the investment, what net annual income must be received to recover the capital investment of the project?

2-3. What is the rate of return on an investment of $10,000 if the company expects to receive $2,000 each year for the next 10 years?

2-4. Determine the equal annual end-of-year payment required each year over the life of these loans to repay them fully during the stated term of the loan.

Loan	Principal	Interest	Term
A	$12,000	8%	3
B	$60,000	12%	10
C	$75,000	10%	30
D	$ 4,000	15%	5

2-5. For each of these mixed streams of cash flows, determine the future value at the end of the final year if deposits are made into an account paying annual interest of 12%, assuming no withdrawals are made during the period.

Year	A	B	C
1	$ 900	$30,000	$1,200
2	$1,000	$25,000	$1,200
3	$1,200	$20,000	$1,000
4		$10,000	$ 900
5		$ 5,000	

2-6. A person borrows $20,000 to be repaid in 8 years with 14% annually compounded interest. The loan may be repaid at the end of any earlier year with no prepayment penalty. What amount would be due if the loan is repaid at the end of year 1? End of year 4?

2-7. Your local heavy machinery company is trying to sell a new line of motor graders. As an incentive, they are offering zero down and 9% compounded daily financing. For the motor grader of interest, you are provided this information:

- Loan amount = $200,000.00
- Length of loan = 48 months
- Monthly payment = $4,979.48

a. Calculate the equivalent monthly periodic and annual interest rate.
b. Calculate the interest portion of the 25th payment.
c. Calculate the loan balance after the 29th payment.

2-8. You plan to buy an apartment on Riverside Drive for $180,000. You put $30,000 initial equity into the apartment, leaving you with a $150,000 mortgage. Develop a mortgage repayment schedule using a spreadsheet for the first year showing the payment, amount of interest in that payment, amount of principal being repaid, and the remaining balance of the loan. The mortgage is for 30 years at a fixed rate of 8.5% APR compounded daily.

2-9. Your company is considering the option of building or leasing a new office facility. The two options are (A) constructing an office complex with other space for rent, and (B) constructing an office to house only your operations. Revenue and expenses are summarized for each of the options:

	A	B
Initial Cost	$1,000,000	$500,000
Annual Maintenance	$50,000	$30,000
Income	$10,000/month	0
Salvage Value	$100,000	$30,000
Life of Facility	10 years	10 years
MARR	18% compounded daily	18% compounded daily

Comparing present worth, which of these two options is more desirable?

2-10. The Port Authorities of New York and New Jersey estimate that the annual net revenues for the George Washington Bridge will total $13 million by the end of this year. At the end of 3 years, tolls are expected to increase by 10%. Revenues will then remain constant for the next 6 years (years 4 through 10). The Port Authorities would like to reinvest this revenue in a comprehensive maintenance and repair program. However, it will take 2 years before plans and specifications can be developed and contracts awarded. What is the annual amount the Port Authorities should expect to spend for a 5-year contract beginning at year 2 with equal payments made at the end of each year? The Port Authorities use a MARR of 7% for all public works projects.

2-11. Using annual, semiannual, and quarterly compounding periods, for each of these calculate the future value if $5,000 is initially deposited:

a. At 12% for 10 years.
b. At 16% for 8 years.
c. At 20% for 6 years.

2-12. For each of these cases, calculate the future value of the annuity at the end of the deposit period, assuming that the annuity cash flows occur at the end of each year. The interest is compounded daily for the annuity.

Case	Amount	APR	Term
A	$ 2,500	8%	10
B	$ 500	12%	6
C	$30,000	20%	5
D	$11,500	9%	8
E	$ 6,000	14%	30

2-13. You are estimating the costs to emplace water pump units in some remote villages in a third world country to decrease the distance people have to travel during the dry season to get water. Each village will require a 20 horsepower unit that will last 4 years. The number of operating hours per year depends on the amount of rainfall during the rainy season.

- Option 1: An $1,800 electric pump requiring a power supply will have a $400 salvage value at the end of 4 years. Electric power costs $1.10 per hour of operation, and maintenance cost is $360 per year.
- Option 2: A $550 gasoline-powered pumping unit will have no salvage value after 4 years. Cost of fuel and oil is $0.35 per operating hour, and estimated labor cost to operate the unit is $1.40 per operating hour.

Determine the minimum number of annual operating hours required to justify purchasing the electric pump if your MARR is 10% (effective annual interest).

2-14. You are considering investing $15,000 for a new file server to handle office e-mail, central document storage, and so on. The computer salesperson offers you a plan in which you can finance 100% of the purchase with terms of monthly payments, at 6% compounded daily, for the next 4 years. The computer will be obsolete at that point and have zero salvage value. Draw a CFD, and determine the monthly payments for this loan. What portion of the 15th payment is interest? What is the pay-off after the 25th payment?

2-15. Develop a payment calculator using Excel. Assume that interest will be compounded daily. Your input should include

- Number of payment periods (months),
- Loan amount, and
- Interest rate expressed as an APR.

Develop a loan amortization table similar to the one shown below. The extra payments column requires a significant amount of logic and can be optional.

Inputs (variables)					
Amount	$400,000.00				
Payment Time (months)	360				
Interest (APR) (%)	6.5				
Formulas (intermediate values)					
i_{eff}					
i_{per}					
Payments					
Payments (via Excel)					
Total Principal Paid via Monthly					
Total Principal Paid via Extra					
Total Principal Paid					
Total Interest Paid					
Total Payment					
End of Period	**Monthly Payment**	**Interest Payment**	**Principal Payment**	**Extra Payment**	**Balance**
0					
1					
2					
3					

2-16. A review of my credit card statement showed that I had a balance of $2,103.82 with a minimum monthly payment of $43.00. Below is information provided by the company with regards to interest rates and so on. How long will it take me to pay off the balance if I only make the minimum payment? How much will I have to pay monthly in order to have a zero balance after 18 months?

Finance Charges

	Average Daily Balances	Daily Periodic Rates	Nominal Annual Percentage Rates	Annual Percentage Rates	Periodic Finance Charges	Trans. Fee Finance Charges
Current Billing Period: 28 Days						
Purchases	$ 0.00	0.03627%	13.24% V	13.24%	$ 0.00	none
Cash Advances	$ 0.00	0.06299%	22.99% F	22.99%	$ 0.00	none
Previous Billing Period: 31 Days						
Purchases	$ 0.00	0.03627%	13.24% V	13.24%	$ 0.00	none

The rates that apply to your account are either fixed (F) or they vary (V) as noted above.

2-17. You have been tasked with fielding an interactive video communications systems. Your job is to provide the Army with the least expensive system from the two alternatives below for the next 5 years:

- Intertactical: This option is an interactive communications system designed to rely on current satellite systems. The Army must spend $10,590,843.42 now (t=0) and $1.7 million this year (t=1), increasing by 13% in subsequent years for 4 additional years (t=2 through 5).
- TacLine: This option provides interactive communications that operate through existing phone lines. The Army must spend $4 million now (t=0) and $3 million this year (t=1), increasing by $500,000 each year thereafter for 4 additional years (t=2 through 5).

a. Draw and label the CFD for the ventures (draw in PowerPoint and cut/copy into your Excel spreadsheet solution).
b. Using an effective annual interest rate of 8%, conduct a present worth analysis for the first venture (Intertactical).
c. Using an effective annual interest rate of 9%, conduct a present worth analysis of the second venture (TacLine).
d. Are these two ventures equivalent? Why or why not?

2-18. Contractors are going to install air conditioners in your building. They have narrowed the choices of air-conditioning systems to two final competitors. Assume that there will be a continuing need for the system and that costs and revenues will continue to repeat at the same amounts.

Air Conditioner System A:

Initial Cost: $200,000.00

Annual Maintenance Costs: $12,000

System Life: 10 years

Air Conditioner System B:

Initial Cost: $240,000.00

Annual Maintenance Costs: $8,000

System Life: 12 years

Our company uses an annual effective interest rate of i = 10% for all capital projects.

a. Draw the two CFDs from the head engineer's perspective.
b. Using equivalent annual cost, recommend the more economical option.
c. By how much could the annual equivalent cost of the recommended alternative increase before you change your recommendation to the other alternative?

2-19. On 2 March 2009, the Los Angeles Dodgers pulled a $45 million, 2-year offer to Manny Ramirez. The issue separating the two sides appeared to be how much the contract was worth in present-day dollars. The Dodgers wanted to pay Ramirez $20 million to start (2009–2010 season) and spread out the remaining $25 million with deferred payments of $10 million until 2010–2011, $10 million more to 2011–2012, and $5 million to 2012–2013. Ramirez's agent stated that the contract was for 2 years with "some deferred compensation" for a "net present value" of $43.5 million, a compromise that Ramirez requested. What is the present value of his contract (NBC Sports, n.d.)?

2-20. Your kids have finally left for college, and you now have the opportunity to buy a black 2018 Harley-Davidson Road King Classic motorcycle. You plan to pay $7,000 down and finance the remaining $12,500 for 5 years at 7.5% APR compounded daily. Using Excel, make the following calculations:

Calculation	Excel Function
Monthly Payment	PMT
Interest Portion of the 12th, 24th, 36th, and 60th Payments	IPMT
Principal Portion of the 12th, 24th, 36th, and 60th Payments	PPMT
Cumulative Interest Paid over the 12th, 24th, 36th, and 60th Payments	CUMIPMT
Cumulative Principal Paid over the 12th, 24th, 36th, and 60th Payments	CUMPRINC

2-21. Battery Park in lower Manhattan has a famous artifact called the Netherlands Monument. The purpose of the monument is to commemorate the arrival of Dutch settlers in 1626. As the story goes, when the settlers arrived, they traded beads worth a mere $24 to unsuspecting native peoples in exchange for the entire island of Manhattan. But who really got the better deal? Today, we estimate the worth of Manhattan to be roughly $500 billion. Answer the following questions:

a. If the Native Americans had sold the beads and invested the money at an interest rate of 10% in a simple interest account, how much would they have in today's dollars?
b. Draw the CFD for the Native Americans in the scenario presented in (a).
c. If the Native Americans sold the beads and invested the money at an interest rate of 10% compounded annually, how much would they have in today's dollars?
d. What is the dollar value of the investment?
e. Draw the CFD for the Native Americans in the scenario presented in (c).
f. What is the compounding factor notation and value?
g. Assuming that the stated present-day value of Manhattan is correct, what value of beads should the early settlers have traded?
h. Create an Excel file that calculates the worth of the investment for every year in which interest is earned. Compare the simple interest versus the compound interest.
i. Develop a plot of years versus value of the investment for both simple and compound interest. Be sure to create a full-scale plot.

2-22. Brooklyn wanted to purchase a used car in excellent condition. She decided on a car with low mileage that cost $20,000. After considering several alternatives, she identified a local lending source that will charge her an interest rate of 6% per annum compounded monthly for a 48-month loan.
a. What will be the size of her monthly payments?
b. What will be the remaining balance on her loan immediately after making her 24th payment?
c. If she chooses to pay off the loan at the time of her 36th payment, how much must she pay?
d. What portion of her 12th payment is interest?
e. What portion of her 12th payment is an equity payment?

2-23. Using factor notations and Excel, solve the following problems:
a. Assuming I deposit my inheritance of $36,000 at 8% APR compounded annually, how much will I have upon retirement in 20 years?
b. I just inherited $100,000 from my uncle's estate. I would like to place that amount in the bank and withdraw $10,000 per year for the next 15 years. Assuming an interest rate of 10% compounded annually, is this enough money?
c. I plan to save $10,000 per year for the next 20 years. At that time, I plan to buy a vacation home on the beach for $375,000. At 6% APR compounded annually, will I have enough money?

2-24. Your sister will begin college in 10 years. Your parents are planning to send her to a university that currently costs $46,000 per year. You expect college costs to grow by 10% per year and that your parents can earn 7%, compounded daily, on market investments.

a. How much would your parents have to save per year (assuming they save a fixed dollar amount per year starting in interest period 1) over the next 10 years to have enough to pay for your sister's 4 years of college at that time? Show your answer with CFD.
b. Suppose your parents wait 2 years before they start saving. How much would they have to save per year over the 8 years starting 2 years from now (starting interest period 3) in order to pay for her college? Show your answer with CFD.
c. Your parents are also concerned about inflation, so they want to save enough money to adjust for inflation. Calculate the average inflation rate from the base price of goods being $100 and the final inflated goods price (in the 13th year), which is predicted to be $140. Include this rate in the increase of tuition cost, and recalculate your answers to (a) and (b).

2-25. Whether to lease or buy a car is a decision every company and individual car buyer must make. Consider the following information for a 2018 Ford F-150 SuperCrew STX 4WD pickup truck.

Lease	Buy
Initial Cost: $1,640	Initial Cost + Tax, Title, etc: $39,955 + $1,700
Monthly Lease Payments: $299	Down Payment: 0
	Salvage Value: $20,000
Length of Lease: 36 Months	Life of New Car: 5 Years

a. You plan to use a loan rate of 3% APR compounded daily and a 60-month car payment plan. Which is the most cost-effective system, buying or leasing?
b. Develop a plot using Excel of trade-in or salvage value versus equivalent annual costs (AEW) to determine the breakeven point for buy versus lease for this Ford F-150.

2-26. Your best friend works in Silicon Valley and has convinced Apple that he has the next great idea for an app for the iPhone to monitor the temperature and weather. Apple will have to pay for startup and equipment today. Startup costs will be $400,000. The equipment cost will be $500,000. Annual cost over the life of the project is $375,000 for product development and $325,000 for counterespionage starting at the end of year 1. At the end of year 5, revenue will be $1 million, and this revenue stream will last until the end of year 10. Apple requires a MARR of 15%. The life of the project is 10 years.

Draw the CFD.

Fill in the appropriate variables. Please show all work.

i:
FV:
NPV:
AEW:

Should Apple invest in your friend's project? Why or why not?

2-27. Your uncle has left you with an inheritance of $5,000, which you want to invest. One bank is offering certificates of deposit (CDs) paying 5% APR compounded annually. A competing bank is offering 6% simple interest. You plan to keep the money in the bank for a minimum of 10 years. Which CD will you plan to buy? If it were only a 5-year CD, would that change your decision?

2-28. Mary purchases a house for $450,000. She makes a down payment of $40,000 at the time of purchase, and the balance is financed at 6% compounded monthly, with monthly payments made over a 10-year period.

a. Draw the CFD that illustrates this loan.
b. What is the effective annual interest rate?
c. What is the periodic interest rate?
d. What is the value of the monthly payment?

2-29. On John's 26th birthday, he deposited $7,500 in a retirement account. Each year thereafter, he deposited $100 less than the previous year. Determine how much was in the account immediately after his 40th deposit with an accounted earned annual interest of 8%.

a. Draw the CFD that illustrates this investment.
b. What is the value of the account after the 40th deposit?

2-30. Dell drafted a 12-year contract to provide a local school district with technology and support. They estimate that initial procurement costs will be $15 million (t=0). Every year (starting at t=1) the firm spends $2 million on labor, supplies/computers, and operations. Dell receives $3.5 million annually from the school district (from t=1 to t=12). The computers have a salvage value of $1.5 million at the end of the contract (t=12), and Dell has a MARR of 15%.

a. Draw the CFD that illustrates this project.
b. Using present or future value criterion, determine if this project is worthwhile for Dell at the proposed annual rate the school district will pay ($3.5 million).

2-31. Assuming the following two alternatives are replaced at the end of their useful life, determine the better alternative using AEW analysis at an interest rate of 8%.

	A	B
First Cost	$4,000	$6,000
Annual Cost	$1,000	$ 500
Annual Benefit	$2,000	$2,200
Useful Life (years)	4	10
Salvage Value	$3,000	$1,000

a. Draw the CFD that illustrates Option A.
b. Draw the CFD that illustrates Option B.
c. What is the annual equivalent worth for Option A?
d. What is the annual equivalent worth for Option B?
e. Which option would you choose and why?

2-32. Suppose the following cash flow represents the base case for a potential investment project. Conduct a sensitivity analysis.

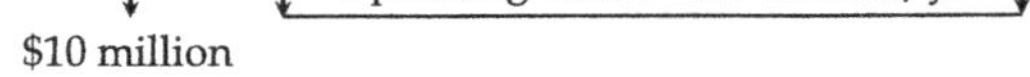

Variable	−30%	−20%	−10%	0	+10%	+20%	+30%
Revenues				20/$8.64 million			
MARR				10%/$8.64 million			
Operating Costs				10/$8.64 million			

a. Complete the above table for the appropriate changes in variable values.

Variable	−30%	−20%	−10%	0	+10%	+20%	+30%
Revenues				20/$8.64 million			
MARR				10%/$8.64 million			
Operating Costs				10/$8.64 million			

b. Find the present value if revenue decreases by 30%, operating costs increase by 20%, or MARR decreases by 20%.

Changing	−30%	−20%	−10%	0	+10%	+20%	+30%
Revenues				$8.64 million			
MARR				$8.64 million			
Operating Costs				$8.64 million			

2-33. Your parents are considering consolidating their debt. They narrowed it down to two options:

- Option 1: AA Credit Card, 16.90% APR, compounding daily
- Option 2: Fisher Bank Loan, 17.00% APR, compounding monthly

Your parents are confused by the compounding and want to know which option really has the lower interest rate. Calculate the annual effective rates of each to compare "apples to apples" and identify the lowest rate for your parents.

2-34. Your parents currently owe $23,790 to four credit card companies and want to consolidate all their debt. They want to know how much their monthly payments will be if they defer payments for the first 9 months, making their first payment at t=10, and make 24 monthly payments to pay off all their debt. Use the lowest rate option from Problem 2-33.

a. Calculate the correct effective rate for this scenario, and draw your CFD.
b. What is the amount of the equal monthly payments?

2-35. You are going to buy a new home and want to conduct an analysis of your likely mortgage. Your 30-year mortgage is for $245,000 at 3% APR, compounded monthly.

a. How much is your monthly mortgage payment?
b. Create an amortization table for the first 2 months to see how each penny of your payment is applied.

c. If you keep the house for 28 years, how is your 336th payment applied:
   i) What new amount goes toward principal?
   ii) What new amount goes toward interest?
   iii) What is the remaining balance?

2-36. Congratulations! You just graduated from college at 22 years old (t=0). You plan to retire at age 65 (t=43) and want to have an annual retirement paycheck of $100,000 with a $6,500 fixed annual increase when you retire. You want to know what lump sum you need to invest now (t=0) in order to meet this goal if you earn 9% compounded monthly on your investments, and you plan to deplete your account by age 85 (t=63).

a. Calculate the correct effective rate for this scenario, and draw your CFD.
b. Solve for the present value.

2-37. You do not have enough money in your savings account to meet your retirement goals right now. You decide that you will invest into your thrift savings plan (TSP) account on a monthly basis for your entire 20-year military career. The TSP account earns 9% compounded monthly.

a. Calculate the correct effective rate for this scenario, and draw your CFD.
b. Determine what monthly investments into your TSP will allow you to reach your retirement goals.

2-38. Apex Corporation requires a chemical finishing process for a product under contract for a period of 6 years. Three options are available. Neither Option 1 nor Option 2 can be repeated after its process life. However, Option 3 will always be available from H&H Chemical Corporation at the same cost during the period that the contract is operative. Here are the options:

- Option 1: Process device A, which costs $100,000, has annual operating and labor costs of $60,000 and a useful service life of 4 years with an estimated salvage value of $10,000.
- Option 2: Process device B, which costs $150,000, has annual operating and labor costs of $50,000 and a useful service life of 6 years with an estimated salvage value of $30,000.
- Option 3: Subcontract out the process at a cost of $100,000 per year.

According to the present-worth criterion, which option would you recommend at i=12% based upon?

a. What is the present worth of Option 1?
b. What is the present worth of Option 2?
c. What is the present worth of Option 3?
d. Which option would you choose? Please explain.

2-39. Jacquline has $2,000 to invest. Usually she would deposit the money in her savings account, which earns 6% interest compounded monthly. However, she is considering three alternative investment opportunities:

- Option 1: Purchase a bond for $2,000. The bond has a face value of $2,000 and pays $100 every 6 months for 3 years, after which time the bond matures.
- Option 2: Buy and hold a stock that grows 11% per year for 3 years.
- Option 3: Make a personal loan of $2,000 to a friend and receive $150 per year for 3 years.

a. Determine the equivalent annual cash flows for each option, and select Jacqueline's best option.
b. What is the equivalent annual cash flows for Option 1?
c. What is the equivalent annual cash flows for Option 2?
d. What is the equivalent annual cash flows for Option 3?
e. What is the best option?

2-40. State University is evaluating several major construction projects. One project is the construction of a new visitor center and purchase of buses for tours to attract visitors and potential students. The Alumni Association has raised some funding, and State U is considering applying for a loan in order to cover the remaining cost of the project. They plan to purchase new buses to increase their tour capacity. Two bus-manufacturing companies have submitted long-term contracts (which is required indefinitely) to sell buses to the visitor center. Each company has agreed to an initial selling price as well as an annual cost for maintenance support. Company A's bus has a service life of 4 years and runs on gasoline. Company B's bus has a service life of 6 years and runs on natural gas. Assume that all costs remain the same over time, and the contract will be renewed indefinitely. The MARR is 6%, compounded annually.

	Company A Bus	Company B Bus
Bus Service Life	4 years	6 years
Initial Purchase Price per Bus	$499,000	$950,000
Fuel Cost per Tour per Bus	$53	$31

a. If the visitor center expects to conduct 335 tours per year, identify the better option for the contract. Show all work.
b. Given the initial purchase cost and annual fuel cost for each company's buses, how many bus tours would the visitor center need to sell each year in order for State U to be indifferent between Company A's buses and Company B's buses? Show all calculations.

2-41. You have just decided to sign for a $200,000 mortgage from the local bank to buy a house near your job. The mortgage has an interest rate of 6.5% APR, compounded quarterly, and you will make monthly payments for 30 years beginning 6 months from now.
a. Draw the CFD for this mortgage, and calculate the appropriate effective interest rate.
b. How much are your monthly payments?
c. Using the tabular method, determine the interest, principal, and remaining balances for the first 3 months.
d. If you decide to pay off the remaining balance of the mortgage after making the payment at the end of the 10th year (following the 120th payment), how much will you owe the mortgage company?

2-42. You would like to purchase a Chevrolet Corvette before you settle down. You are trying to decide between a 2010 model with 50,000 miles or a 2015 model with 10,000 miles. Regardless of your choice, you plan to keep the car until it has 150,000 miles. Use the data below to compare each model. Use a 6% annual interest rate.

	2010 Corvette	2015 Corvette
Initial Cost	$30,000	$45,000
Lifespan (years)	5	8
Annual Fuel/Maintenance	$ 2,700	$ 1,750
Resale Value	$10,000	$18,000

a. Draw the CFD that illustrates the 2010 model.
b. Draw the CFD that illustrates the 2015 model.
c. Which option would you choose and why? Show all calculations below, and provide a detailed explanation for your decision.

2-43. You are going to buy a new car following your recent promotion and want to trade in your old car too. Assume the following:
- The purchase price of the new vehicle is $50,000.
- You have saved $10,000 for a down payment.
- Your old vehicle has a trade-in value of $5,000.

- You can secure a 5-year loan with an interest rate of 4.8% APR, compounded monthly, for the remainder of the purchase price.
- Your first payment is deferred until the sixth month.

a. Draw the CFD for this scenario.
b. What is the effective periodic interest rate?
c. Given the scenario above, how much is your monthly car payment?

2-44. Develop a loan calculator in Excel that is flexible enough to calculate monthly payments on different interest rates for your graduation loan of $45,000 for a new car. Then plot monthly payment rates with interest rates of 0.25%, 1%, 5%, and 18% APR, compounded daily, for a 5-year loan. Write a paragraph explaining whether you should take the loan, why or why not, and, if yes, the amount of your monthly payment. Use your loan calculator, and explain how much disposable income you will have as an entry-level engineer making $65,000. Should you keep the clunker you inherited from your parents instead?

2-45. The city of Chicago is trying to determine whether to break a major infrastructure renewal project (roads, drainage, utilities, steam lines, and IT) for a major road in the downtown business area into one of two contracts. The one hard-to-quantify piece of information is the long-term cost of lost business. Conduct an NPV sensitivity analysis varying cost per lost business. Present this study in a plot that can be used by key leaders to determine if the additional expense for two contracts is justified.

	Two Construction Contracts	One Construction Contract
Interest Rate	0.05	0.05
Award Fee (millions)	$16	$5
Length of Contract (months)	14	30
Cost per Month (millions)	$2.80	$1.50
Business Lost per Month	1.5	2
Cost per Lost Business	Unknown	Unknown

2-46. You are considering two investment options to save for a house you wish to purchase in 20 years upon retirement:
- Option A: Invest $20,000 immediately (one-time investment).
- Option B: Save $2,000 per year starting in year 1 and ending in year 20.

For both options, you have talked to your financial advisor and are confident you can achieve a 6.5% average annual return on all investments you make. What will be the value of each investment in 20 years, and which option would you select? (Draw complete CFDs for both options, and show all work.)

2-47. Using factor notations, solve the following problems:

a. Assuming I deposit my signing bonus of $36,000 at 9% APR, compounded annually, how much will I have upon retirement in 20 years?
b. I just inherited $250,000 from my uncle's estate. I would like to place that amount in the bank and withdraw $35,000 per year for the next 15 years. Assuming an interest rate of 12% compounded annually, is this enough money?
c. I plan to save $9,000 per year for the next 20 years. At that time, I plan to buy a vacation home on the beach for $400,000. At 7% APR, compounded annually, will I have enough money?
d. I am currently 20 years old and plan to retire at 65. What annual contributions must I deposit into an account to have $1,600,000 in my retirement account at age 65, if I earn a return of 6% compounded annually?
e. I just won the lottery and decide to take annual payments of $8,752,000 each year over 30 years. If I can earn 4% APR, compounded annually, did I make a wise choice as opposed to taking the lump sum payment of $187,541,000?

2-48. You recently bought a 55-inch, curved, Ultra HD TV at Better Buy because the salesperson offered you a 0% interest rate with no payments for 12 months. You end up financing $2,000 at 0%. After you get home, you read the fine print, and it says that full compound interest will be charged if the balance is not paid in full by the end of month 12. You also see that the loan APR is 24% with interest compounded monthly (if not paid in full).

a. What is the effective periodic interest rate, i_{per}, you should use for monthly interest (assuming you do not pay off the loan in 12 months or less)?
b. Calculate the effective annual percentage rate, i_a, on this 24% APR, compounded monthly.
c. If you do not make any payments for a year, how much will be required to pay off the $2,000 loan after 12 months have passed? (Do not include interest beyond 12 months.)

2-49. One of the biggest baseball card transactions was made in 1991 when Wayne Gretzky and Bruce McNall, the owner of the Los Angeles Kings, bought a T206 Honus Wagner baseball card for $451,000.

According to Daniel Roberts (2016), another T206 Honus Wagner card sold for $3.12 million in 2016. The American Tobacco Co. originally produced and sold 50 of these cards from 1909 to 1911. The Wagner has reached such mythical proportions that it is often simply called "The Card." The last rumored transaction for one of these cards before 1991 was at a collectibles market in 1933, where it sold for $50. Consider,

a. If the collector who sold a T206 Honus Wagner baseball card in 1933 invested the money at an interest rate of 10% in a simple interest account, how much would the collector have in today's dollars? (Draw a complete CFD, and show all work.)
b. If the collector sold the same Honus Wagner card in 1933 and invested the money at an interest of 10%, compounded annually, how much would the collector have in today's dollars? What is the dollar value of the investment? (Draw a complete CFD, and show all work.)
c. What is the compounding factor notation and value for (b) above?
d. Assuming that Wayne Gretzky and Bruce McNall were correct about the present-day value of the Honus Wagner card in 1991, what should the collector have sold it for in 1933?
e. If the collector invested the $50 in 1933, what compound annual rate of return would he need to get so that his investment was worth $451,000 in 1991?
f. Create an Excel plot that calculates the worth of the investment for every year in which interest is earned. Compare the simple interest versus compound interest for each year of the investments made in (a) and (c) above. Develop a plot of years versus dollar value of the investment for both simple and compound interest. Be sure to create a full-scale plot that maximizes the values on the axis.

2-50. Your business is planning to renovate a research laboratory. The plan currently under review requires an investment estimated at $1.5 million. The planning horizon is 10 years, and the estimated salvage value at the end of this period is $600,000. Modernization should result in annual savings of $550,000 from operating and maintenance expenses. A MARR of 12% is used by the laboratory.

a. Determine the NPW for the baseline case defined above.
b. Calculate the NPWs for the range of –15% to +15% for the investment costs. All other variables remain at their base values.
c. Calculate the NPWs for a range of –30% to +30% for the planning horizon. All other variables remain at their base values.
d. Calculate the NPWs for a range of –12.5% to +12.5% in annual savings. All other variables remain at their base values.

Bibliography

Farr, John V., 2011, *Systems Life Cycle Costing: Economic Analysis, Estimation, and Management*, CRC Press, Boca Raton, FL.

Lang, H. J., and Merino, D. N., 1993, *The Selection Process for Capital Projects*, John Wiley and Sons, New York, NY.

Newnan, D. G., Eschenbach, T. G., and Lavelle, J. P., 2008, *Engineering Economic Analysis*, 9th Edition, Oxford University Press, New York, NY.

Park, C. S., 2012, *Fundamentals of Engineering Economics*, 3rd Edition, Pearson/Prentice Hall, Upper Saddle River, NJ.

Park, W. R., and Jackson, D. E., 1984, *Cost Engineering Analysis: A Guide to Economic Evaluation of Engineering Projects*, 2nd Edition, John Wiley and Sons, New York, NY.

Sage, A. P., 1983, *Economic Systems Analysis: Microeconomics for Systems Engineering, Engineering Management, and Project Selection*, North-Holland, New York, NY.

References

NBC Sports, n.d., accessed March 2009.

O'Connor, John J., and Robertson, Edmund F., September 2001, "MacTutor History of Mathematics Archive, History Topic the Number e," accessed 11 April 2018 at http://www-history.mcs.st-and.ac.uk/PrintHT/e.html.

Roberts, Daniel, October 2, 2016, "Honus Wagner Baseball Card Sells for New Record $3.12 Million," *Yahoo! Finance*, accessed 1 August 2018 at https://finance.yahoo.com/news/honus-wagner-baseball-card-sells-for-new-record-3-2-million-050323471.html.

chapter three

Advanced economic analysis of alternatives

3.1 Introduction

This chapter is broken into three distinct elements. The first section summarizes the "three worths." When making capital decisions, one can use net present value (NPV)[1], net future value, and/or annual equivalent worth (AEW) to make choices among alternatives. These three analysis results can then be applied to detailed cash flows over time using after-tax analysis. In the second section, the importance of rate of return, breakeven time, and meaningful sensitivity analysis to conducting sound analysis and data interpretation is discussed. The third section involves using the effects of taxes, inflation and deflation, and depreciation of capital assets. This is often referred to as after-tax analysis. After-tax analysis is vital for the selection and optimization of projects and program portfolios. After-tax analysis consists mainly of developing income and cash flow statements.

In order to remain competitive in the marketplace, firms must continually select new projects for investment. Often, due to high initial costs, the decision can make or break a company. As a result, considerable effort has gone into developing methods to select the most appropriate and profitable project for a firm. None of the methods presented is a panacea, and all have weaknesses. However, each of the following methods provides a helpful decision support tool for decision makers.

3.2 Net present value

The NPV or net present worth (NPW) is the current worth of the cash flows over the life cycle at time 0. The cash flows are discounted over the course of the life cycle at the minimum attractive or acceptable rate of return (MARR) or some other interest rate that reflects the cost of capital. If there is a large disparity between the normal discount rate (the current value of money) and the MARR, then a compromise interest rate must be

[1] The terms "net present worth," "present value," and "net present value" are used interchangeably to describe the time value of money at year 0.

EXAMPLE 3.1

Given the following cash flows at a MARR of 12% compounded annually, determine the NPV.

Year	Project A
0	–$76,000
1	$35,560
2	$37,360
3	$31,850
4	$34,400

$$NPV_A = 34{,}400(1+.12)^{-4} + 31{,}850(1+.12)^{-3} + 37{,}360(1+.12)^{-2} + 35{,}560(1+.12)^{-1} + (-76{,}000)$$

$$NPV_A = \$30{,}065.19 > 0, \text{ Thus Accept Project}$$

assumed. After solving for the NPV of a single project, the following decision rules are applied:

$$NPV > 0, \text{ Accept}$$

$$NPV = 0, \text{ Indifferent}$$

$$NPV < 0, \text{ Reject}$$

When comparing projects based solely on NPV, a firm should choose the project with the largest NPV as long as it is positive. Examples 3.1 and 3.2 demonstrate the concept of NPV.

EXAMPLE 3.2

Given the following cash flows for two projects, determine which provides the best return at a MARR of 12% compounded annually.

Year	Project A	Project B
0	–$76,000	–$95,000
1	$35,560	$43,000
2	$37,360	$56,000
3	$31,850	$40,000
4	$34,400	$20,000

$NPV_B = 20{,}000(1 + .12)^{-4} + 40{,}000(1 + .12)^{-3} + 56{,}000(1 + .12)^{-2} + 43{,}000(1 + .12)^{-1} + (-95{,}000) = \$29{,}217.29$ which is > 0. We then compare with Project A, which has an NPV = \$30,065.19 or

$$\$29{,}217.29 < \$30{,}065.19$$

Thus, we would select Project A since $NPV_B < NPV_A$.

3.2.1 *NPV in Excel*

The =NPV function in Excel can be used to calculate the NPV of a project or a group of projects. Note: Excel discounts every value entered in the function and assumes the first value occurs at the end of time period 1. Thus, the initial investment should not be included in the function as shown below. Excel also requires the user to enter a rate (discount rate). The MARR should be entered for this value. The "Excel Traditional View" and "Excel Equation View (Ctrl + ')" are shown for a NPV problem in Example 3.3.

EXAMPLE 3.3

Conduct an NPV analysis using Excel for the following cash flows.

Year	Project A	Project B
0	-\$76,000	-\$95,000
1	\$35,560	\$43,000
2	\$37,360	\$56,000
3	\$31,850	\$40,000
4	\$34,400	\$20,000
NPV	\$30,065.19	\$29,217.29

Solution

Shown below is the Excel solution for a MARR of 12% used to determine the NPV of these cash flows.

Year	Project A	Project B
0	-76,000	-95,000
1	35,560	43,000
2	37,360	56,000
3	31,850	40,000
4	34,400	20,000
NPV	=NPV(0.12,D20:D23)+D19	=NPV(0.12,E20:E23)+E19

(continued)

Note how year 0 must be added separately to the Excel NPV command. In the above example, the amounts in cells D19 and E19 represent the cash flows at year 0.

3.3 Annual equivalent worth

3.3.1 Introduction

The AEW is a uniform flow of benefits less costs at equally spaced time periods over the life cycle of the project. It is a measure of the net return on a project on an annualized or amortized basis. It is an easy-to-understand method to report cash flows or dollars/time period. More importantly, it allows for a direct comparison of unequal time periods. AEW can also be calculated simply by converting the project NPV to an annuity or uniform cash flow. In fact, the most direct method for solving these problems by hand is to first find the NPV of a project and then calculate the annuity. When comparing multiple projects, the project with the larger NPV will always have the larger AEW as well.

Referring to Example 3.1, the NPV = \$30,065.19. We can now solve for the AEW or A given P

$$A = P\left[\frac{i(1+i)^n}{(1+i)^n - 1}\right] \tag{3.1}$$

$$A = \$30,065.19\left[\frac{.12(1+.12)^4}{(1+i)^4 - 1}\right] = \$9,898.49$$

or using the PMT function in Excel (see Table 2.5) = –PMT(0.12,4,30065.79) = \$9,898.69.

Example 3.4 demonstrates how to use the NPV function in Excel for multiple cash flows and then convert the NPV to an AEW. Note that the cash flow in year 0 should not be included in the NPV calculation.

3.4 Capital budgeting: selection of multiple projects

One of the basic tenets of investment is to maximize return while minimizing the risk associated with the investment choice. In the financial markets, this tenet drives the countless financial instruments that have been constructed and continue to be constructed as you read this text. An additional constraint that we must add to this discussion is the idea that every investor has a limit to the amount that he or she can invest. Even if we consider that certain investors have immense resources and

EXAMPLE 3.4

Use the PMT function with an interest rate of 12% to determine AEW for the cash flows shown below in Projects A and B.

Excel Traditional View

Year	Project A	Project B
0	-$76,000	-$95,000
1	$35,560	$43,000
2	$37,360	$56,000
3	$31,850	$40,000
4	$34,400	$20,000
NPV	$30,065	$29,217
AEW	$9,898	$9,619

Excel Equation View

Year	Project A	Project B
0	-76,000	-95,000
1	35,560	43,000
2	37,360	56,000
3	31,850	40,000
4	34,400	20,000
NPV	=NPV(0.12,D20:D23)+D19	=NPV(0.12,E20:E23)+E19
AEW	=PMT(0.12,4,-D24)	=PMT(0.12,4,-E24)

Note that Project A has a greater NPV and also a greater AEW as we would expect because of the same number of years (i.e., n=4).

the capability to borrow even more, there exists a limit to the amount that can be invested. Previously, we have discussed methods used for selecting a single project from a group of candidates. We used methods such as internal rate of return (IRR) and NPW to inform our selection decision.

Now we will expand this discussion by including the scenario facing many firms: choosing multiple projects to pursue simultaneously. In reality, large organizations maintain a portfolio of projects that occur simultaneously. The selection of which projects to include in this portfolio is based on the concept of portfolio theory. While portfolio theory describes the process for solving for the optimal mix of assets that simultaneously minimizes risk while maximizing return, our analysis will differ slightly as the firm is faced with deciding which project to invest in. Unlike in the application of portfolio theory for finance, we assume that a firm does not have the option to partially invest in a project. The choice to invest in a project is viewed as binary (yes or no). A firm would never choose to build 40% of a bridge. The firm would build 100% or nothing because only a complete project will earn a return. This varies from financial portfolio theory in that it assumes that an investor can purchase any portion of an asset.

3.5 Breakeven analysis

Breakeven analysis has many applications in engineering economy and is often used to describe the internal rate of return—the price of one or more variable when one decision is more economically viable than another, or simply the amount of goods that need to be produced in order to make a profit. Conducting meaningful sensitivity of the effects of the input variables on the output used to make decision is important to conducting meaningful and defensible analysis. Example 3.5 illustrates the concept of breakeven analysis.

EXAMPLE 3.5

Your friend is considering entering the pizza business. Her MARR is 10%. She is considering two different pizza ovens:

	Model 1	Model 2
Cost of Oven	$95,000	$35,000
Life of Oven	5 years	4 years
Salvage Value at End of Life	$10,000	$5,000
Annual Profit	$2.00/pizza	$1.00/pizza

Two assumptions: (1) There is a continuing requirement for the service of the selected alternative (she will continue to sell pizzas for many years to come), and (2) each alternative will be replaced by an identical asset that has the same costs and performance.

a. If your friend projects that she will sell 12,000 pizzas annually, which model should she buy?

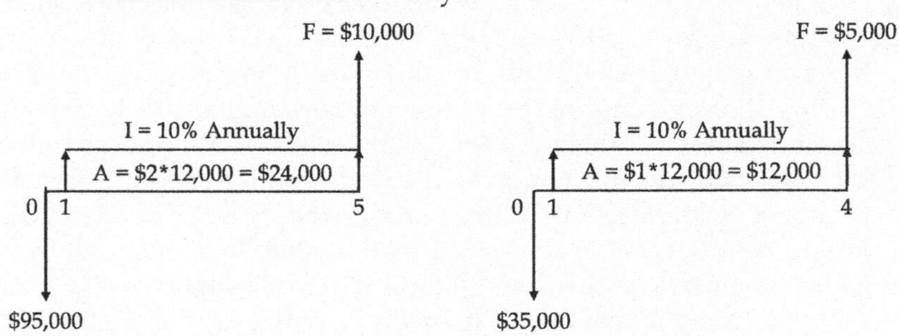

AE_1 = −95K(A/P,.10,5) + A_1 + 10K(A/F,.10,5) = −95K(.2638) + 24k + 10K(.1638) = $577 per Year or = PMT(0.1,5,95000) + 24000 − PMT(0.1,5,,10000) = $577.21

AE_2 = −35K(A/P,.10,4) + A_2 + 5K(A/F,.10,4) = −35K(.3155) + 12k + 5K(.2155) = $2,035 per Year or = PMT(0.1,4,35000) + 12000 − PMT(0.1,4,,5000) = $2,035.88

Note that Model 2 has a higher annual equivalent cost, $2,036 > $577, therefore, select Model 2.

b. At what level of pizza sales would your friend be indifferent between the two models?

Let $AE_1 = AE_2$ and let x = Number of Pizzas Sold per Year
PMT(0.1,5,95000) + 2*X − PMT(0.1,5,,10000) = PMT(0.1,4,35000) + x − PMT(0.1,4,,5000) = − $25,060.76 + 2X + $1,637.97 = − $11,041.48 + x + $2,035.88

x = 13,458 pizzas

3.6 *After-tax cash flow analysis*

To conduct meaningful analysis, the effects of taxes, inflation and deflation, and depreciation of capital assets must be addressed. This is often referred to as after-tax analysis. After-tax analysis is vital for the selection and optimization of projects and program portfolios. After-tax analysis consists mainly of developing income and cash flow statements. A fundamental understanding of their effects is also important when accountants and business managers are involved in the planning and execution of programs.

When first considering conducting analysis of a project, we must investigate:

Cash Outflows

- Procurement costs,
- Operations and support,
- Disposal costs,
- Interest and repayment of borrowed funds, and
- Income tax.

Cash Inflows

- Borrowed funds,
- Revenue of cost avoidance/savings, and
- Salvage value.

3.6.1 Depreciation

Depreciation is simply the systematic allocation of cost of a capital expenditure item over its useful life. In reality, depreciation is nothing more than an accounting charge that reduces the overall value of an asset due to its depletion for income tax purposes. There are many methods to calculate depreciation. To demonstrate depreciation, we will present the straight-line and Modified Accelerated Cost Recovery System (MACRS)[2] methods. The MACRS methods is the most common for tax depreciation because, as the name implies, it allows for accelerated depreciation.

Straight-line depreciation is simply

$$D_n = \frac{I - S}{N} \tag{3.2}$$

where

$D_n \equiv$ depreciation allowance in year n
$I \equiv$ cost base
$S \equiv$ salvage value
$N \equiv$ estimated useful life of the asset

Note that the cost base includes both the actual cost and the cost to put the asset into operation. The book value of the asset at year n can be expressed as

$$B_n = I - \sum_{n=1}^{N} D_n \tag{3.3}$$

The MACRS is used both for tax purposes and for internal accounting. It allows for the recovery of more costs early in the life of the investment, as shown in Table 3.1. Table 3.2 contains the MACRS classes and amount of time they can be depreciated. Depending on the type of asset, the IRS allows for different cost recovery periods. A yearly depreciation rate is then multiplied by the cost base to determine the annual depreciation amount. Example 3.6 shows how the

[2] See IRS Publication 946 for the details of MACRS depreciation at https://www.irs.gov/publications/p946.

Table 3.1 MACRS depreciation table[3]

Recovery Year	3-Year	5-Year	7-Year	10-Year	15-Year	20-Year
1	33.33	20	14.29	10	5	3.75
2	44.45	32	24.49	18	9.5	7.219
3	14.81	19.2	17.49	14.4	8.55	6.677
4	7.41	11.52	12.49	11.52	7.7	6.177
5		11.52	8.93	9.22	6.93	5.713
6		5.76	8.92	7.37	6.23	5.285
7			8.93	6.55	5.90	4.888
8			4.46	6.55	5.9	4.522
9				6.56	5.91	4.462
10				6.55	5.9	4.461
11				3.28	5.91	4.462
12					5.9	4.461
13					5.91	4.462
14					5.9	4.461
15					5.91	4.462
16					2.95	4.461
17						4.462
18						4.461
19						4.462
20						4.461
21						2.231

MACRS allows for earlier depreciation of assets when compared to the straight-line method. With the exception of this example, the details of MACRS deprecations are not presented because those rates are subject to change by the IRS.

Internally developed software is amortized on a straight-line basis over 5 years (or shorter if you can show it is appropriate). Software acquired as part of a business acquisition can be amortized over 15 years. Purchased software is generally amortizable over 3 years. Depreciation can play a role in choosing whether to develop in-house or procure commercially because software can be amortized over different time horizons depending on whether it is an off-the-shelf item or is internally developed.

[3] See IRS Publications 534 and 946 at https://www.irs.ustreas.gov.

Table 3.2 Property descriptions and MACRS classes

Property Class	Personal Property (all property except real estate)
3-year property	Special handling devices for food and beverage manufacture Special tools for the manufacture of finished plastic products, fabricated metal products, and motor vehicles Property with asset depreciation range (ADR) class life of 4 years or less
5-year property	Information systems; computers, peripherals Aircraft and parts (of non-air-transport companies) Computers Petroleum drilling equipment Property with ADR class life of more than 4 years and less than 10 years Certain geothermal, solar, and wind energy properties
7-year property	All other property not assigned to another class Office furniture, fixtures, and equipment Property with ADR class life of more than 10 years and less than 16 years
10-year property	Assets used in petroleum refining and certain food products Vessels and water transportation equipment Property with ADR class life of 16 years or more and less than 20 years
15-year property	Telephone distribution plants Municipal sewage treatment plants Property with ADR class life of 20 years or more and less than 25 years
20-year property	Municipal sewers Property with ADR class life of 25 years or more
Property Class	Real Property (real estate)
27.5-year property	Residential rental property (does not include hotels and motels)
39-year property	Non-residential real property

EXAMPLE 3.6

Your small consulting company is evaluating a circuit board testing machine. The device costs $35,000, and the maker estimates that it will have a salvage value of $6,000 after 5 years of use. Determine the annual depreciation using both the straight-line and the MACRS methods.

Solution

a. Straight-line

$$D_n = \frac{I-S}{n} = \frac{35{,}000-6{,}000}{5} = \$5{,}800 \text{ per year}$$

n	Depreciation Amount D_n	Book Value B_n
0	-	35,000
1	5,800	29,200
2	5,800	23,400
3	5,800	17,600
4	5,800	11,800
5	5,800	6,000

b. MACRS

Under IRS guidelines, a circuit board testing machine would be classed as high-tech equipment, which has a 5-year recovery period (see Tables 3.1 and 3.2). Thus, the equipment could be depreciated according to this schedule.

n	Depreciation Percent
1	20
2	32
3	19.2
5	11.52
5	11.52
6	5.76

(continued)

n	Depreciation Amount D_n	Book Value B_n
0	-	35,000
1	.2(35,000)	28,000
2	.32(35,000)	16,800
3	.192(35,000)	10,080
4	.1152(35,000)	6,048
5	.1152(35,000)	2,016
6	.0576(35,000)	0

Note that a 5-year recovery period is depreciated over 6 years. This is based on the assumption that the equipment will be sold during the sixth year. Also note that if a piece of capital equipment is sold before it can be depreciated fully, one-half the normal amount can be depreciated during that year.

3.6.2 *Corporate income taxes*

Until 2018, corporate taxes in the United States, like personal taxes, were based on a progressive structure as shown in Table 3.3. Example 3.7 demonstrates how corporate rates are calculated.

Since January 1, 2018, the corporate tax rate in the United States has been a flat 21% based on the passage of the Tax Cuts and Jobs Act on December 20, 2017. Note that during the 1950s, the corporate tax accounted for one-third of all tax revenues, which dropped to 9% in 2016 (Jacobson, 2014).

Table 3.3 Corporate tax structure for year 2017[4]

Taxable Income	Tax Rate	Tax
\$0 – \$50,000	15%	X * .15
\$50,001 – \$75,000	25%	\$7,500 + [(X – 50,000)*.25]
\$75,001 – \$100,000	34%	\$13,750 + [(X – 75,000)*.34]
\$100,001 – \$335,000	39%	\$22,250 + [(X – 100,000)*.39]
\$335,001 – \$10,000,000	34%	\$113,900 + [(X – 335,000)*.34]
\$10,000,001 – \$15,000,000	35%	\$3,400,000 + [(X – 10,000,000)*.35]
\$15,000,001 – \$18,333,333	38%	\$5,150,500 + [(X – 15,000,000)*.38]
Over \$18,333,333	35%	\$6,416,667 + [(X – 18,333,333)*.35]

[4] See IRS Publication 542 at https://www.irs.gov/pub/irs-pdf/p542.pdf for the details of the corporate tax structure. At the time this book was being published, the Tax Cuts and Jobs Act of 2017 was being signed into law by President Trump.

EXAMPLE 3.7

A small software company has this balance sheet:

Gross Income:	$27,000,000
Expenses:	
Salaries	$10,000,000
Depreciation	$ 3,000,000
Leases	$ 2,500,000
Taxable Income	$11,500,000

Determine how much tax the company must pay using the 2017 rates.

Solution

From Table 3.3

Tax = $50,000*.15 + $25,000*.25 + $25,000*.34 + $235,000*.39
+ $9,665,000*.34 + $1,500,000 * .35
= $3,925,000
or $3,400,000 + [($11,500,000 – $10,000,000)*.35] = $3,925,000

3.7 *Income and cash flow statements*

Discussions of depreciation, income taxes, and inflation have been presented for the sole purpose of developing net cash flow schedules and conducting after-tax analysis. In evaluating capital projects, income and cash flow statements similar to those presented in Example 3.8 must be used to compare alternatives. Table 3.4 contains the standard elements

EXAMPLE 3.8

To expand your IT support service, you are considering the benefits of acquiring a two new small hybrid utility vehicles not only to save money but also to advertise your eco-friendly business. You estimate that incremental cash flows from this investment include annual revenues of $8,000 and annual operating costs of $3,500 for the next 6 years. You have two options for acquiring these vehicles:

Option 1: Purchase the vehicles new for $20,000 cash. You can then salvage the vehicles for $2,000 at the end of 6 years.

Option 2: Lease the both vehicles for an initial payment of $3,000 and six annual end-of-year payments of $1,500. At the end of 6 years, you will return the vehicles.

(continued)

Your company uses 12% for the cost of capital. You can use the MACRS tax depreciation schedule shown in Tables 3.1 and 3.2 and a marginal tax rate of 34%. All cash flows given are actual dollar amounts. Should you lease or purchase the vehicles? Ignore the effects of inflation.

Option 1

Income Statement

Year	0	1	2	3	4	5	6
Revenue		$ 8,000.00	$ 8,000.00	$ 8,000.00	$ 8,000.00	$ 8,000.00	$ 8,000.00
Expenses							
Costs		$ (3,500.00)	$ (3,500.00)	$ (3,500.00)	$ (3,500.00)	$ (3,500.00)	$ (3,500.00)
Depreciation		$ (6,666.00)	$ (8,890.00)	$ (2,962.00)	$ (1,482.00)	$ -	$ -
Taxable Income		$ (2,166.00)	$ (4,390.00)	$ 1,538.00	$ 3,018.00	$ 4,500.00	$ 4,500.00
Income Tax		$ 736.44	$ 1,492.60	$ (522.92)	$ (1,026.12)	$ (1,530.00)	$ (1,530.00)

Cash Flow Statement

Net Income		$ (1,429.56)	$ (2,897.40)	$ 1,015.08	$ 1,991.88	$ 2,970.00	$ 2,970.00
Depreciation		$ 6,666.00	$ 8,890.00	$ 2,962.00	$ 1,482.00	$ -	$ -
Investment/Salvage Activites	$ (20,000.00)						$ 1,320.00
Net Cash Flow	$ (20,000.00)	$ 5,236.44	$ 5,992.60	$ 3,977.08	$ 3,473.88	$ 2,970.00	$ 4,290.00
PV of Net Cash Flow	$ (20,000.00)	$4,675.39	$4,777.26	$2,830.81	$2,207.71	$1,685.26	$2,173.45
Cum. PV Net Cash Flow	$ (20,000.00)	$ (15,324.61)	$ (10,547.34)	$ (7,716.54)	$ (5,508.82)	$ (3,823.56)	$ (1,650.12)

NPW	$ (1,650.12)
AEW	$401.35
IRR	8.8%
Disc PB	NA

Discount Rate/WACC	12%

Depreciation							
	Time	1	2	3	4	5	6
	Factor	33.33%	44.45%	14.81%	7.41%	0.00%	0.00%

Option 2

Income Statement

Year	0	1	2	3	4	5	6
Revenue		$ 8,000.00	$ 8,000.00	$ 8,000.00	$ 8,000.00	$ 8,000.00	$ 8,000.00
Expenses							
Costs		$ (1,500.00)	$ (1,500.00)	$ (1,500.00)	$ (1,500.00)	$ (1,500.00)	$ (1,500.00)
Operating Expenses		$ (3,500.00)	$ (3,500.00)	$ (3,500.00)	$ (3,500.00)	$ (3,500.00)	$ (3,500.00)
Depreciation		$ -	$ -	$ -	$ -	$ -	$ -
Taxable Income		$ 3,000.00	$ 3,000.00	$ 3,000.00	$ 3,000.00	$ 3,000.00	$ 3,000.00
Income Tax		$ (1,020.00)	$ (1,020.00)	$ (1,020.00)	$ (1,020.00)	$ (1,020.00)	$ (1,020.00)

Cash Flow Statement

Net Income		$ 1,980.00	$ 1,980.00	$ 1,980.00	$ 1,980.00	$ 1,980.00	$ 1,980.00
Depreciation		$ -	$ -	$ -	$ -	$ -	$ -
Investment/Salvage Activites	$ (3,000.00)						$ -
Net Cash Flow	$ (3,000.00)	$ 1,980.00	$ 1,980.00	$ 1,980.00	$ 1,980.00	$ 1,980.00	$ 1,980.00
PV of Net Cash Flow	$ (3,000.00)	$1,767.86	$1,578.44	$1,409.32	$1,258.33	$1,123.51	$1,003.13
Cum. PV Net Cash Flow	$ (3,000.00)	$ (1,232.14)	$ 346.30	$ 1,755.63	$ 3,013.95	$ 4,137.46	$ 5,140.59

NPW	$ 5,140.59
AEW	$1,250.32
IRR	62.4%
Disc PB	1.78

Discount Rate/WACC	12%

Note when using Excel to calculate NPW, make sure you separate the year 0 costs and simply add them to the NPV calculation for years 1 through 6. Also note a salvage activity of $1,320 in year 6 of Option 1. The vehicle is fully depreciated and can be sold for $2,000. However, since the asset is fully depreciated, we must pay tax on the capital gain.

By all metrics, Option 2 or leasing the vehicle is the best option.

of an income statement and cash flow statement used in after-tax cash flow analysis. After-tax analysis consists of developing a project cash flow statement and discounting the cash flow's typically used NPW. In their simplest form after tax analysis consists of an income statement and a cash flow statement.

Table 3.4 Standard form for income and cash flow statement

Income Statement						
End of Year	0	1	2	3	4	5
Revenue						
Expenses:						
Labor						
Material						
Overhead						
Debt Interest						
Depreciation						
Taxable Income						
Income Taxes						
Net Income						

Cash Flow Statement						
Operating Activities:						
Net Income						
Depreciation						
Investment Activities:						
Investment						
Salvage Value						
Gains Tax						
Working Capital						
Financing Activities:						
Borrowed Funds						
Principal Repayment						
Net Cash Flow						
Cumulative Net Present Cash Flow						

NPW	
AEW	
IRR	
Discounted Payback	

Discount Rate	

Depreciation	Time	1	2	3	4	5
	Factor					

Loan Repayment	Time	1	2	3	4	5
	Payment					
	Interest					
	Principal					
	Balance					

Companies typically finance projects with a mixture of borrowing or debt and internal cash or equity. Many companies have policies regarding the ratio of the total debt to the total investment in a project, or the debt ratio. Since interest is tax deductible, the debt ratio can affect the viability of a project, especially for companies and individuals in the higher tax brackets.

Often, especially early in a project, expenses may be greater than revenues, producing a negative taxable income. This does not mean that the government will give you a refund or tax credit. This does mean that the negative income must be offset by other projects in the company's portfolio. Remember that after-tax analysis is for a project and often does not reflect the financial health of the total company. Also, the corporate income tax rate for the company is still applicable no matter the financials of an individual project. Example 3.8 presents a simple after tax analysis consisting of an income and cash flow statements.

3.8 *Expected value*

The expected value of the NPW is often used as a measure of probability distribution. Each possible value is multiplied by an associated probability to determine the expected value. Since the probabilities are weights, they must sum to 1. Typically, these probabilities can be classified as objective or subjective, where

- Objective probabilities are based on objective data (historical data) and assume the same trends and characteristics of the past will prevail in the future, and
- Subjective probabilities are assigned usually based on subject-matter expert opinion.

Mathematically, we define expected value as

$$E[NPW] = NPW_1 p_1 + NPW_2 p_2 + \ldots + NPW_j p_j \tag{3.4}$$

If E[NPW] > 0, then we pursue the project as a feasible option. Example 3.9 demonstrates this concept.

EXAMPLE 3.9

Consider three possible investment returns on a new product. The income is based on your competitor's response and the amount of sales. An $8 million initial investment is required. Your MARR is 12%. You expect the income to be uniform for 4 years. Based on marketing results, you believe that the following probabilities and income are appropriate:

	Light	Moderate	High
Demand	$1.3 million	$2.5 million	$4 million
P[NPW]	.2	.4	.4

NPW(High) = – $8M + $4M (P/A,.12,4) = –$8M + $4M(3.0373)
= $4.1492M
NPW (Moderate) = –$.4068M
NPW (Light) = –$4.0515M

E[NPW] = (–4.0515)(.2) + (–.4068)(.4) + (4.1492)(.4) = $.6867M

Thus, based on expected value, you should invest in this project.

3.9 *Sensitivity analysis*

Sensitivity analysis is the study of how inputs, variations, and assumptions affect the output of a mathematical model. Excel has made sensitivity analysis not only easy but also an important component of all economic analysis. Sensitivity analysis allows us to

- Identify the key input elements, which can allow for more effort quantifying the value of the most important inputs,
- Develop a visual presentation of the effects of various inputs on the output, and
- Ask "what if" to determine the amount of change in a data point that might change the output of the analysis.

Sensitivity analysis is important in building confidence for a model. It plays an important role in model validation and verification. Example 3.10 and Figure 3.1 demonstrate this concept.

EXAMPLE 3.10

The economics of hybrid cars is confusing at best. The benefits of driving a hybrid are a function of miles driven per year, cost of fuel, maintenance, salvage value, insurance, and so on. Considering all things equal, conduct a sensitivity analysis based solely on initial costs and miles driven annually. Use the following information:

	Prius	Corolla
Sales Price	$22,000	$15,350
Mileage Estimates	48 mpg	38 mpg

The cars will be driven roughly 15,000 miles per year (mpy).

Develop a plot of annual costs as a function of gasoline prices. Assume the full price of the cars will be financed at 8% for 36 months, which will produce payments for the Prius of $681.04 and for the Corolla of $444.99.

(continued)

Solution

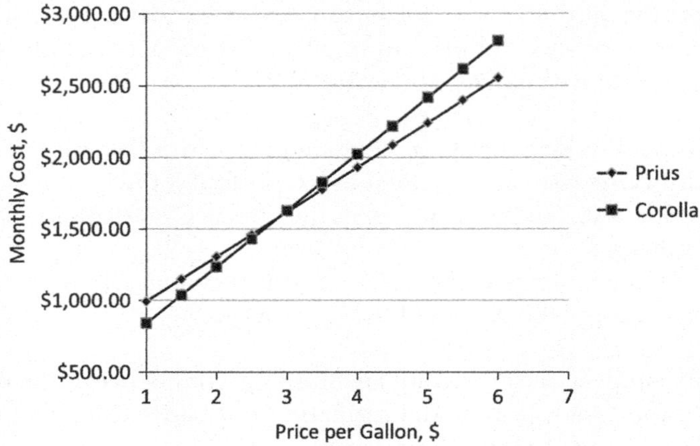

Figure 3.1 Sensitivity analysis for hybrid versus a traditional car

3.10 *Summary*

The engineering profession is no different from any other business in that costs are a key business driver. From analysis of alternatives for large projects to personal finance, understanding the economic analysis of alternatives governs most of our decisions. Engineers cannot replace professional accountants, business administrators, and lawyers; however, turnkey projects (financing, designing, building, operating, and retiring) and smart business practices are driven by a basic understanding of the time value of money from a life cycle cost perspective. For any engineer, a basic understanding of economic analysis is essential in conducting meaningful life cycle cost analysis.

QUESTIONS

3-1. Consider the hybrid problem in Example 3.10. Most alternative energy solutions can cost significantly more than the current fossil fuel solution. What other non-economic factors should be included when (a) buying a hybrid car, (b) investing in wind and solar energy, (c) investing in green construction, and (d) investing in nuclear versus fossil fuel.

3-2. Do a search on bonus depreciation and provide an example.

3-3. Three methods for selecting mutually exclusive projects have been presented thus far: net future worth, net present worth (net present value), and annual equivalent worth. Describe the pros and cons of each of these methods. Pick the one you think is the best, and justify your answer.

PROBLEMS

3-1. The so-called Rule of 72 is used to quickly estimate the number of years required for an investment to double. The equation can be expressed as

$$n_{double} = 72 / i$$

where i is expressed as a percentage (i.e., 10% would be expressed as 10). Develop a plot of time versus interest to double your investment. Overlay the actual formula or $2 = 1^*(1+i)^n$, and comment on the accuracy of this rule.

3-2. Your business purchases new computer-aided drafting workstations at the beginning of the year for $20,000. The estimated salvage value after 5 years is $4,000.

a. Calculate the straight-line depreciation amount for each of years 1 through 5.

b. Determine the MACRS depreciation schedule for this equipment (see Tables 3.1 and 3.2) using this table:

End of Year	0	1	2	3	4	5	6
Depreciation %							
Depreciation Amount (D_n)							
Book Value (B_n)							

c. If you actually sell the equipment at the end of 7 years for $4,000, determine the after-tax (net) cash flows from this investment if the marginal tax rate is 34%.

3-3. Trainor's Tracks was awarded a government contract to provide the next-generation tracks for the Army's newest track vehicle. In order to meet the Army's demand, Trainor's Tracks must purchase a special tool to manufacture the tracks. The cost associated with this tool is $758,000 and is depreciable on a 3-year MACRS scale.

Complete the MACRS depreciation table for the special tool purchase. Show all work.

Year	Depreciation Percentage	Depreciation Amount	Book Value

a. Trainor's Tracks may sell the special tool in year 2. Determine the predicted gains or loss savings.
b. In 2014, Trainor's Tracks' revenues totaled $890,000, while labor costs were $195,000 and operating costs were $375,000. Calculate Trainor's Tracks' net income for 2017 using a tax rate of 39%.
c. What was Trainor's Tracks' effective tax rate for 2017?

3-4. An organization is considering the installation of a new computer control system. The initial investment in the system is $125,000, and the project life is 5 years. The salvage value of the computer control system is $50,000, the system depreciates using a 7-year MACRS method, and the income tax rate is 40%. If the book value at the end of year 5 is $33,475, answer the following questions:
a. What is the value of the gain on the asset?
b. What is the gains tax that needs to appear in the cash flow statement?

3-5. Your consulting company is planning to purchase a new server for knowledge management. The initial cost of the server amounts to $500,000 with a salvage value of $30,000 at the end of its useful life, which is 5 years. Prepare a table in which the depreciation expense and the book value are compared for straight-line and MACRS depreciation methods.

3-6. A firm is considering purchasing a machine that costs $65,000. It will be used for 6 years, and the salvage value at that time is expected to be zero. The machine will save $35,000 per year in labor but will incur $12,000 in operating and maintenance costs each year. The machine will be depreciated according to 5-year MACRS. The firm's tax rate is 40%, and its after-tax MARR is 15%. Should the machine be bought, based on your analysis of the first 3 years of cash flows? Use the template provided in Table 3.4 to conduct your analysis.

3-7. A manufacturing company is considering acquiring a new injection-molding machine at a cost of $120,000. Because of the rapid change in product mix, the need for this particular machine is expected to last only 8 years, after which time the machine is expected to have a salvage value of $10,000. The annual operating cost is estimated to be $11,000. The addition of the machine to the current production facility is expected to generate annual revenue of $48,000. The firm has only $70,000 available from its equity funds, so it must borrow the additional $50,000 required at an interest rate of 10% per year with repayment of principal and interest in eight equal annual amounts. The applicable marginal income tax rate for the firm is 40%. Use a cost of capital of 4.2% for this project. Determine the NPW, AEW, IRR, and discount payback period for this project. Use the template provided in Table 3.4 as a starting point for your income/cash flow statement.

3-8. On 2 January 2016, the Allen Flour Company purchased a new machine at a cost of $82,000. Installation costs for the machine were $3,000. The machine was expected to have a useful life of 10 years, with a salvage value of $3,000. The company uses straight-line depreciation for financial reporting. On 3 January 2013, the machine broke down, and an extraordinary repair had to be made to the machine at a cost of $8,000. The repair extended the machine's life to 13 years but left the salvage value unchanged. On 2 January 2016, an improvement was made to the machine in the amount of $5,000 that increased the machine's productivity and increased the salvage value (to $6,000) but did not affect the remaining useful life.

a. What is the depreciation expense for 31 December 31, 2016?
b. What is the depreciation expense for 31 December 31, 2019?
c. What is the depreciation expense for 31 December 31, 2022?

3-9. An automobile-manufacturing company is considering purchasing an industrial robot to do spot welding, which is currently done by skilled labor. The initial cost of the robot is $210,000, and the annual labor savings are projected to be $150,000. If purchased, the robot will be depreciated under MACRS as a 5-year recovery property. The robot will be used for 7 years, at the end of which time the firm expects to sell it for $60,000. The company's marginal tax rate is based on the company's revenue of $12 million over the project period. Determine the net after-tax cash flows for each period over the project life. Assume MARR is 15%. To finance the industrial robot, the company will borrow the entire amount from the local bank, and the loan will be paid off at an interest rate of 10%. Use this problem narrative to create a model in Excel. Your model must include a calculation for NPW, AEW, IRR, and cumulative net present cash flow.

a. What is the NPW? Explain the meaning of this calculated NPW.
b. What is the AEW? Explain the meaning of this calculated AEW.
c. What is the IRR? Explain the meaning of this calculated IRR.

3-10. You are buying a new computer-aided drafting and design system for your business that costs $100,000 today. To use this system fully, you must invest an additional $25,000 in training costs. You finance $80,000 of the total investment cost at an effective annual interest rate of 8%, payable in five annual payments. The manufacturer has guaranteed you a salvage value of $20,000 for the system at the end of 5 years. The incremental cash flows generated with this system include $80,000 in annual revenues and $20,000 in annual non-capital expenses. Your MARR is 12%. Use the 7 years MACRS tax depreciation and a marginal tax rate of 39%. Use the template provided in Table 3.4 as a starting point for your income/cash flow statement.

3-11. Creamy Dairy is a family-owned dairy located in upstate New York. They are trying to decide how to process their corn into feed for their livestock. Because they are dairy farmers (and good ones at that), financial choices have a mystical quality about them, so they could use your help. The farm is currently earning $300,000 per year, and an improvement could provide additional revenue. Each of the options will result in an additional (above the current dairy earnings) yearly pre-tax cash flow of $100,000 for 15 years starting at time 1:

- Option 1: Subcontract the feed process at a cost of $75,000 per year starting at time 1.
- Option 2: Upgrade the current feed facilities at a cost of $200,000 for equipment and yearly maintenance cost of $25,000.
- Option 3: Build a new facility for $300,000 with no maintenance required for 10 years then $15,000 for the remaining useful life.

Creamy Dairy uses the cost of capital of 4%. Assume no short-term debt, a market rate of 8%, risk-free rate of 2.5%, and marginal tax rate of 35%. Create an Excel model (use the blank template provided in Table 3.4), and determine the NPW/AEW of the three options.

3-12. Yours Inc. is bidding on the Army's Computer/Radio Subsystem (C/RS), which consists of four separate components: a lightweight notebook computer, a soldier radio, a squad radio, and a GPS receiver. To support your bid, you must develop a thorough actual dollar after-tax cash flow analysis. The Yours Inc. board of trustees has approved $1 million for the initial investment. The bank has agreed to a 3-year loan at 8% APR, compounded daily to finance the remainder of the debt. Yours Inc. policy for establishing a MARR for all investment projects is 19%. The corporation's marginal tax rate, from all sources, is 35%. This includes federal, state, and local taxes. The corporation will receive a tax credit for any years showing a negative taxable income at the marginal total tax rate. The tax credit will be treated as a positive cash flow in the year assessed. The production costs for each element are as follows:

	Soldier Radio	Squad Radio	Computer	GPS
Initial Capital Investment	$2,225,000	$1,645,750	$1,500,000	$6,545,000
Annual O&M Costs	$1,765,400	$2,565,000	$2,325,000	$1,965,850
Material Costs per Component	$275	$355	$375	$335

We expect that the annual operation and maintenance (O&M) costs will increase at a rate of 4% annually, exclusive of inflation. Assume an inflation rate of 3%. Material costs are estimated to increase at a rate of 3% annually, exclusive of inflation. The annual increase in both O&M and material costs will begin at time 0, which will affect the expenses at the end of the first year. Administrative costs are a fixed annual cost that is associated with all operations but should be adjusted for inflation. It is estimated that administrative costs will remain at $1,875,500 annually. Annual administrative costs represent the administrative costs for production of a C/RS that includes a squad radio, soldier radio, computer, and GPS.

If Yours Inc. is awarded the contract, they will be required to produce 10,000 C/RS per year for the next 3 years. In order to execute this contract, Yours Inc. must purchase some new capital equipment for each alternative. The cost of the new equipment is represented by summing up the initial investment costs of each alternative. Assume that this equipment has a 3-year MACRS recovery period. Upon completion of the contract, Yours Inc. will sell all equipment purchased for this project. Yours Inc. estimates they can sell (salvage) the equipment for 30% of the original purchase price. The current existing equipment has already been fully depreciated. The current existing equipment must be kept upon completion of the new contract in order to meet current obligations. Construct an after-tax cash flow worksheet using actual dollars to determine your bid price.

Bibliography

Griffis, Fletcher H., and Farr, John V., 1999, *Construction Planning for Engineers*, McGraw-Hill, New York, NY.

Lang, H. J., and Merino, D. N., 1993, *The Selection Process for Capital Projects*, John Wiley and Sons, New York, NY.

Newnan, D. G., Eschenbach, T. G., and Lavelle, J. P., 2008, *Engineering Economic Analysis*, 9th Edition, Oxford University Press, New York, NY.

Park, C. S., 2012, *Fundamentals of Engineering Economics*, 3rd Edition, Pearson/Prentice Hall, Upper Saddle River, NJ.

Park, C. S., 2016, *Contemporary Engineering Economics*, 6th Edition, Pearson/Prentice Hall, Upper Saddle River, NJ.

Park, W. R., and Jackson, D. E., 1984, *Cost Engineering Analysis: A Guide to Economic Evaluation of Engineering Projects*, 2nd Edition, John Wiley and Sons, New York, NY.

Sage, A. P., 1983, *Economic Systems Analysis: Microeconomics for Systems Engineering, Engineering Management, and Project Selection*, North-Holland, New York, NY.

Sullivan, William G., Wicks, Elin M., and Koelling, C. Patrick, 2008, *Engineering Economy*, 14th Edition, Prentice Hall, Upper Saddle River, NJ.

References

Jacobson, L., August 28, 2014, "Bernie Sanders Says Tax Share Paid by Corporations Has Fallen from 33% to 9% since 1952," accessed 10 April 2018 at http://www.politifact.com/truth-o-meter/statements/2014/aug/28/bernie-s/bernie-sanders-says-tax-share-paid-corporations-ha/.

chapter four

The basic theory of interest

4.1 Introduction

The purpose of this chapter is to answer the question, "What is the best interest rate for analysis?" This seemingly simple question is, actually, quite challenging to solve. Many organizations address this problem by using subjective or heuristic techniques. However, a more rigorous method and philosophy can be leveraged to arrive at a useful rate valuable for specific analysis.

In the previous chapters, topics covered included the time value of money and techniques for choosing between competing alternatives. Making the best decisions becomes the chief concern as firms or individuals face financial options. However, the time value of money strongly influences all choices. With most engineering economy texts, the concept of interest is used extensively without practical consideration of its origin. Often the estimation of cash flows is well understood and can be replicated in the "real world" with ease. Many economic factors are easily translated to a model or ledger, such as costing or demand forecasting. However, the same is often not true of interest.

This chapter will explore the topic of interest from a pragmatic standpoint. Primary questions such as "What interest rate do I use for the project?" or "Does the firm's interest rate change over time?" are addressed through the concept of cost of capital. Finding and applying the proper interest rate ensures consistent and valuable analysis of alternatives.

4.2 Interest as time value of money

At the most basic level, the idea of interest is quite simple and familiar. Interest is the expression of the change of value from a principal amount over a given period. For example, if a 1-year investment of $100 (the principal) 1 year hence yields $110, the interest (value of providing control of the principal) is said to be 10%. The interest concept is, at its core, based on the value received for the transfer of principal control over varying time horizons. An investment refers to the transfer of principal control for an investment.

From a practical standpoint, many things affect the value demanded from various investments. These include the perceived investment quality, investors' preferences, and risk tolerances. However, opportunity cost and uncertainty are the primary drivers of interest rates.

For a principal owner seeking investment opportunities, there are many alternatives that yield potential value. The various conditions of these options inform what opportunities are available. When applied to a single investment, the principal cannot, by definition, be used for another purpose. In kind, if an investment has a high degree of uncertainty, a higher interest rate will be demanded. Typically, interest rates for publicly available investments, like fixed-income securities, are set by the "market" and are said to be market interest rates. However, it is crucial for the reader to understand that, for practical purposes, all interest rates are market rates.

Interest is expressed in two quantitative forms: simple and compound. The mechanics of these approaches are covered in Chapter 2. The difference between the two concepts has to do with the treatment of reinvested value. Many times, quick-turn or "on the spot" analysis is done using simple interest, which was presented in Equation 2.1, and compound interest, presented in Equation 2.2.

In each of these cases, the future value of money is determined by the same three parameters: the initial principal, the number of time periods, and the interest rate. The value for the interest rate has the same characteristics in either scenario. Value is positively correlated with interest value. In general, the closer the interest rate is to zero (assuming non-negative-only values) the closer the future value is to the present value. However, with these necessary calculations, a few assumptions are made but seldom stated.

The first assumption in a standard value calculation is that of an ideal bank. An ideal bank implies that an investor can assume the same rate for initial investments and deposits. Also, most textbook problems include the notion of a constant ideal bank. This refers to the constant application of rates regardless of scenario or duration. These assumptions are required to achieve equivalence calculations, which are statements about future and present values. Two values (one present and one future) are said to be equivalent at a given constant interest rate. However, in practice, rates can vary significantly over both long- and short-term horizons. Also, compounding is not done cleanly as assumed in Equations 2.1 and 2.2. Instead, the lens of compound interest is used to view real-world investment performance.

Investment performance can be interpreted as an equivalent compound annual growth rate (as if money had been invested in a constant ideal bank). This value is often referred to as the calculated compound annual growth ratio (CAGR). This value is derived using Equation 2.14. Example 4.1 shows how to calculate CAGR.

EXAMPLE 4.1

Apple Inc. stock has performed exceptionally well in recent years. In November 2000, the split and dividend* adjusted stock price was $1, and in November 2017, it was $174. What is the CAGR for Apple stock over this 17-year period?

Using the compound interest formula (see Equation 2.2), we have

$$174 = 1(1+)^{i17}$$

Rearranging gives us

$$i = \left(\frac{174}{1}\right)^{\frac{1}{17}} - 1$$
$$= .35$$

*Over the course of this period, an investment in Apple common stock would have yielded an impressive equivalent constant ideal bank investment of about 35%. Note that stock prices are adjusted for splits and dividends, which are discounted using the returns implied by these events. As such, these values are lower than actual recorded values.

Tools like the CAGR are excellent for retroactively evaluating how well an investment performed but provide little information regarding what interest rate should be used when projects are forward-looking. To answer the forward-looking concern, a better understanding of what drives interest rates at the decision-maker level is needed. As shown in Figure 4.1, the value of the interest is driven by two competing forces: purchasing power and earning power.

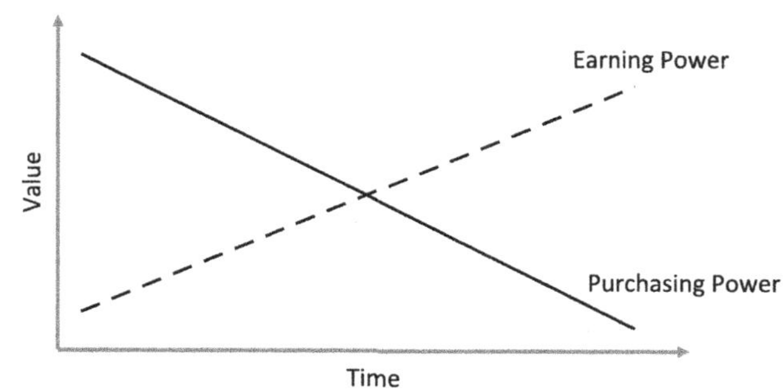

Figure 4.1 Time value of money: earning versus purchasing power

Two factors influence the principal investment amounts in the context of time: time value of money and purchasing power. Current values always have lower earning power in the present (assuming positive interest). However, current purchasing power of money is always at its highest due to inflation. The balance of these forces meets at different points for various individuals and firms, which ultimately determines the interest rate used for capital budgeting decisions. The idea of earning power is intuitive, if not obvious: the more you invest, the more your money grows. However, purchasing power is not as straightforward. Why does the same dollar buy less and less over time? The next section addresses this topic.

4.3 *Inflation*

Inflation is the phenomenon that explains the change in purchasing power of a given currency over time. In general, healthy economies have an inflation rate of about 2%. Figure 4.2 shows the rate of change over time in the US dollar.

The inflation rate is, in fact, a difficult thing to measure. In the United States, this rate is estimated by tracking the price of a basket of goods that are commonly purchased by consumers or employers. The most common tool used to determine inflation is the consumer price index (CPI). The CPI includes some of the following categories:

- Food and beverages,
- Housing,
- Apparel,

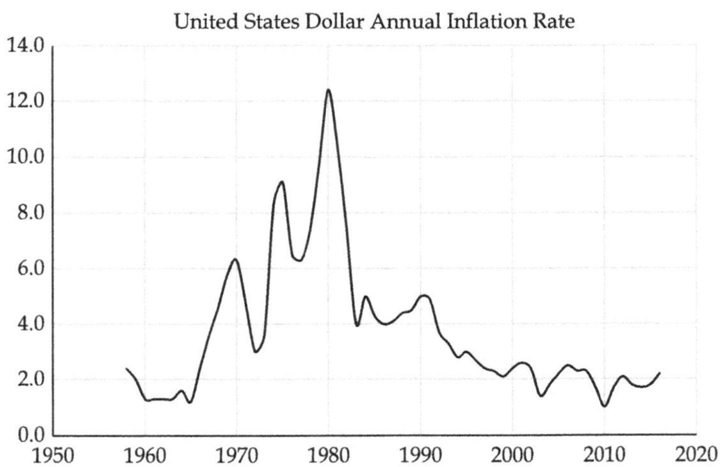

Figure 4.2 US inflation rate 1957–2016

- Transportation,
- Medical care,
- Recreation,
- Education and communication, and
- Other goods and services.

The CPI is a measure of the average change over time in the prices paid by urban consumers for a market basket of consumer goods and services (Bureau of Labor Statistics, n.d.). The CPI is frequently called a cost-of-living index, but this is probably not correct because consumers tend to adjust their spending habits based on price increases and decreases. This information is directly observable from market conditions and does not need to be estimated for its use in modeling and analysis.

Inflation is a convenient concept when exploring interest rates. The existence of the change in purchasing power means that capital (principal values discussed earlier) will not be idle. If inflation is positive, purchasing power decreases. However, if inflation is substantial, there will be no incentive to save.

Extremely risk-adverse investors may not want to part with their principal. So, if the principal value is entirely stable (meaning that purchasing power does not change), many investors would choose not to invest but instead stockpile value. This is the famous "money under the mattress" concept whereby investors do not seek an advantage in risking capital. To encourage investment, modern capitalist systems attempt to regulate inflation through a central bank. The job of the central bank is to ensure that inflation is positive but not too high. This is a "Goldilocks" approach to managing an economy. A modest amount of loss in purchasing power will encourage investment, which, in turn, results in economic growth.

However, the existence of inflation has a significant impact on reasonable interest rates. Most discussion of money across time involves adjusting for inflation. This adjustment is so standard that it has its own semantics. Adjusted past values for comparison with today's values are referred to as constant or real dollars. It is common to hear in conversation something like the following: "The house cost $5,000 in 1950, which is equivalent to $100,000 today." Example 4.2 demonstrates this concept.

The use of constant dollars allows for a familiar reference point. Values that have not been adjusted for inflation are considered actual or nominal dollars. Nominal values are essential for modeling and analysis in capital budgeting. In later sections, topics of individual interest rates are explored. When discounting future values, it is important that those values are expressed in nominal terms.

EXAMPLE 4.2

Inflation rates are calculated using the changing price of a basket of goods. The most common measure of this basket is the consumer price index. The following table represents the changes in this basket of goods between January 2000 and December 2016. Calculate the inflation rate over this period.

Month	CPI
2000 Jan	169.3
2000 Feb	170
2000 Mar	171
2000 Apr	170.9
......	......
2016 Jun	239.842
2016 Jul	239.898
2016 Aug	240.389
2016 Sep	241.006
2016 Oct	241.694
2016 Nov	242.199
2016 Dec	242.821

$$i = \frac{242.821 - 169.3}{169.3} = \sim .43$$

Over the given period of 16 years, there was a change in purchasing power of 43%. The interpretation of this change is that what \$1 could buy at the beginning of 2000 would take \$1.43 at the end of 2016.

The presence of inflation as a factor implies that some adjustments need to be made to better understand the interest rates in Equations 2.1 and 2.2. The rate at which real dollars change over time is referred to as the real interest rate. Real rates remove inflation from the change in the value of money. If i is the nominal interest rate and f is the inflation rate, then it follows that

$$1 + i_0 = \frac{1+i}{1+f} \tag{4.1}$$

where i_0 is the real interest rate or the interest rate adjusted for inflation. Some manipulation yields

$$i_0 = \frac{i-f}{1+f} \tag{4.2}$$

Example 4.3 demonstrates this concept.

EXAMPLE 4.3

Consider an investment of $100 over a 1-year period. The inflation rate is 5%, and the nominal interest rate is 10%. What is the change in value over this period in real terms?

$$P = \$100$$

$$f = 5\%$$

$$i = 10\%$$

$$\dot{i}_0 = \frac{(0.1 - .05)}{1.1}$$

$$= 0.045$$

With a real interest rate of ~4.5%, the change in value is

$$100 * 0.045 \cong 4.5$$

This change in value is a significant departure from the $10 earned from the nominal rate alone. Alternatively, this exercise can be done in reverse to uncover the real interest rate from changes in investment value.

Changes in purchasing power over time have significant impacts on the practical use of interest rates. As with consumer goods, enterprises have to consider how market forces will affect their ability to invest funds or purchase materials.

4.3.1 *Determining inflation and deflation*

Two terms are used to describe inflation: average inflation rate and general inflation rate. The average inflation rate is used for a specific market item or sector and is calculated by

$$\bar{f} = \left(\frac{CPI_n}{CPI_0}\right)^{\frac{1}{n}} - 1 \tag{4.3}$$

Interest rates may or may not be adjusted for inflation. The inflation-free interest rate or i′ is often referred to as a real interest rate. The market interest rate or simply i includes the cost of capital and inflation. The market interest rate can be related to the inflation-free interest rate using

$$i = i' + \bar{f} + i'\bar{f} \tag{4.4}$$

Two types of economic analysis can be conducted that account for the effects of inflation. An actual dollar analysis consists of cash flows that include inflation. For this type of analysis, the cash flows are adjusted by applying the general inflation rate to the base year dollar amounts. Also, you can conduct a constant dollar analysis. This type of analysis assumes purchasing power is independent of time. Real interest rates should be used for this type of analysis.

4.4 Market interest rates

Market forces set all interest rates. As mentioned in the opening section, the time value of money is affected by uncertainty and opportunity costs. Presenting alternatives to broad market participants, opportunity costs are, mostly, localized, and a risk-and-reward construct sets rates. In general, the lower the risk, the smaller the interest rate (or rate of return) and vice versa. However, at the firm level, opportunities that are unique to the circumstances of the organization may yield interest rates that are either higher or lower than publicly available market investments. As a rule, firms perceive their internal projects as having a higher return than public market investments. If firms could not get better returns internally, they would use their capital for publicly traded investment vehicles and not go to the bother of running a project.

In the public realm, market interest rates are set by and derived from secondary markets. These types of markets trade securities, assets, or derivatives after they have been issued to a primary set of investors. The most famous secondary markets include stock and bond exchanges like the New York Stock Exchange.

One of the most crucial interest rates in financial modeling is the risk-free rate. This is the rate than investors expect to receive if they are unwilling to take any risk. From a practical standpoint, a purely riskless investment vehicle does not exist. However, several real-world opportunities are available that are very close to being riskless. The majority of financial analysis considers sovereign debt from first-tier countries to be, mostly, risk-free. For this reason, we will now consider debt issued by the United States.

4.4.1 United States treasury debt

The debt issued by the United States is often referred to as "T-bills" or "Treasuries" and is currently considered to be among the safest investments in the world (Luenberger, 1997). When debt is issued to an investor, it typically comes with a stream of smaller payments and one balloon payment upon maturity. The interest rate when purchasing a debt instrument is referred to as the yield to maturity (YTM). Specific calculations and examples will be explored in later sections. However, the use of Treasuries

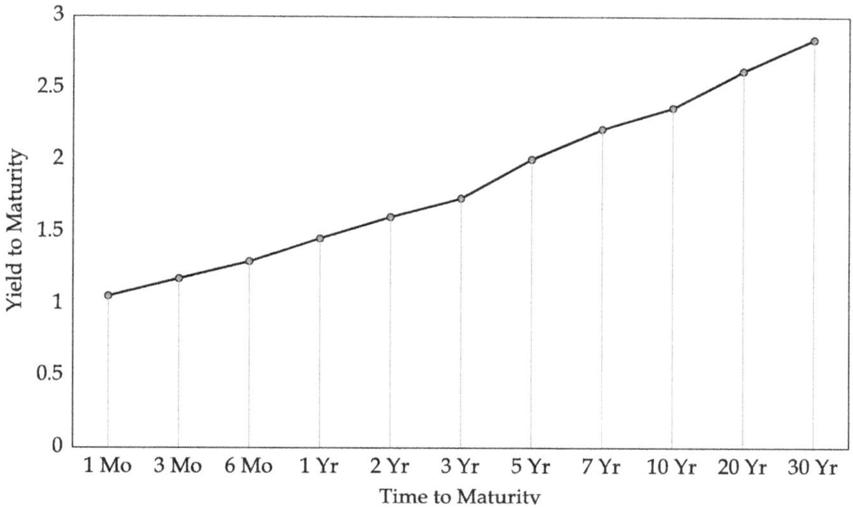

Figure 4.3 US Treasuries YTM in November 2017

as a riskless asset gives the first indication of what interest rate a project planner should use.

The first primary consideration in interest rates is "How long is my project?" Debt instruments offer a useful analog for this question. Specifically, Treasuries are issued at a broad range of maturities. The collection of observed interest rates at given maturities is called spot rates. The spot rate is used to construct the yield curve, which is the interpolation of spot rates to create a continuous function. Figure 4.3 shows the yield curve for Treasuries in November 2017.

One important, and fundamental, observation of the yield curve is that investments with longer time horizons demand higher yields from the market. This is true for projects as well. If the principal is tied up, then opportunity costs exert upward pressure on the required interest rate. For virtually the same asset, a risk-free investment has different interest rates for different time horizons. This gives our first determination of what interest rate to use for an economic analysis, or a "floor." One should never use an interest rate for analysis that is less than the risk-free rate. In the following section, the concept of individual interest rate is explored.

4.4.2 Debt

Debt interest rates are determined based on how much it costs firms to issues bonds or to take on loans. Often debt financing is the least expensive way to finance projects (lower cost). Debt holders have many legal rights to a company before stockholders do; however, debt financing is often the least expensive way to raise capital. As discussed with Treasuries, companies will also have a yield curve based on issued debt YTM. These interest rates are set by the market and have some risk premium over the risk-free rate. Typically, when calculating the cost of debt, we use a 10-year average bond yield (or similar horizon). Debt (bonds) are typically categorized by the quality and will be given a grade by third-party rating agencies. Table 4.1 gives the ratings for the top three rating agencies.

Table 4.1 Ratings of the top three credit rating agencies (from Wikipedia, n.d., and Standard & Poor's, n.d.)

Moody's	Standard & Poor's	Fitch	Credit Worthiness
Aaa	AAA	AAA	An obligor has EXTREMELY STRONG capacity to meet its financial commitments.
Aa1 Aa2 Aa3	AA+ AA AA–	AA+ AA AA–	An obligor has VERY STRONG capacity to meet its financial commitments. It differs from the highest-rated obligors only to a small degree.
A1 A2 A3	A+ A A–	A+ A A–	An obligor has STRONG capacity to meet its financial commitments but is somewhat more susceptible to the adverse effects of changes in circumstances and economic conditions than obligors in higher-rated categories.
Baa1 Baa2 Baa3	BBB+ BBB BBB–	BBB+ BBB BBB–	An obligor has ADEQUATE capacity to meet its financial commitments. However, adverse economic conditions or changing circumstances are more likely to lead to a weakened capacity of the obligor to meet its financial commitments.
Ba1 Ba2 Ba3	BB+ BB BB–	BB+ BB BB–	An obligor is LESS VULNERABLE in the near term than other lower-rated obligors. However, it faces major ongoing uncertainties and exposure to adverse business, financial, or economic conditions which could lead to the obligor's inadequate capacity to meet its financial commitments.

(continued)

Table 4.1 (*continued*)

Moody's	Standard & Poor's	Fitch	Credit Worthiness
B1 B2 B3	B+ B B–	B+ B B–	An obligor is MORE VULNERABLE than the obligors rated 'BB,' but the obligor currently can meet its financial commitments. Adverse business, financial, or economic conditions will likely impair the obligor's capacity or willingness to meet its financial commitments.
Caa	CCC	CCC	An obligor is CURRENTLY VULNERABLE, and is dependent upon favorable business, financial, and economic conditions to meet its financial commitments.
Ca	CC	CC	An obligor is CURRENTLY HIGHLY-VULNERABLE.
	C	C	The obligor is CURRENTLY HIGHLY-VULNERABLE for nonpayment. May be used where a bankruptcy petition has been filed.
C	D	D	An obligor has failed to pay one or more of its financial obligations (rated or unrated) when it became due.
e, p	pr	Expected	Preliminary ratings may be assigned to obligations pending receipt of final documentation and legal opinions. The final rating may differ from the preliminary rating.
WR			Rating withdrawn for reasons including: debt maturity, calls, puts, conversions, etc., or business reasons (e.g. change in the size of a debt issue), or the issuer defaults.
Unsolicited	Unsolicited		This rating was initiated by the ratings agency and not requested by the issuer.

(*continued*)

Table 4.1 (*continued*)

Moody's	Standard & Poor's	Fitch	Credit Worthiness
	SD	RD	This rating is assigned when the agency believes that the obligor has selectively defaulted on a specific issue or class of obligations but it will continue to meet its payment obligations on other issues or classes of obligations in a timely manner.
NR	NR	NR	No rating has been requested, or there is insufficient information on which to base a rating.

Companies or governments can issue a large range of debt, so to estimate the after-tax cost of debt (interest rate of debt), we use

$$i_d = \left(\frac{c_s}{c_d}\right)k_s(1-t_m) + \left(\frac{c_b}{c_d}\right)k_b(1-t_m) \tag{4.5}$$

where

c_s = amount of short-term loan, \$
c_b = total amount of bond financing, \$
c_d = total debt capital, \$ = $c_s + c_b$
k_b = before-tax interest on a bond, %
k_s = before-tax interest rate on a loan, %
t_m = marginal tax rate

Taxes are typically included in the debt calculations because, from a firm's perspective, payments on bonds are tax deductible. The before-tax interest rate on a bond is often referred to as the YTM.

Buying and selling bonds has its own terminology:

- Market value: Price paid by an investor to purchase the bond on the market.
- Face value: Value printed on the front of the bond (typically \$100 or \$1,000). If the market value equals the face value, the bond is called "par."
- Coupon: Periodic payment based on the coupon rate (a percentage of face value). Most bonds have a semiannual coupon payment that is calculated at half of the value of the coupon rate times the face value (coupon payment (A) = (coupon rate * face value)/2). Bonds also make one final payment of the face value (in addition to the final coupon payment).
- Maturity: Time until the last coupon payment.

EXAMPLE 4.4

If a firm issues a 10-year bond with a face value of $1,000 and a coupon rate of 5% and the yield to maturity is 8%, what is the market value of the bond?

The coupon payments are made semiannually for 10 years at a rate of 5%, so A = .05*$1,000/2 = $25. Because the maturity is 10 years, we have 20 interest periods (semiannual periods). Because the YTM = 8% and it is a nominal rate, the periodic interest rate is 8%/2 = 4%.

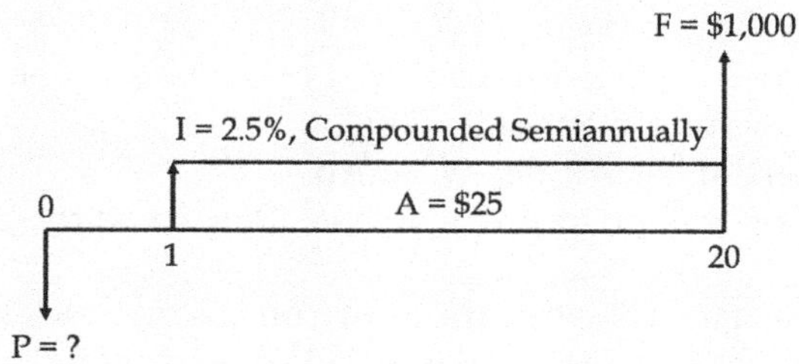

Market value is the *present value* of the stream of coupon payments and the final face value payment, so

$$\begin{aligned}\text{Market Value} &= A\left[P/A, 4\%, 20\right] + F\left[P/F, 4\%, 20\right]\\ &= \$25 * (13.59) + \$1,000 * (.46)\\ &= \$339 + \$456\\ &= \$796.16\end{aligned}$$

- Yield to maturity: The return on investment an investor will receive if he or she holds the bond to maturity (i.e., all the coupon payments and the final face value payment).
- Current yield: The return on investment an investor receives for only holding the bond for the next year (i.e., only the next year's coupon payment). Note that the yields are annual percentage rates or nominal interest rates.

Example 4.4 demonstrates the characteristics and how to calculate the market values of a bond.

4.4.3 *Equity*

Issued equity has an interest rate as well. There are many ways to estimate the return on a share of common stock, including retained earnings, information from financial statements, historical averages based on the previous year's performance, and rate of return expectation on a risky asset using some risk premium. All are appropriate. However, we will focus on historical averages based on the previous year's performance. Also, we will use a 10-year performance window because capital projects, broadly speaking, last about 10 years (adjustments in horizon can be easily made).

When using historical averages, investors' expected return from a common stock is thought to have three components. These factors are expressed as the expected return of stock, also expressed as i_e:

$$i_e = i_f + \beta_{stock}\left(i_m - i_f\right) \tag{4.6}$$

i_f = risk-free real return, which is typically based on a 10-year US Treasury bill yield[1]
i_m = historical market rate of return
β_{stock} = market risk premium, which is often called the β factor

This approach to evaluating the return on a firm's stock is referred to as the capital asset pricing model. A significant amount of financial information is available free of charge on the Internet. For example, as of November 2017, Yahoo Finance and Morningstar are excellent sources of quality data. Both of these sources have benefits and drawbacks, so we will use their strengths for specific data points. Yahoo Finance is useful as a source to obtain historical stock prices and β_{stock}, whereas the Morningstar site is a valuable reference for corporate bond information. The Morningstar bond information has a significant amount of data about the debt a company has issued, including the capital structure. Example 4.5 demonstrates how to calculate β_{stock}.

4.4.4 Minimum attractive rate of return

Decision makers within the US government can stop reading the chapter now. The appropriate interest rate for this class of analysis is to use the yield from the project-appropriate maturity on the Treasury yield curve. The reason is that, as an organization, the cost of capital for the US government is equal to the yield paid on the debt it issues; this is the same for all sovereign governments. This is how all individuals and groups should determine their interest rate for analysis. Interest rates used for economic analysis are equivalent to the cost of capital. The cost of capital is the market-determined interest rate demanded to transfer capital to an entity. In the case of business, this is derived from issued financial assets, and for an individual, it is the opportunity cost of other market instruments (such as stocks or bonds).

1 See http://www.treasury.gov/resource-center/data-chart-center/interest-rates/Pages/TextView.aspx?data=yield for yields.

EXAMPLE 4.5

Estimating β_{stock} is quite straightforward. Beta is merely the slope of the best-fit line between market returns and the returns of the given stock. For example, if I wanted to estimate the beta of Google, I would make a scatter plot of the last 60 monthly returns of Google versus the returns of the market as a whole (we typically use the S&P 500 as our market proxy). Such a chart as shown in Figure 4.4 with a best-fit trend line would look as follows:[2]

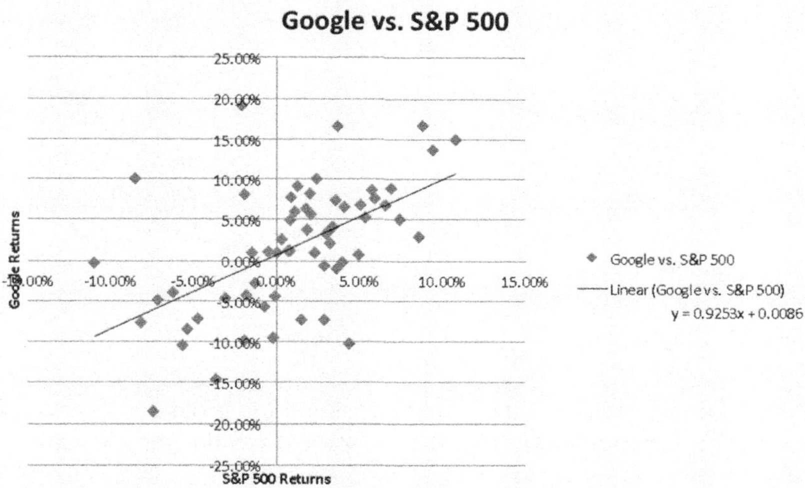

Figure 4.4 Google versus S&P 500 returns

You can see from the chart that the trend line between the market returns and Google has a slope of 0.9253, which means that in general Google moves with the market.

[2] The best-fit line is, in fact, the result of a single factor linear regression.

In business and engineering, the minimum acceptable or attractive rate of return (MARR) or hurdle rate is the minimum rate of return on a project a manager or company is willing to accept before starting a project, given its risk and the opportunity cost of forgoing other projects (Park, 2012). MARR usually consists of the cost of capital. The MARR then becomes the most appropriate interest rate to use for all internal economic analysis.

The cost of capital is often expressed as the weighted average cost of capital (WACC) and is best defined as the rate that a company is expected to pay, on average, to obtain debt and equity financing. Simply, WACC is the

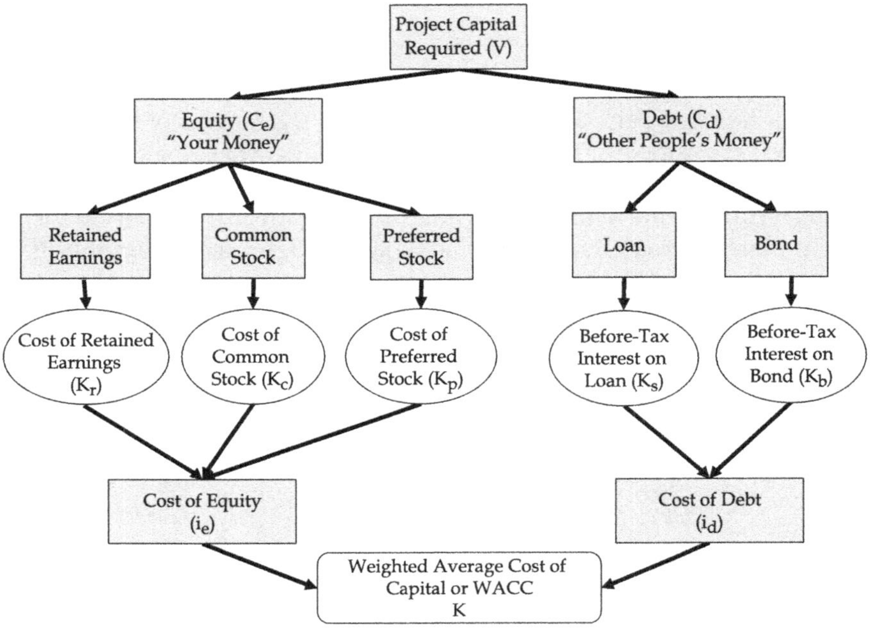

Figure 4.5 Graphical depiction of the WACC components

cost to get finances from issuing financial assets and is defined as the cost of debt times percent debt plus the cost of capital times the percent capital, or

$$k = \frac{i_d c_d}{v} + \frac{i_e c_e}{v} \tag{4.7}$$

where

i_d = cost of debt
c_d = total debt capital
c_e = total equity capital
$V = c_d + c_e$
i_e = average equity interest rate per pay period, taking into account all equity sources
i_d = after-tax average borrowing interest rate per period, taking into account all debt sources

This is shown graphically in Figure 4.5.

4.5 *Summary*

Systematic exploration of interest rates is the best way for firms to understand the true cost of capital. Specifically, the use of the WACC is valuable to understand how the market has set a firm's interest rate through the demand for its fund-raising efforts. We believe that there is no more appropriate way to determine an interest rate for use in analysis. Firms should use the WACC as the MARR.

Often, in practice, some firms take the WACC and add some arbitrary value to arrive at a MARR. However, injecting a subjectively selected rate can negatively affect proper project selection. It is always best practice to use market-derived information unless a compelling reason requires deviation. The interpretation of market prices through the lens of compounding is the best summary of the basic theory of interest.

Using Figure 4.6 with market asset prices will significantly aid in calculating an appropriate interest rate.

Lastly, we have included Table 4.2 as a summary of the different terms and definitions used for interest rates in engineering economics.

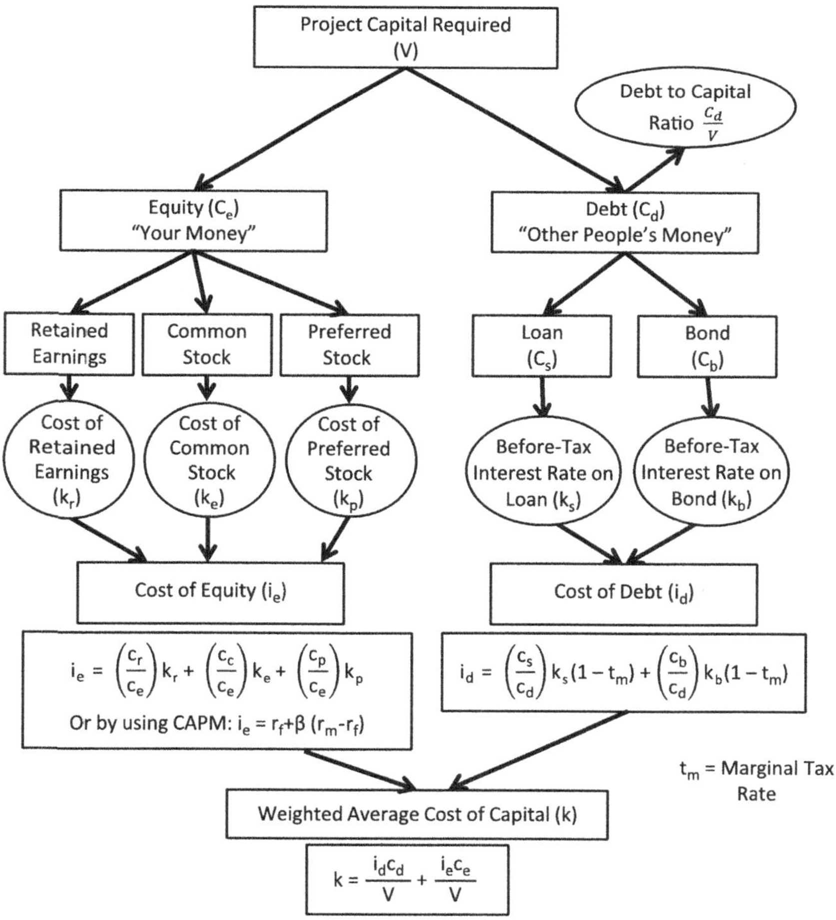

Figure 4.6 Cost of capital flow chart

Table 4.2 Definition of interest rate variables used in engineering economics

Interest Rate Variable	Definition
i or r	Often called the cost of money. This is typically expressed as a percentage of the principal paid by the borrower to a lender for the use of assets. Unless otherwise defined or implied, this is referred to as the nominal interest rate and is expressed as an annual percentage rate.
I	I is often used instead of i. I is often used to express interest rates as a percentage, and i is used for fractions. For example, if i = .025 then I = 2.5%.
i_f	The risk-free rate of return. This is the interest rate received by an investor from an asset that has no risk.
IRR	The internal rate of return is the interest rate that makes the net present value of all cash flows from a particular project equal to zero.
i_a	The effective annual interest rate truly represents the interest rate for 1 year.
i_d	After-tax borrowing interest rate per period, taking into account all equity sources.
i_{eff}	Average equity interest rate per period, taking into account all debt sources.
i_{per}	The periodic interest is the true interest charged by the lender and must be in the same units as the payment periods. For example, monthly payments will require a monthly periodic interest rate and must be based on the effective annual interest rate.
MARR	The minimum acceptable or attractive rate of return or hurdle rate is the minimum rate of return on a project a manager or company is willing to accept before starting a project, given its risk and the opportunity cost of forgoing other projects. This also represents the rate at which a firm can invest the money from its investment pool.
WACC	The weighted average cost of capital represents the costs to raise capital based on the weighted cost of debt (i_d) and the cost of equity (i_e).

QUESTIONS

4-1. Consider Figure 4.5, the flow chart for the WACC calculations. If a firm has not issued any debt or equity, how might the concepts in the WACC be used to arrive at a usable interest rate?

4-2. Search the Dow Jones industrial average companies and compare their total equity versus debt issued. What is the balance? If companies can get much lower rates for debt, why do you think they issue equity?

PROBLEMS

4-1. Consider an investment that requires a principal of $500 and promises to return $600 in 8 years. What is the CAGR of this proposal?

4-2. You are given an investment with a current market value of $700 with a continuous compounding rate of return of 10%. What will the value be in 6 years?

4-3. Consider an investment in the following cost and revenue structure. With an annual inflation rate of 4%, what is the real CAGR?

Time	Value
0	($100)
1	$ 30
2	$120

4-4. Consider the following table of cash flows with an inflation rate of 5% and a nominal interest rate of 10%. What is the real value of the proposal?

Time	Value
0	($200)
1	$ 40
2	$220

4-5. Find the United States' CPI values from 2005 to 2015. Calculate the annual inflation rate over this period. Are there any anomalies?

4-6. A 10-year bond has a face value of $1,000, a coupon rate of 6%, and a YTM of 8%. What is the current value of the bond?

4-7. A 10-year bond with a face value of $1,000 and a coupon rate of 5% is issued. It has a current market value of $900. What is the YTM of the bond?

4-8. Consider the following table. This information represents market values for a company's stock prices over five time periods. Use this information to estimate the equity rate of return using the capital asset pricing model. Assume a risk-free rate of 2%.

Time	Stock Price	Market Price
0	57.96	1,257.64
1	67.67	1,257.60
2	81.15	1,426.19
3	105.86	1,848.36
4	127.51	2,058.90
5	151.67	2,043.94
6	152.09	2,238.83
7	172.91	2,599.03

4-9. A firm has decided to invest in a new piece of machinery. The initial cost of the machine is $3 million. The debt-to-capital ratio for this project is 33.3% (one-third). To finance this project, the chief financial officer has asked you to determine the cost of equity given the following parameters:

Source of Equity Capital	Amount	Fraction of Total Equity
Retained Earnings	$ 500,000	
New Common Stock	$1,000,000	
Preferred Stock	$ 500,000	

Note: the remainder of the total cost will be financed using debt financing.

Supplemental information:

- Current common stock price = $40/share
- Previous year's annual cash dividend = $3
- Estimated growth rate of annual cash dividends = 5%
- Flotation costs for issuing common stock = 10%
- Issuing price for preferred stock = $90
- Fixed annual dividend for preferred stock = $12
- Flotation costs for issuing preferred stock = 6%
- Average market return = 13.25%
- Risk-free interest rate = 3.2%
- The contractor's β_{stock} = 1.3

Determine the cost of equity to finance this alternative, and calculate the number of shares required to raise $1 million.

Bibliography

Luenberger, David, 1997, *Investment Science*, Oxford University Press, Oxford, United Kingdom.

References

Bureau of Labor Statistics, n.d., "Is the CPI a Cost-of-Living Index?" accessed 18 January 2014 at http://www.bls.gov/cpi/cpifaq.htm#Question_9.

Park, C. S., 2012, *Fundamentals of Engineering Economics*, 3rd Edition, Pearson/Prentice Hall, Upper Saddle River, NJ.

Standard and Poor's, n.d., "S&P Global Ratings," accessed 19 April 2018 at https://www.standardandpoors.com/en_US/web/guest/article/-/view/sourceId/504352.

Wikipedia, n.d., "Bond Credit Rating," accessed 10 April 2018 at https://en.wikipedia.org/wiki/Bond_credit_rating.

chapter five

Simulation-based costing

5.1 Introduction

Systems and enterprises at the most basic level are an integrated composition of elements or subsystems governed by processes that provide a capability to satisfy a stated need (i.e., requirements) or objective. Simulation is an ideal way to analyze these systems because we can model the characteristics of the elements and replicate the processes with logic. Figure 5.1 shows the domains in which engineers use simulation to represent and model their systems. To develop a system or enterprise successfully, you must

- First define the problem that exists,
- Identify the mission requirements (or business drivers) of the organization(s) needing the problem to be solved,
- Evaluate high-level concept of operations (CONOPS) for solving the problem,
- Select the concept that makes the most sense in light of the product or mission requirements,
- Develop an operational concept around the selected concept,
- Create architectures and derived requirements for the subsystems, components, and configuration items consistent with the decomposition of the system,
- Design the integration, test, and evaluation process for the parts of the system,
- Conduct the integration and test process for the parts of the system,
- Manufacture and assemble the parts of the system,
- Deploy the system, and train people to operate and maintain it,
- Iterate and refine the system, and
- Finally, retire the system.

Simulation can play a key role during each of these phases to assess risk for operational analysis and life cycle cost (LCC). Simulation can be used to prototype the systems, evaluate CONOPS, develop corresponding training systems, and determine the cost and associated risk.

Both buyers and developers incorporate modeling and simulation (M&S) into all phases of the development of new products, covering the entire life cycle from concept development to retirement. The military

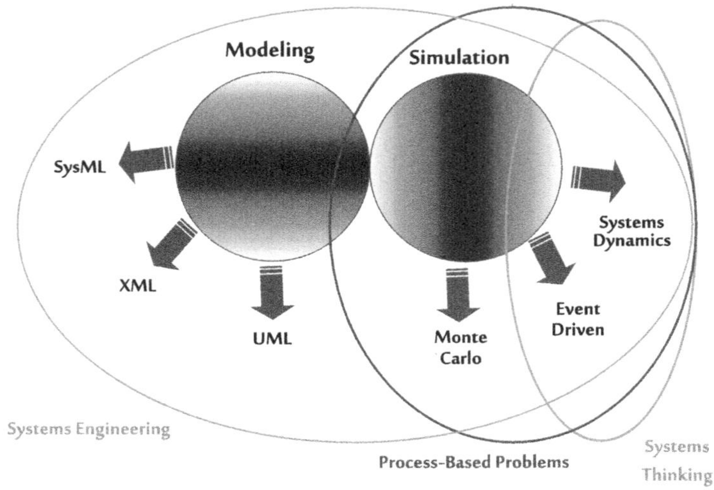

Figure 5.1 Types of simulation used in modeling systems

(see Department of Defense, n.d.) has used M&S for many years because it can provide a realistic and cheaper way to train decision makers and to conduct analysis of complex material and force structure. Specifically, M&S can be used (1) to evaluate requirements for new systems and equipment; (2) to evaluate and conduct research, development, and analysis activities; (3) to develop digitized prototypes and avoid the building of costly full-scale mockups; and (4) to plan for efficient production and sustainment of the new systems and equipment. Because M&S has become a catchall term for all aspects of computer-based analysis and is very domain dependent, we need to start with some basic terms and points of discussion:

- A *model* is a physical, mathematical, or logical representation of a system, entity, phenomenon, or process.
- A *simulation* is the implementation of a model over time. It is a technique that can be used for design, testing, analysis, or training. A simulation moves or changes over time. You can see the model(s) in the simulation moving—whether it shows military units moving across a battlefield or engine parts moving in a simulated car engine. M&S provides virtual duplication of products and processes and represents those products or processes in readily available and operationally valid environments. Use of such M&S can reduce the LCC by conducting trade space studies.

Typically, we categorize simulations into three classes (Department of Defense, n.d.): virtual, constructive, and live:

- Virtual simulations represent systems both physically and electronically. Think of a video game or a cockpit mockup used to train pilots—these are virtual simulations.

- Constructive simulations represent systems and their employment through the use of extensive, complex mathematical and decision-based modules and statistical techniques. A constructive simulation is a computer program. The user inputs data to cause an event to occur and then gets the results. For example, a military user may input data on a military unit telling it to move and to engage an enemy target. The constructive simulation determines the speed of movement, the effect of the engagement with the enemy, and any battle damage that may occur. Results can be provided digitally or visually, depending on the type of simulation used.
- Live simulations are simulated operations conducted by real people using real equipment. Military training events using real equipment are live simulations.

For simulation-based costing (SBC) analysis, constructive simulations are the primary analysis tool. Simulation is important for cost analysis because

- The system can be virtually prototyped,
- CONOPS can be used to conduct "what if" or trade space studies, and
- When used jointly, prototyping with CONOPS, can be used to assess cost versus risks for a LCC estimate.

5.1.1 Ways to study a system

Obviously, building a prototype and testing it in actual field trials is the desired—but certainly not the most cost-effective—way to develop a new system. "What if" analysis and large enterprise-wide solutions can only be conducted in a synthetic environment. From the simplest component to modern airplanes, detailed models are developed and simulated under a wide variety of conditions. Only then should prototypes be built and some actual tests conducted. Before the advent of cheap computing, most engineering analysis was conducted using physical models. Today, few physical models are used in the design and testing even below the system level. However, they are still used extensively to develop input and understand fundamental behavior that occurs below the system level. Figure 5.2 presents the many ways to model the behavior of a system.

5.1.2 Advantages and disadvantages of simulations

There are many advantages to conducting simulations:

- Once the model is explained, most people can understand it and accept its results as legitimate representations of the system under consideration. A simulation is more "intuitive" than a closed-form mathematical equation.

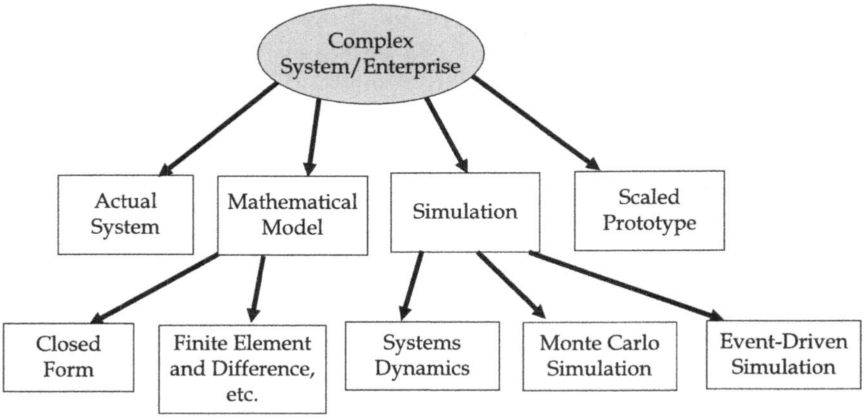

Figure 5.2 Ways to model a system

- Simulation can be used for complex, real-world situations or conditions that are not included in analytical models.
- We can simulate extended periods of time in a short period of time on a computer.
- It is much less expensive to build something in a computer language and experiment with the model than it is to construct the physical system for experimentation.
- Simulation allows for easier "what if" analysis and variations on the existing model (sensitivity analysis).
- Simulations are relatively straightforward and can be developed with minimum cost.
- Simulations are easier to apply than analytical methods.
- Simulations provide greater flexibility in representing the real system with fewer simplifying assumptions.
- Simulation precludes loss of life and damage to the environment.
- Models can be used repeatedly.

Simulations also have disadvantages:

- They do not lead to fundamental understanding. We observe outcomes of a process but may not understand why the outcomes are as they are.
- Simulation is an abused analytical tool that is often used inappropriately.
- The best M&S languages can be expensive and require considerable time investment to learn the software.
- Simulation models do not provide optimal solutions.
- Only the conditions that are included in the model can be examined.
- You may not discover fundamental relationships that are sometimes illuminated by analytical models.

When done correctly, proper cost estimation can provide benefits that greatly surpass the costs to perform it. We propose the process shown in Figure 5.3 as a way to use SBC in cost analysis. Different problems and requirements require different sets of methods, processes, and tools to produce defensible and rational analysis. Simulation-based costing is but

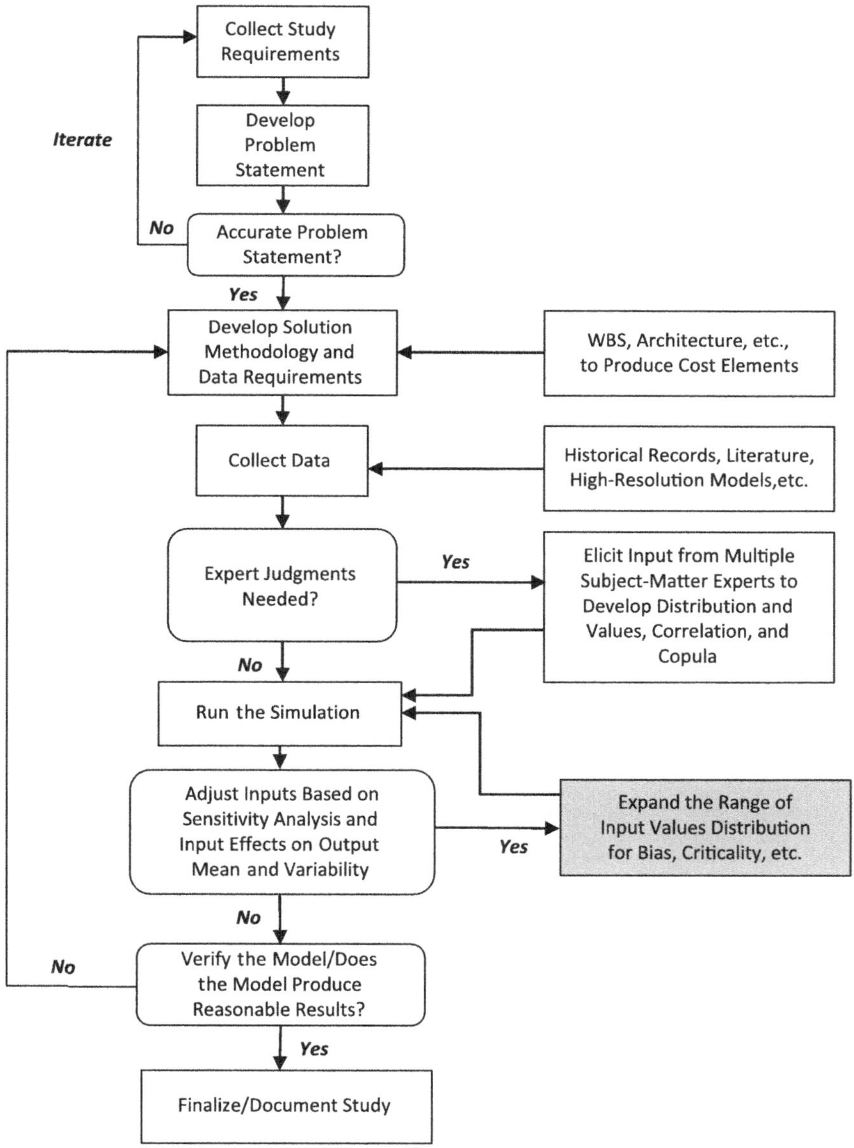

Figure 5.3 Use of SBC in conducting total ownership cost analysis (Farr et al., 2016)

one of the techniques that we can use. However, its ability to assign risk makes it very useful for early life cycle cost analysis.

5.2 Review of probability and statistics

5.2.1 Introduction

Many of the simulation techniques presented in this text require some knowledge of probability and statistics. The reason for this section is not to present the material in depth but to introduce or review the concepts of probability theory and elements of statistics.

In probability, properties of the population are assumed to be known and can be modeled mathematically. The term "probability" refers to the study of randomness and uncertainty. Probability can be used to answer questions about the sample. Statistics is used to determine properties from a sample that can be applied to the general population. These relationships are shown in Figure 5.4.

5.2.2 Random variables

A random variable is a special type of function critically important to the science of mathematical statistics and thus simulation. The concept of a random variable allows us to relate the exponential outcomes to a numerical function describing the outcome. It is a function that

- Has as its domain the sample space,
- Assigns one and only one number to each point in the sample space (integer, 0 or 1, or real), and
- Has value only through an experiment.

The value assumed by a random variable associated with an experiment depends on the outcome of the experiment. As a convention, this text

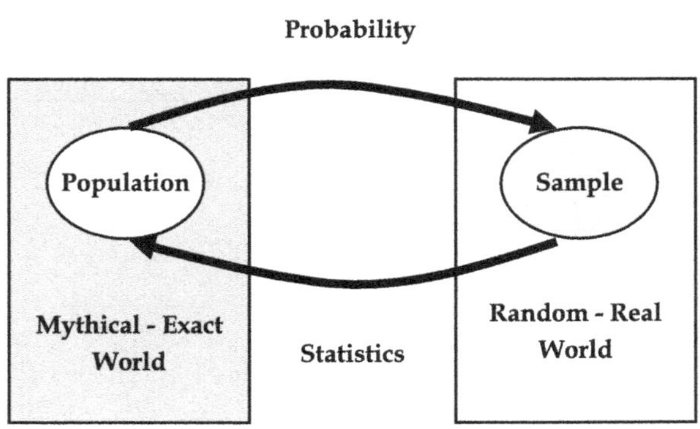

Figure 5.4 The relationship between probability and statistics

EXAMPLE 5.1

You are trying to determine the probability of a contractor filing more than two claims for a given contract. You analyze the last five jobs, and the number of claims are:

Job	1	2	3	4	5
Claims	0	4	2	5	2

Let X be the number of claims. There are four possible outcomes: 0, 2, 4, and 5. Then

$$p(0) = P(X = 0) = \frac{1}{5} = 0.2$$

$$p(2) = P(X = 2) = \frac{2}{5} = 0.4$$

$$p(4) = P(X = 4) = \frac{1}{5} = 0.2$$

$$p(5) = P(X = 5) = \frac{1}{5} = 0.2$$

These values specify the probability distribution function. In simple terms, for every possible value of x, the probability distribution specifies the probability of observing that value when the experiment is performed. Thus,

$$P(X \geq 2) = 1 - [P(X = 0) + P(X = 1) + P(X = 2)] = 1 - [0.2 + 0 + 0.4] = 0.4$$

will use a capital letter (usually X or Y) to represent a random variable and its lowercase version to represent a specific numerical value of the random variable. Example 5.1 presents a simple probability problem. The probability that X is greater than a given value is an important element of risk.

5.2.3 Probability density functions

The probability distribution (or density) function (PDF) provides a complete description of a random variable. A PDF is an important part of engineering planning when trying to model complex processes (e.g., construction). Typically, the expected value and some measure of scatter are used to summarize the important characteristics of a PDF.

Let X be a random variable such that $P(X = x) = f(x)$ is known for each x in the sample space. The term f(x) is the PDF of X. It describes the probability for X over the entire sample space. There are two types of random variables, the discrete and the continuous, depending on whether the results are countable or infinite.

1. The discrete type of random variable: Let X be a special type of random variable defined on the set of real numbers in a way that for any finite interval there exists a finite number of values. Let f(x) be a function such that

$$\sum_{i=1}^{n} f(x_i) = 1 \tag{5.1}$$

Example 5.2 illustrates how to develop a discrete PDF.

EXAMPLE 5.2

As part of a major land development project, 125 construction sites are assigned a 1–5 rating, with 5 being the most desirable. The rating is based on view, trees, constructability, access, and lot size. Based on a study of the area, we obtained the values in this table:

Lot Rating, x	1	2	3	4	5
Number	10	35	20	27	33
Probability, p(x)	0.08	0.28	0.16	0.216	0.264

The numbers 1 through 5 in these tables are values of the random variable. A PDF of discrete random variable is defined for every number x by p(x) = P(X = x) or

$$p(1) = P(X = 1) = 0.08$$
$$p(2) = P(X = 2) = 0.28$$
$$p(3) = P(X = 3) = 0.16$$
$$p(4) = P(X = 4) = 0.216$$
$$p(5) = P(X = 5) = 0.264$$

and equivalent description, and the most widely used for a PDF is

$$f(x) = \begin{Bmatrix} 0.08 & \text{if } x = 1, \\ 0.28 & x = 2, \\ 0.16 & x = 3, \\ 0.216 & x = 4, \\ 0.264 & x = 5, \\ 0 & \text{otherwise} \end{Bmatrix}$$

Figure 5.5 is a graph of the relative frequency histograms.

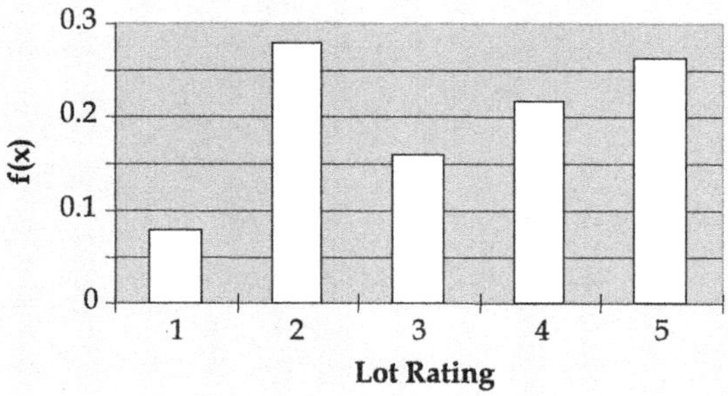

Figure 5.5 Probability histogram

2. The continuous type of random variable: A random variable is said to be continuous when the value constitutes an infinite set between two numbers, say a and b. Let X be a random variable such that

$$P(a \leq X \leq b) = \int_a^b f(x)dx \tag{5.2}$$

and

1. $f(x) \geq 0$ for all x,
2. $a < b$,
3. f(x) has, at most, a finite number of discontinuities on every finite interval, and
4. $\int_a^b f(x)dx = 1$

then X is said to be a continuous random variable. Example 5.3 illustrates how to develop a continuous PDF.

Whether the random variable is continuous or discrete, the function f(x) completely defines its probability properties. The function f(x) is called the probability distribution function of x.

5.2.4 Cumulative distribution functions

For some fixed value of x, we often wish to compute the probability that the observed value of X will be at most x. Consider the random variable X. Take

EXAMPLE 5.3

Let the random variable X have the PDF

$$f(x) = \begin{cases} 2x & 0 \le x \le 1 \\ 0 & \text{otherwise} \end{cases}$$

Find P(1/2 ≤ X ≤ 3/4) and P(–1/2 ≤ X ≤ 1/2)

$$P(1/2 \le X \le 3/4) = \int_{1/2}^{3/4} 2x\,dx = \left[3/4^2 - 1/2^2\right] = 5/16$$

$$P(-1/2 \le X \le 1/2) = \int_{-1/2}^{0} 0\,dx + \int_{0}^{1/2} 2x\,dx = 1/4$$

x to be a real number, and consider the unbounded set, ∞ to x, including the point x itself. Let F(x) = P(X ≤ x). The function F(x) is called the cumulative distribution function (CDF) of the random variable X. Since F(x) = P(X ≤ x), then with f(x) the PDF,

$$F(x) = \sum_{n=-\infty}^{x} f(n) \tag{5.3}$$

for the discrete type of random variable.

For the continuous type of random variable, the CDF can be expressed as

$$F(x) = \int_{-\infty}^{x} f(u)\,du \tag{5.4}$$

In Figure 5.6, for each x, F(x) is the area under the PDF curve to the left of x.

Note here that F(x) is a function of the point x, and

$$\frac{dF(x)}{dx} = f(x) \tag{5.5}$$

The probability p(a ≤ X ≤ b) = F(b) – F(a) except in rare functions involving discontinuities. Example 5.4 is an example of discrete CDF.

Conceptually, a simulation is simple. As shown in Figure 5.7, for all data input (e.g., behavioral properties, cost, time), an expected value

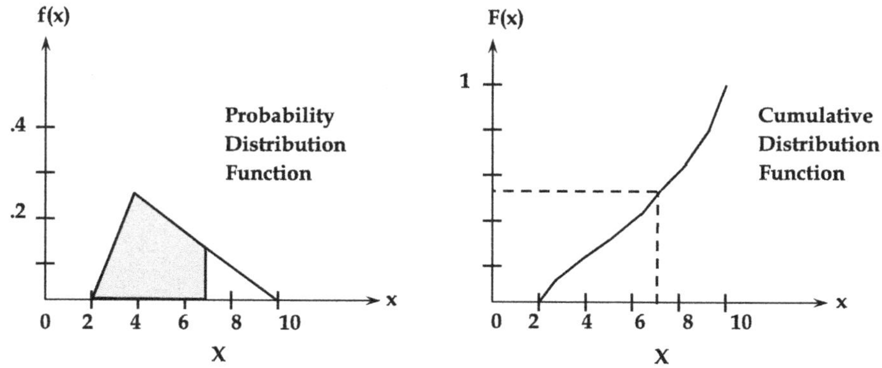

Figure 5.6 A PDF and associated CDF

along with a probabilistic representation of its behavior is developed. An inverse transformation of that function is performed and converted to a mathematical function where a random number drawn from a uniform distribution between 0 and 1 is inputted into the function, and the output is a value representative of distribution. This equation is referred to as a

EXAMPLE 5.4

The PDF for the number of contractors that will bid on a computer upgrade for a major federal client

X	1	2	3	4
p(x)	.4	.3	.2	.1

Thus,

$$P(X \le 1) = 0.4 = p(1)$$
$$P(X \le 2) = 0.7 = p(1) + p(2)$$
$$P(X \le 3) = 0.9 = p(1) + p(2) + p(3)$$
$$P(X \le 4) = 1.0 = p(1) + p(2) + p(3) + p(4)$$

The CDF is

$$F(x) = \begin{cases} 0 & \text{otherwise} \\ .4 & x < 1 \\ .7 & x < 2 \\ .9 & x < 3 \\ 1.0 & x \le 4 \end{cases}$$

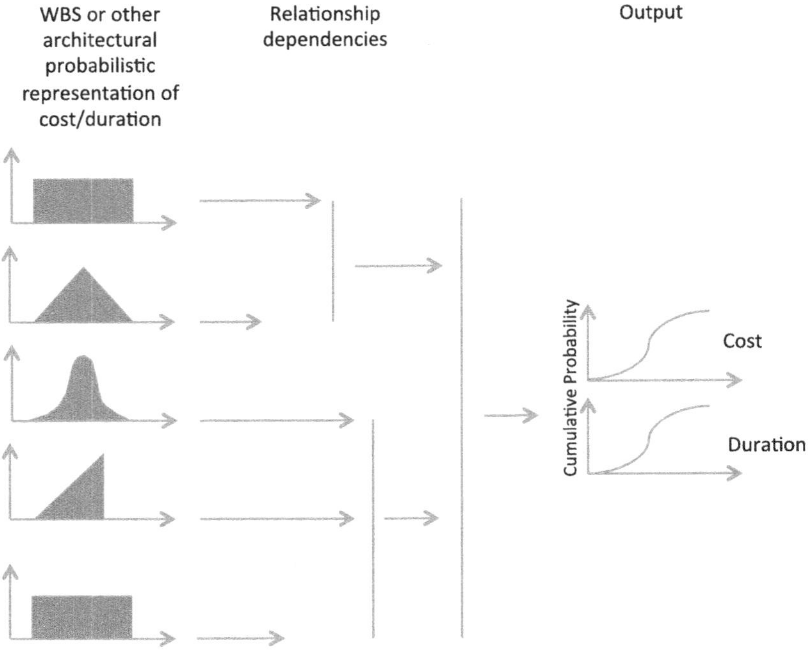

Figure 5.7 Conceptual representation of a simple simulation (modified from NASA, 2015)

process generator, which was developed using the Inverse Transformation Method (ITM). For each process generator, a random number is input many times (thousands). Then, based on the central limit theorem, which says that the distribution of a sum or average of many small random quantities is close to normal[1] under a wide range of conditions and distributions representative of the behavior of the system or enterprise can be modeled. Note that these process generators can be discrete or continuous.

5.3 Discrete process generators

Discrete process generators (DPGs) associate an outcome with a random number generated from a uniform distribution between 0 and 1 or U(0,1). We use U(0, 1) because this is the only random number distribution most software applications can approximate. Other random number generators in software applications are nothing more than functions of the U(0, 1) distribution. You will use a DPG when you wish to simulate an experiment that has a finite set of possible outcomes.

1 This is true even if the quantities are not independent (as long as they are not too strongly associated).

Table 5.1 Game outcome generator

Outcome	Probability	CDF	Number Range
Go Forward 1	.50	.50	$0 \le r < .50$
Go Back 1	.25	.75	$50 \le r < .75$
Go Forward 2	.20	.95	$.75 \le r < .95$
Go Back to Start	.05	1.0	$.95 \le r \le 1.0$

Before we discuss the actual development of a DPG, we will discuss the concept. Understanding DPGs is easy if you think of it in terms of a spinner that comes with a board game. Assume we have a game with four outcomes, and each outcome has a different chance of occurring. The outcomes and their associated probabilities are given in Table 5.1.

If we were to represent these outcomes and their probabilities on a game spinner, the spinner would look like Figure 5.8. The area of the circle allocated to each outcome is equal to the associated probability for each outcome. Therefore, if we were to spin the spinner, 50% of the area where it could end up would indicate the outcome "Go Forward 1." Each of the outcomes has a portion of the circle that is equal to its probability of occurrence.

We could take this spinner analogy one step further and associate it with random numbers. You could also use 100 equal-sized pieces of paper; number them from 0 to 99 and toss them into a hat. Then, you could randomly draw a piece of paper. The number on the paper would

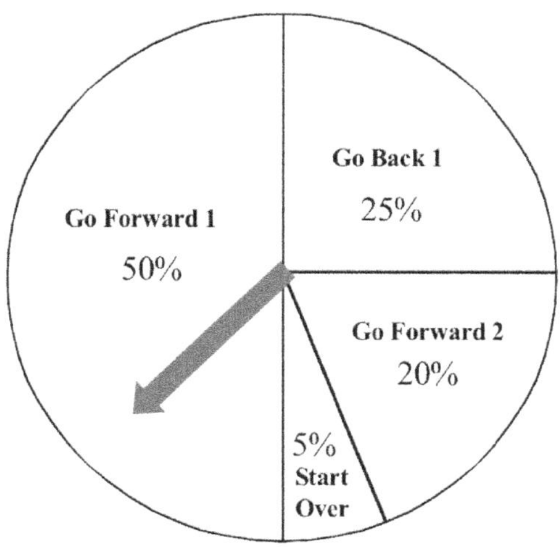

Figure 5.8 Spinner for game

correspond to one of the possible outcomes. Because 50% of the area is associated with "Go Forward 1," we could let the numbers 0 through 49 indicate the outcome of moving forward one space. The area associated with "Go Back 1" represents 25% of the circle, so we should let 25 of the numbers indicate the outcome of going back one. But we cannot use 0 through 24 because those numbers are included in "Go Forward 1." Therefore, we will use the next 25 numbers, 50 through 74. We continue this way until all the numbered pieces of paper are associated with an outcome on the board. For example, if you pulled the number 64 out of the hat, you know you should "Go Back 1." Likewise, drawing an 83 means you would "Go Forward 2." Table 5.1 shows which numbers correspond with each outcome.

Developing a DPG is very similar to drawing numbers out of a hat—it's just a lot faster and more useful in simulation. The difference is that, in the that example, we had only 100 numbers to draw. In a U(0, 1) distribution, there are infinite numbers between 0 and 1 that can be drawn. We can handle that, however, when we set up the DPG. There are four steps to developing a DPG:

1. Classify and count outcomes. From a sample set of historical data, identify all possible, relevant outcomes of the process you wish to model. Then, based on the historical data, determine how many times each outcome occurred or was observed.
2. Determine the relative frequency of occurrence. Divide the number of times each outcome was observed by the total number of observations. This is the probability mass function for the experiment.
3. Determine the CDF.
4. Determine random number ranges. Identify the range of numbers, based on the CDF, that will associate a U(0, 1) random number with a specific outcome from the observed event. The CDF is used to set the lower and upper bounds for each range.

These four steps, when carried out in sequence, define a DPG that will return appropriate outcomes on the experiment based on a series of random numbers from a U(0, 1) generator. Each of these steps is demonstrated in the following scenario:

> *You wish to simulate the business at a local call center so you can determine how many employees of certain types should be hired and trained. You observed the order counter during the lunch rush one day. One of the pieces of information you collected was the type of information ordered by each customer. The results of your observations are provided in Table 5.2.*

Table 5.2 Call center information demand

Shipping Status	Payment Questions	Shipping Status	Payment Questions	Order Status
Order Status	Shipping Status	Order Status	Order Status	Shipping Status
Technical Support	Order Status	Technical Support	Order Status	Order Status
Payment Questions	Technical Support	Shipping Status	Technical Support	Shipping Status
Technical Support	Shipping Status	Technical Support	Payment Questions	Technical Support
Payment Questions	Technical Support	Request for Supervisor	Technical Support	Payment Questions

We can see that Table 5.3 consists of a finite set of outcomes and that we could use it to develop a DPG in order to simulate the ordering process. The first step in developing the DPG is to classify and count outcomes. We see there are only five different choices the customers can make. We identify and classify them by their type: order status, shipping status, payment questions, technical support, and request for supervisor. We count the number of times each outcome was observed and record it as shown in Table 5.3.

The second step is to determine relative frequency of occurrence. Divide each value by 30, the total number of observations, and record the resulting decimal value in the table.

Step three is to determine the CDF. Add each of the previous probabilities to derive a CDF value for each outcome in the experiment. We do this so we can avoid duplicate outcomes for some random numbers, as discussed in the spinner example.

Fourth and finally, determine random number ranges. By developing a range from the CDF calculated above, and comparing a U(0, 1) random number "r" to the set of ranges, we can simulate random information request orders that follow the probability distribution from the original observed data. We develop each range by allowing each outcome's CDF

Table 5.3 Discrete process generator

Information Requested	Times Ordered	Frequency	CDF	Random Number Range				
Order Status	7	.2333	.2333	0	≤	r_1	<	.2333
Shipping Status	7	.2333	.4666	.2333	≤	r_1	<	.4666
Payment Questions	6	.2000	.6666	.4666	≤	r_1	<	.6666
Technical Support	9	.3000	.9666	.6666	≤	r_1	<	.9666
Request for Supervisor	1	.0333	1.000	.9666	≤	r_1	≤	1

Table 5.4 Information request DPG being used

Random Number Generated	Information Request Simulated
0.1531	Order Status
0.6937	Technical Support
0.8508	Technical Support
0.6297	Payment Questions
0.3118	Shipping Status

value to be the upper bound of its range and the previous outcome's CDF value to be the lower bound for the range.

Notice only the lower bound value has the "equality" associated with it. This is for two reasons. First, we cannot let the value .2333 represent both the *Order Status* and the *Shipping Status* outcomes, so we will let any value just up to but not including .2333 represent the order status. Second, while the random number generator can produce the value 0, it cannot produce the value 1. Therefore, we do not want to erroneously bias the DPG against the *Request for Supervisor* frequency by inadvertently decreasing its range.

We use the DPG by generating a series of random numbers. Each of these random numbers represents a customer's information request. Therefore, from five different random numbers, we could generate five requests. Table 5.4 shows an application of the constructed information requested DPG. Just the DPG alone does not generate sufficient information to qualify as a simulation. Generally, a DPG will be used in conjunction with other process generators in order to create a model that returns interesting information.

A graphical representation of the DPG and how the information requested selection is related to the random number is presented in Figure 5.9. The random number r receives the value .50; its location is identified on the *r* axis (vertical axis). We extend a line from the axis to the CDF function. Where it touches the CDF, we drop a line and identify which information request was "generated" via the information request DPG and the U(0, 1) random number.

5.4 *Continuous process generators*

The DPGs you learned in the previous section are extremely useful and cover many applications, and you will find yourself using them frequently in simulation. But not all processes are discrete. Some have continuous results; the sample space is infinite. For example, the quantity of oil pumped from a well in one hour could be 34 barrels, it could be 34.12 barrels, it could be 34.1275 barrels, and so on. In fact, it could be any number of barrels between 0 and some reasonable number. There are an infinite

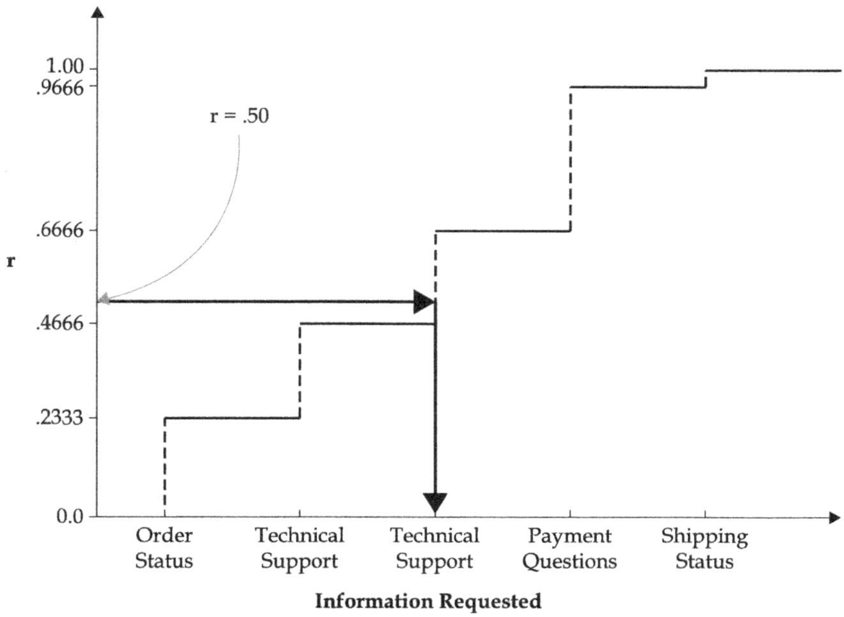

Figure 5.9 Graphical representation of information request DPG

number of possibilities, and therefore the process must be simulated with a continuous process generator.

There are a large number of possible distributions in the probability and statistics literature that can be used to describe element costs. However, in practice, only a few are often employed. The following distributions are among the most often used:

- Empirical distribution is based on sampling from historical data.
- Uniform distribution can be used to describe situations in which all outcomes between a low and high bound are equally likely.
- Triangular distribution is used to describe a function with high and low bounds but with a mode or most common peak; note that ramp functions are a special case of triangular distribution where the expected value is equal to the maximum (ramp up) or minimum value (ramp down).
- Normal distribution is a symmetric bell-shaped distribution centered on the mean with dispersion determined by variance.

Figure 5.10 shows a graphical display of the uniform, triangular, and normal distributions. Their empirical distribution is not shown as its form depends on the set of data from which it is derived.

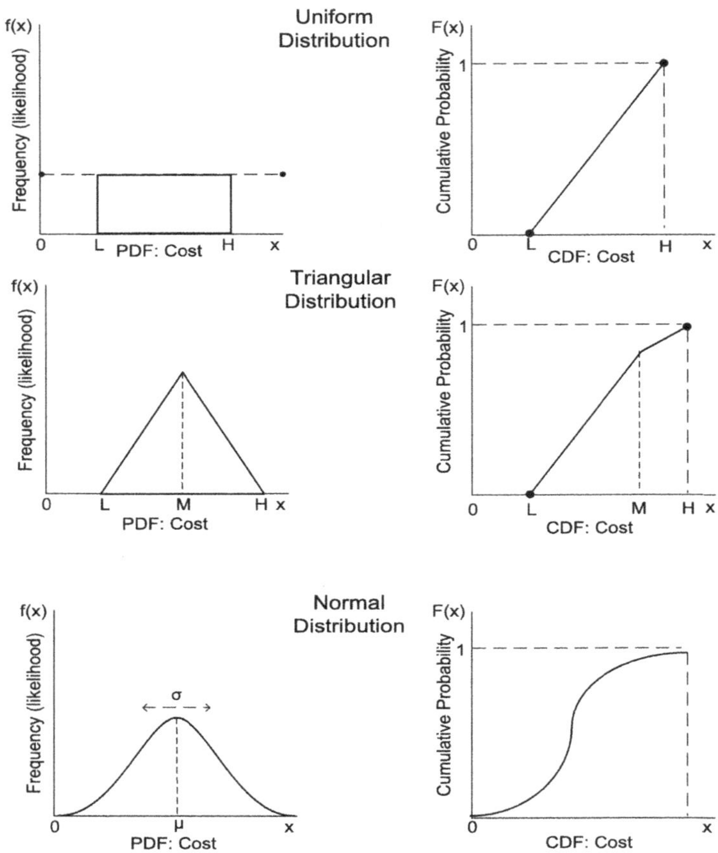

Figure 5.10 Common cost estimation distributions

Table 5.5 describes the four common distributions of interest. These distributions will be referred to throughout the rest of the chapter.

5.5 Probability and statistics summary

Obviously, a single discussion doesn't begin to develop the probabilistic and statistical concepts needed by an engineer. Statistical analysis must be persuasive in our profession. A better understanding of the randomness that affects complex systems, combined with better analytical tools, allows us the ability to incorporate statistical methods in the design process.

DPGs allow us to simulate events with limited outcomes and control each outcome's probability of happening. The four steps explained in detail throughout this reading allow you to successfully create DPGs

Table 5.5 Common probability distributions in cost estimation

Distribution	Empirical (discrete)	Uniform (continuous)	Triangular (continuous)	Normal (continuous)
Parameters	None	High (H), Low (L)	High (H), Mode (M), Low (L)	μ, σ
PDF	Data Driven	$f(x)=\begin{cases}\frac{1}{L-H} & \text{for } x\in[L,H]\\ 0 & \text{otherwize}\end{cases}$	$f(x)=\begin{cases}0 & \text{for } x<a\\ \frac{2(x-L)}{(H-L)(M-L)} & \text{for } a\le x\le c\\ \frac{2(b-x)}{(H-L)(M-L)} & \text{for } c\le x\le b\\ 0 & \text{for } x<a\end{cases}$	$f(x)=\frac{1}{\sigma\sqrt{2\pi}}e^{-\frac{(x-\mu)^2}{2\sigma^2}}$
CDF	Data Driven	$F(x)=\begin{cases}0 & \text{for} < L\\ \frac{x-L}{H-L} & \text{for } x\in[L,H]\\ 0 & \text{for} > H\end{cases}$	$F(x)=\begin{cases}0 & \text{for } x<a\\ \frac{(x-L)^2}{(H-L)(M-L)} & \text{for } a\le x\le c\\ 1-\frac{(H-x)^2}{(H-L)(H-C)} & \text{for } c\le x\le b\\ 0 & \text{for } x<a\end{cases}$	*standard normal N(0,1) $\Phi(x)=\frac{1}{\sqrt{2\pi}}\int_{-\infty}^{x}e^{\frac{-t^2}{2}}$
Mean	$\frac{\sum_{i=1}^{n}x_i}{n}$	$\frac{1}{2}(H+L)$	$\frac{L+M+H}{3}$	μ
Standard Deviation	$\sqrt{\frac{\sum_{i=1}^{n}(x_i-\overline{x})^2}{n}}$	$\sqrt{\frac{1}{12}(H-L)^2}$	$\sqrt{\frac{H^2+M^2+L^2-..HM-LM-LH}{18}}$	σ
Pros	Retains fidelity of historical data	Only two pieces of information required to estimate	Only three pieces of information required to estimate	Simple to implement and describe
Cons	Does not include values that have not already occurred	Few real-world variables actually follow this distribution	Does not include values outside of its stated range	Fails to predict 'tail' events

regardless of the discrete event being modeled. The concept and development of DPGs presented in this reading integrate seamlessly with spreadsheet simulations.

5.6 Simulation in practice

5.6.1 Introduction to simulation in practice

Almost any process associated with making a decision can be modeled using simulations: business processes, inventory, forecasting, project delivery, combat models, reliability, and so on. In fact, given the world of modern simulation software, everything from biological systems to complex processes associated with a large human, hardware, or software system of systems can be replicated in a virtual environment. The challenges are developing the appropriate level of resolution, mathematically describing the entity behavior, and understanding and modeling the processes and interactions associated with entities in the system. Figure 5.11 shows a visual representation of a simple logistics simulation where airplanes unload logistics packages at an airport.

5.6.2 Building complex simulations

Creating a simulation is no different than any other software project requiring well-defined requirements, appropriate design, configuration management, and a structured verification, validation, and accreditation approach.

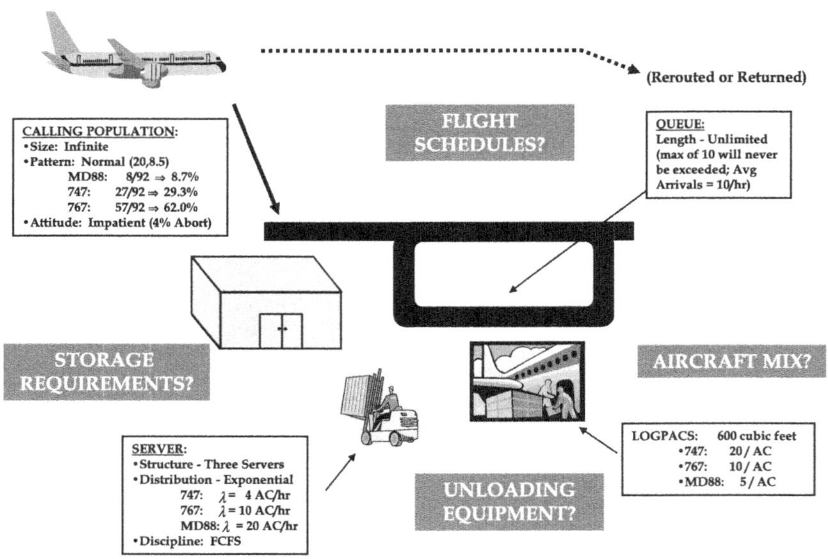

Figure 5.11 Visual representation of a simple logistics simulation

Referring to Figure 5.11, the first step in the development of any system is to determine the customer's requirements. Translating the customer's needs into meaningful requirements is a huge challenge for simulations because they are often developer driven. Once a need for a simulation has been identified, stakeholder requirements must be developed. Many companies accomplish this using a formal process called "voice of the customer." Whether or not a formal process is used, it is important to have a clear understanding of who the stakeholders are and the requirements of each before system design begins. Simple good requirements practices should be followed, including no abstract language, no statement of "how to," and no unquantifiable requirements. Too many simulations are built without knowing the problem, the customer, and the consumer. A simple simulation is shown in Example 5.5.

EXAMPLE 5.5

Use the following random number streams:

Stream 1: .11, .36, .45, .98, .78, .67, .51, .23, .15, .89
Stream 2: .91, .39, .72, .14, .42, .78, .55, .38, .60, .71

And the following information from the Petroco Service Station:

Time between Arrivals	Probability
1	.15
2	.3
3	.4
4	.15
Total	1.0

Construct a Monte Carlo simulation using the tables below.

First, we must construct a table and determine the random number ranges.

Time between Arrivals	Probability f(x)	Cumulative Probability F(x)	Random Number Ranges
1	.15	.15	$.00 \le x < .149$
2	.3	.45	$.15 \le x < .449$
3	.4	.85	$.45 \le x < .849$
4	.15	1.0	$.85 \le x \le 1.0$
Total	1.0		

(continued)

Now we can simulate the arrival of cars using the first stream of random numbers.

Arrival	Random Number	Time between Arrivals
1	.11	1
2	.36	2
3	.45	3
4	.98	4
5	.78	3
6	.67	3
7	.51	3
8	.23	2
9	.15	2
10	.89	4
Average		2.7
Std Dev		0.95

Next, we simulate the arrival of cars using the second stream of random numbers.

Arrival	Random Number	Time Between Arrivals
1	.91	4
2	.39	2
3	.72	3
4	.14	1
5	.42	2
6	.78	3
7	.55	3
8	.38	2
9	.60	3
10	.01	1
Average		2.4
Std Dev		0.97

Is there a difference in the results? If yes, why?

Yes, there is a difference in the results. Ten random numbers are not a large enough sample to have a true set of random numbers. This variation is expected for 10 values.

There is not a great deal of difference in the averages of the 10 interarrival times generated in this simulation (2.7 minutes vs. 2.4). Different random number streams can, however, produce great differences in simulations. Thus, we must run the simulation many times to eliminate any random number biases.

During system design, when key decisions are made about the architecture, many bad decisions are made. Most of the LCC commitments are made at this juncture. Simulations are often the victim of bad software language choice and inappropriate levels of resolution for the processes and entities. Most simulations are void of formal techniques to help prioritize system design, such as quality function deployment. Fabrication and coding has become the least of the challenges in developing a simulation. Modern computer-aided software engineering has made mundane yet critical implementation issues such as configuration management easy to address.

5.7 Using readiness levels for model input

Probably the biggest criticism of simulations is the lack of data. High-resolution simulations, surveys, interviews, and so on can all be used for developing process generators. However, early during the concept development phase, often technologies are immature and not well defined. Readiness level is one of many tools that can be used to assess the variability of the input.

Numerous readiness levels exist in the literature:

- Technology readiness levels (TRLs),
- Systems readiness levels,
- Integration readiness levels,
- Cost readiness levels,
- Manufacturing readiness levels, and
- Commercialization readiness levels.

TRL can be defined as a systematic metric or measurement system that supports assessments of the maturity of a particular technology and the consistent comparison of maturity between different types of technology (Mankins, 1995). The TRL is the accepted standard in government and industry as a measure of technology maturity. The TRLs range from 1 to 9, with 9 satisfying the highest degree of readiness (ready to market) and 1 the lowest (development of the concept only). TRLs can be used to assess the maturity of a technology (both hardware and software) and, in turn, the risks that might be associated with it. TRL might also serve as an indicator of technology obsolescence and replacement (Ganguly et al., 2010). Readers are referred to Mankins (1995) for more information on TRLs along with their detailed definitions.

Although TRL serves as an important evaluation scale to assess the readiness of a technology, a metric that provides a description of integration would greatly increase the robustness of the TRL metrics. Sauser et al., 2009), Gove (2007), and Gove and colleagues (2008) created an integration readiness level to measure integration maturity on a scale similar to TRL to provide a system-level readiness assessment. TRLs and integration readiness levels can be used to form a basis for a logical and defensible means to input variation for SBC.

NASA's relative risk-weighting methodology is used to translate elicited risk information based on given work breakdown structure (WBS) elements into cost impacts and scenarios by constructing triangular WBS cost-risk distribution analysis. The cost estimators and systems engineers collaborate to perform cost-risk assessment by evaluating cost-risk drivers and producing cost-risk scenarios within four suggested fundamental program categories: technology, design and engineering, complexity, and schedule (NASA, 2015). The results of risk scores for each WBS element risk scenario may be developed by adapting the analytic hierarchy process to derive weights for both the risk drivers' categories and the rating scales' intensities (NASA, 2015); therefore, the risk-weighted cost impacts within a range of triangular distribution (pessimistic, reference or "most likely," and optimistic) can be achieved by multiplying the risk factors and likelihood of system costs (NASA, 2015).

We assess and evaluate the appropriate TRL for each technology within a system and in which the risk values for each technology can be derived and translated based on its defined TRL indication. As mentioned, the risk values are assessed and mapped "reversely" from their associated readiness levels, in which the improved TRL-weighted risk rating scale may be viewed as the opposite of the TRL scale we often use, as shown in Table 5.6. However, it is likely that a system may be at a lower risk level when there is a higher TRL value representing the maturity state of a given technology.

As an example, Table 5.7 shows the risk levels of Technology 1, 2, 3, and n, which can be cross-referenced based on the assumed TRL values

Table 5.6 Improved TRL-weighted risk rating

TRL	Risk Description	TRL-Risk Level
1	Extra High	8
2	Very High	7
3	High	6
4	Moderate High	5
5	Moderate	4
6	Moderate Low	3
7	Low	2
8	Very Low	1
9	Extra Low	0 ~ 1

Table 5.7 Weighted TRL cost-risks

	Technology 1	Technology 2	Technology 3	Technology n	Weighted TRL Risk
Cost-Weighted Risk Contribution	0.05	0.23	0.44	0.28	1.00
Pessimistic TRL Risk Level	6	4	4	4	0.45
Most Likely TRL Risk Level	5	3	3	3	0.34
Optimistic TRL Risk Level	4	2	2	2	0.23

above. Also, in Table 5.7, the row of cost-weighted risk is equal to the percentage of cost-risk weight for a given technology in a system, and it sums to 100%.

Based on Table 5.7, the weighted technology cost-risk factor values are calculated for three different likelihood parameters, in which both low and high ends of the weighted TRL cost-risk factors within a triangular technology cost distribution can be calculated based on the formulas as follows:

$$\text{Weighted Low End Cost-Risk Factor} = \frac{\text{Weighted Optimistic TRL Risk Factor}}{\text{Weighted Most Likely TRL Risk Factor}} = \frac{0.23}{0.34} = 0.68$$

$$\text{Weighted High End Cost-Risk Factor} = \frac{\text{Weighted Pessimistic TRL Risk Factor}}{\text{Weighted Most Likely TRL Risk Factor}} = \frac{0.45}{0.34} = 1.32$$

Figure 5.12 shows a graphic representation of the results of our triangular technology cost distribution by applying weighted TRL cost-risk factors. Note that cost and time can be used interchangeably.

Therefore, based on the example above, as we hypothetically assume a system has an overall (most likely) TRL score of 5 and the technology cost equals $200,000, as well as with an assumption of 0.68 and 1.32 as its optimistic and pessimistic TRL cost-risk factors in respective order, we can conclude that the technology costs in a triangular cost distribution range from $136,000 to $264,000 with a mean of $200,000.

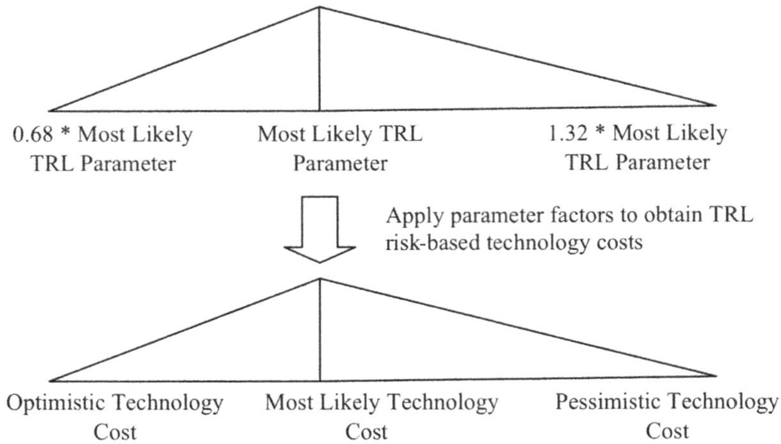

Figure 5.12 Graphic representation of weighted TRL risk-based costs in a triangular distribution (modified from NASA, 2015)

5.8 Simulation using spreadsheets

5.8.1 Introduction

Teaching simulation can be difficult. So, we took the approach in this book to teach the mathematical underpinnings, introduce you to simulation using manual techniques, expand the complexity of what we can model by using spreadsheets, and finally, using a spreadsheet add-in, model realistic systems. Many spreadsheet add-ins exist, including @Risk, Crystal Ball, Extend, and even some freeware packages. Hopefully, this hierarchal approach will allow modelers to build on basic skills before advancing to more complex simulations. Spreadsheets have a distinct advantage in that most students know how to use the built-in capabilities of tools such as Excel. Spreadsheets provide an ideal platform on which to learn the basics of simulation methodology.

Conducting simulations in Excel has some challenges:

- It's difficult to run many iterations.
- Many continuous process generators do not have built-in Excel functions.
- You must use the graphing capabilities to post process results.
- Developing the logic for queuing systems is very difficult without writing Visual Basic code.

Example 5.6 is a simple simulation example illustrating the concepts of discrete process generators. This basic single server queuing model would produce a wealth of information about the behavior of the system.

EXAMPLE 5.6

As a newly assigned operations engineer, you have developed a self-service supply center for your factory. You have one supply clerk to process the paperwork for the customers. You did a queuing analysis on your store but found the mathematical solution did not accurately model the real-world system; therefore, you decide to manually simulate the store's operation. You spend one day taking data and arrive at the following distribution of arrival intervals and service times:

Distribution of Arrival Intervals (Arrivals at the Check-Out Counter)

Arrival Interval (minutes)	Probability P(x)	Cumulative Probability	Random Number Range (r_1)
1	.45	.45	$0 \le r_1 < .45$
2	.35	.8	$.45 \le r_1 < .8$
3	.15	.95	$.8 \le r_1 < .95$
4	.05	1.0	$.95 \le r_1 \le 1.0$

Distribution of Service Times

Service Time (minutes)	Probability P(y)	Cumulative Probability	Random Number Range (r_2)
1.0	.35	.35	$0 \le r_2 < .35$
2.0	.55	.9	$.35 \le r_2 < .9$
3.0	.10	1.0	$.9 \le r_2 \le 1.0$

Below is a manual simulation of the process at the self-service supply center store.

1 Cust #	2 r_1	3 Inter-arrival Time	4 Total Clock Time	5 At Cashier Time	6 Wait Time	7 Queue Length after Entry	8 r_2	9 Service Time	10 Depart Clock	11 Time in System
1	—	—	0	0	0	0	.06	1	1	1
2	.93	3	3	3	0	0	.47	2	5	2
3	.72	2	5	5	0	0	.30	1	6	1
4	.83	3	8	8	0	0	.97	3	11	3
5	.35	1	9	11	2	1	.96	3	14	5
6	.17	1	10	14	4	2	.20	1	15	5
7	.82	3	13	15	2	2	.16	1	16	3
8	.63	2	15	16	1	1	.50	2	18	3
9	.68	2	17	18	1	1	.03	1	19	2
10	.98	4	21	21	0	0	.07	1	22	1
					$\Sigma = 10$					$\Sigma = 26$

(continued)

At Cashier Time for Entity 2 = MAX (Total Clock Time 2, Depart Clock 1)

Depart Clock = At Cashier Time + Service Time

Time in System = Service Time + Wait Time

You can determine the average waiting time, average queue length, and average time in the system and many other characteristics using these results:

Average Waiting Time (Wq): 10/10 = 1 minute
Average Queue Length (Lq): Little's Law: $Lq = \lambda Wq >$ (10/21) (10/10) = .4762 cust
Average Time in System (Ws): 26/10 = 2.6 minutes
Average Wait (Wq) = 10/10 = 1 Average Queue Length (Lq) = 7/10 = .70
Average Time in the System = 26/10 = 2.6

5.9 *Building systems simulations*

5.9.1 *Introduction*

Simulation has not only become accepted; analysts have also come to the realization that simulation is the only means to study complex systems. Tables 5.8 and 5.9 list some of the key mistakes made in simulation development and domain applications of the tools, respectively. Example 5.7 presents a simple example of risk versus schedule and how M&S can be used to incorporate risk into a decision.

5.9.2 *Using expert elicitation for data development*

There is a large body of work concerning subject-matter expert (SME) elicitation in the operations research community and in particular within the defense domain where problems are typically new and have minimal related empirical data (see Marks et al., 2013). Galway (2007) recommends the following procedure to elicit subjective probabilities for cost-risk analysis:

- Use multiple independent experts.
- Ask an expert to provide, at a minimum, upper, lower, and most-likely values for cost elements under consideration.
- Fit a triangle distribution to these three numbers, using the upper and lower values to bound 90% of the probability (where reasonable, to counteract known biases in elicitation).

Table 5.8 Key mistakes made in simulation development

Steps in the Systems Engineering Process	Mistakes Made in Development
Gather and Define Requirements	No formal systems engineering process and tools are used The customer is not clearly defined
System Design	No formal process or prioritization tools used No formal requirements management tools used Programmers not only wrote code but also developed the meta model and design
Detailed Design	Various levels of resolution of the entities and processes
Code	Wrong languages are used Did not investigate off-the-shelf solutions or prior subroutines for the simulation Did not follow processes Poor documentation
Software Testing (Validation)	No integrated plan beyond simply running the model
Integration (Verification)	Limited knowledge on how to look at the system No way to test the synergisms of various modules
Operational Acceptance (Accreditation)	Verification and validation is often not documented nor conducted by an independent group Accreditation is nothing more than a "rubber stamp"

- In addition to the upper, lower, and most-likely values, elicit at least two more percentiles (perhaps the 25th and 75th).
- Provide feedback to the expert about the results of the elicitation, including the final range of non-zero probabilities, the median estimated cost, the probability that the final cost will exceed the most-likely cost, and so on. Carefully document the process and the results and archive the data obtained for future retrospective studies.

Table 5.9 Typical M&S problems

Systems Engineering	Process-Based Problems	Systems Thinking
Hardware and Software Integration	Inventory	Social Problems
Interface Design	Decision and Risk Analysis	Governance
	Project Management	Strategic Investments
	Human Behavior Performance	Enterprise Management

EXAMPLE 5.7

You need to assess the time risk for building a new house. You need to sign a lease while the new home is being built. Below is a network of the building process.

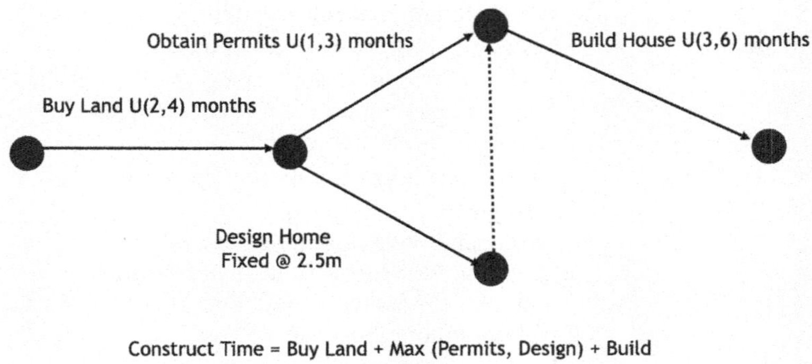

A model was built in @Risk and is shown below. The bigger question is, do you sign a lease for 50% probability or 80%? What is your risk tolerance?

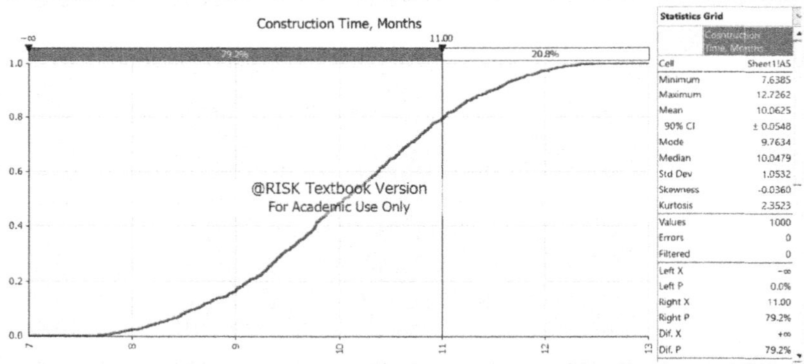

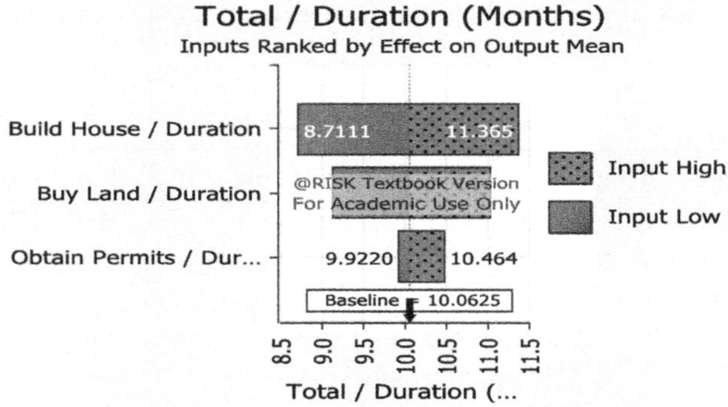

Developing subjective probability estimates involves identifying the right experts, ensuring that there is no bias in their estimates, and then developing probability distributions that are appropriate for the analysis (Farr et al., 2016).

5.9.3 *Adjusting for bias*

Biases are important and can affect subjective probability estimates. Contractors, owners, academicians, and others interject their own bias into data solicitations. Farr and colleagues (2016) developed a methodology for adjusting for bias. Bias categories are not equal, and more research to determine the percentages is required. It is clear that this approach to assessing the biases of given stakeholders is required to improve the quality of information that is elicited for cost considerations. Kahneman's (2011) book, *Thinking, Fast and Slow*, is an excellent reference on bias in decision making.

5.10 *Sensitivity analysis*

Sensitivity analysis involves varying different input values, processes, or assumptions to determine their effects for comparison with the results of the basic analysis. Sensitivity analyses conducted on major unknowns for each feasible alternative can provide a range of costs and benefits that may provide better and more defensible analysis than a single estimate. It is not sufficient to present the decision maker with a set of alternatives whose costs and benefits are based on most-likely factors and assumptions. Sensitivity analysis and SBC provide other information needed for any decision.

5.11 *Summary*

Modeling and simulation continue to be key tools in assessing cost risk throughout the life cycle, especially early in programs. M&S allows program managers to quickly develop CONOPS. Further along in the life cycle, M&S can be used for detailed design. M&S can be used in a distributed collaborative environment that supports authoritative information exchange and rapid refinement of the design or concept and over the system life cycle to respond to changing circumstances, such as technological advances and changing threats, tactics, or doctrine. During the early phases of defining the needed M&S environment, the engineering team must also establish the metrics, such as cost, to be used for evaluating candidate concepts. The specifics of the M&S environment can then be filled in such that meaningful measures of merit can be extracted from the simulations and used to focus further rounds of simulation. For the modern engineer, M&S should be applied throughout a system's life cycle in support of product development activities.

Unfortunately, M&S has been slow to be adapted by most cost engineers and estimators, perhaps due to concerns about the complexity of the analysis, a lack of data, or difficulties with management; it is an underutilized technique.

QUESTIONS

5-1. What was your preconceived definition of modeling and simulation before you started this class?

5-2. Can you think of problems that cannot be modeled using a simulation? (Hint: a simulation only gives you an answer for what you have modeled.)

5-3. Simulations are essential models for processes with a mathematical representation of the entity behavior. Which is more difficult to model: the process or the entity behavior?

PROBLEMS

5-1. Local college students drink a lot of coffee at the recently remodeled Café Java. The manager has gathered data on arrival, coffee demand, and service time using a single cashier. The relative frequency for the demand of coffee size for the 8:00 to 9:00 a.m. hours (peak consumption period) of operation is shown in the table below. Develop a discrete process generator that you can use in simulating operations.

Demand	Frequency
0	8
Small	10
Medium	22
Large	10

5-2. The restaurant manager next contacted one of her best coffee customers, Dr. Farr, to develop the other data needed for the simulation. Based on his extensive experience as a patron, Dr. Farr determined that the interarrival time of students was, on average, 2 minutes (closely approximating a Poisson process). Also, the service times were uniformly distributed, with a minimum time of 10 seconds and a maximum time of 55 seconds. Develop the continuous process generators needed for the simulation.

a. Arrival time process generator
b. Service time process generator

5-3. Use the following random number streams:

Stream 1: .11, .36, .45, .98, .78, .67, .51, .23, .15, .89
Stream 2: .91, .39, .72, .14, .42, .78, .55, .38, .60, .71

and the following information from the Petroco Service Station:

Time between Arrivals	Probability
1	.15
2	.3
3	.4
4	.15
Total	1.0

Construct a Monte Carlo simulation using the tables below.

a. Construct a table and determine the random number ranges.

Time between Arrivals	Probability f(x)	Cumulative Probability F(x)	Random Number Ranges
1			
2			
3			
4			
Total			

b. Simulate the arrival of cars using the first stream of random numbers.

Arrival	Random Number	Time between Arrivals
1		
2		
3		
4		
5		
6		
7		
8		
9		
10		
Average		
Std Dev		

c. Simulate the arrival of cars using the second stream of random numbers.

Arrival	Random Number	Time between Arrivals
1	.91	
2	.39	
3	.72	
4	.14	
5	.42	
6	.78	
7	.55	
8	.38	
9	.60	
10	.01	
Average		
Std Dev		

d. Is there a difference in the results? If yes, why?

5-4. Consider a continuous random variable with the following PDF, which represents service time:

$$f(x)\begin{cases} \frac{1}{3}, & \text{for } 2 \le x \le 5 \\ 0 & \text{otherwise} \end{cases}$$

Below is a sketch of the PDF:

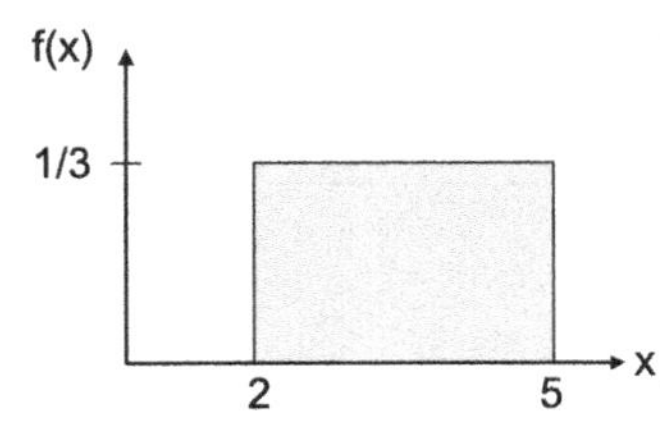

Develop a process generator for these service times using the ITM.

5-5. Consider a continuous random variable with the following PDF, which represents service time:

$$f(x) = \begin{cases} 2 - \frac{x}{2}, & \text{for } 2 \le x \le 4 \\ 0 & \text{otherwise} \end{cases}$$

Below is a sketch of the PDF:

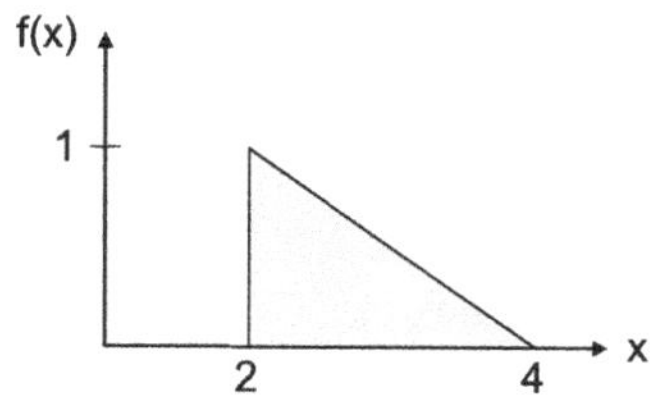

Develop a process generator for these service times using the ITM.

5-6. The operations manager is considering a refinement to the simulation model to study another possible change in inventory policy. The number of days the customer has to wait from the time they place the order until the time the products arrive at the site is not constant. The lead-time (a random variable X) is described by the following probability:

$$f(x) = \begin{cases} 8x - 4 & \frac{1}{2} \le x \le 1 \text{ weeks} \\ 0 & \text{otherwise} \end{cases}$$

a. Graph the function.
b. Use the ITM to derive the continuous process generator for the lead-time to receive the products.

5-7. Given the continuous probability distribution below, develop a process generator by using the ITM technique.

$$f(x) = \begin{cases} \frac{x}{18}, & 0 \le x \le 6 \\ 0, & \text{otherwise} \end{cases}$$

Graph the function, and use the ITM to derive the continuous process generator.

5-8. As part of a business process reengineering study, you collected the following times for calls received at your IT support desk. You would like to build a simulation to investigate bottlenecks and the return on investment of new technology. Your IT department provided you with the following interarrival times in minutes.

a. Using @Risk, fit a continuous process distribution to the data for use in a simulation.

Interarrival Times (mins)	1.108568755	0.732264692
0.588814749	0.547610383	0.242657492
0.397494588	1.538859923	0.352512396
1.673683718	2.719185953	0.112583034
0.084404664	3.242295179	0.115828188
0.230179077	0.118464846	3.459217465
0.587504092	0.583542146	0.711006448
1.37812423	0.389833717	0.610918746
0.327181262	0.746621891	0.478771272
0.535358812	0.658668746	0.291238526
1.064741685	0.03681801	0.018324774
2.761979873	0.187436268	0.621802042
0.56351202	0.382481155	0.817091787
3.29436124	0.990546057	0.290123951
0.366030742	0.771284964	0.335684761
0.073726122	1.694452517	3.786707383
1.780062427	0.562536755	0.46974402
0.544917667	0.039699342	0.926288378
0.182478972	0.669752309	0.189516512
3.550765671	1.590598732	0.737572538
0.592049529	0.162172788	0.662148766
1.932861976	0.109502818	0.835056714
0.331807512	0.874628264	2.907609454
0.39967384	0.25855252	1.18275267
0.523997406	0.330339127	0.067442005
1.413733145	0.069066114	0.403250281
0.492974166	0.365648322	0.043291821
0.909592149	3.288387525	1.49132871
0.471302103	2.225006862	1.309829203
0.722365606	0.632092026	0.22914525
1.887436095	1.475870936	0.891470352
0.197749859	1.172587523	0.576193114
1.82730058	3.223637925	0.867377053
	1.374153606	2.127784231

b. Using a histogram,[2] develop a discrete process generator representative of this data.
c. Under what circumstances is using a discrete process generator preferable to a continuous process generator? Give an example.

5-9. You run a @Risk simulation of project completion time as shown below:

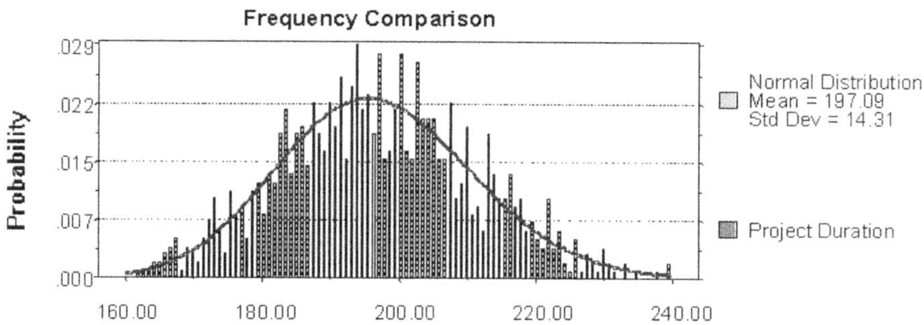

a. Your company bid this contract on completing the project within 185 days. What is the probability of completing this project in less than 185 days? (Show your work.)
b. You would like to develop a chart for the vice president of engineering to justify additional resources. Develop any chart that would be of interest using the information provided, and explain the chart in simple terms.

5-10. Using the ITM, determine the continuous process generators that will generate random variables with the following triangle distribution:

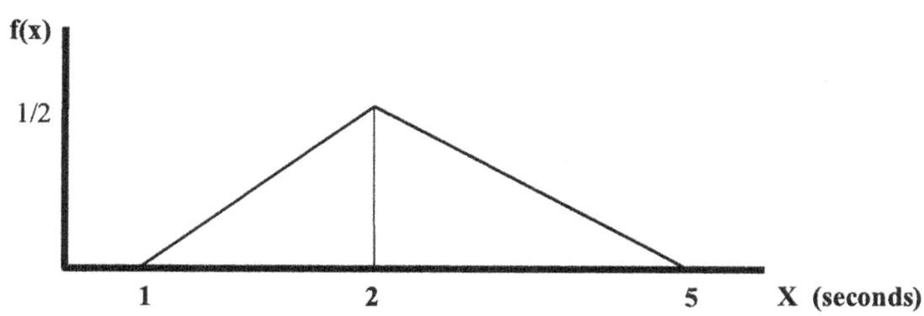

The sketch of this PDF:

Determine the PDF equation and the process generator for these interarrival times.

[2] Typically, we use the n^1/2 rule for determining bin sizes for histograms.

5-11. In order to verify the proposed service time process generator, you take a stopwatch and actually time 30 transactions.

a. Using Excel and bin sizes of 5 and 15, plot histograms using the data shown below. (Attach on a separate piece of paper.) Qualitatively, using these plots, does the uniform process generator (10,55) accurately reflect your observations?

42	12	28
28	21	20
46	30	32
45	42	26
13	49	27
38	20	40
11	45	25
47	22	10
34	45	50
27	38	43

b. What technique would you use to quantitatively assess (or hypothesize) whether this data behaves as a uniform distribution varying from 10 to 55?

5-12. The manager of Café Java is trying to determine whether to hire another cashier for the morning rush hour. Does she need one? Justify your answer using the simulation results obtained from five customers. Use the following random number streams:

First Set: .25, .95, .47, .02, .17
Second Set: .70, .89, .02, .46, .49

Assume that the interarrival times are exponentially distributed with a mean of $1/\lambda = 1.5$ minutes/customer. The ordering, preparation, and collecting money time (i.e., total service time) is an upward ramp function with a minimum of 15 seconds and a maximum of 1 minute.

a. Develop the process generators.
b. Simulate the arrival of five students by hand as the first step in designing a spreadsheet to study the problem in depth. (Hint: this is a basic single-server queuing system.)
c. Determine the average waiting time, average queue length, and average time in the systems.
d. Based on these limited results, is another cashier needed? (Justify your answer.)

e. How and why would you incorporate the demand for coffee into the simulation?

5-13. The commercial off-the-shelf program manager for the Seawolf class attack submarine has asked you to examine their proposed inventory policy for passive sonar geophones (PSGs). The inventory policy must support the Navy's diverse missions. The proposed inventory policy is to order 30 PSGs whenever the inventory drops below 16 at the east coast depot.

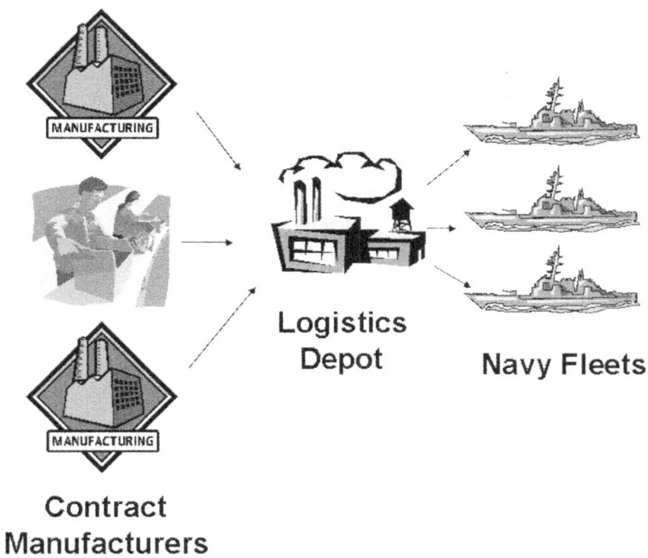

This new policy and your analysis is restricted to the following conditions:

- The Navy is prohibited from having more than one order in the system at any one time. They have to wait until the initial order of 30 arrives at the depot and the inventory level drops below 16 before a special order is placed.
- The Navy operates its depots 10 hours per day, 7 days a week. The days on the system clock are in terms of working days. Since the depot operates 10 hours/day, 1/10 of a day is actually 1 working hour. This accounting method prevents orders from being processed when the depot is not open.
- The depot can place and receive an order any day of the week.
- Assume a starting inventory of 21 PSGs.

The program manager feels that the following three process generators can be used to accurately model the behavior of this supply chain:

Process Generator #1

Continuous process generator for the interarrival time between demand orders:

$$x = -\frac{5}{6}\ln(r_1)$$

where

x = days between arrival of the PSGs
r_1 = random number

Process Generator #2

The PSG demand from the depot (demands per order)

Number of PSGs	Number of Occurrences
1	6
2	5
3	9
4	30
5	25
6	25

where

r_2 = random number
total count = 100

Number of PSGs	P(x)	Cum Prob	Random Number Range
1	.06	.06	$0 \le r_2 < .06$
2	.05	.11	$.06 \le r_2 < .11$
3	.09	.20	$.11 \le r_2 < .20$
4	.30	.50	$.20 \le r_2 < .50$
5	.25	.75	$.50 \le r_2 < .75$
6	.25	1.00	$.75 \le r_2 \le 1.00$

Process Generator #3

Continuous process generator for order lead-time:

$$Y = -12.5\ln(r_3)$$

where

Y = # of days lead-time and r_3 is a random number

Use these process generators to simulate the proposed inventory policy of ordering 30 PSGs whenever the number of geophones in inventory initially drops below 16. Use the general format in the following table to record the times, demands, inventory levels, and reorder points. Note that some of the random numbers in the table may not be needed. The clock time must be recorded in the table for each event in the simulation. Process 11 customers through the system in your simulation. Assume that the first order arrives in the system at a clock time of 0 minutes. The blank line between order arrivals allows you to insert receipt of an order when it occurs. The following column heading will help you set up this manual simulation.

r_1	Days between Orders	Clock Time	r_2	PSG Demand	r_3	Lead Time (Days)	Inventory Level

What is the average demand of PSGs per order?

5-14. Big Red Inc. rents trucks on a weekly basis. Trucks are picked up and dropped off at one of five locations:
- Enid, Oklahoma,
- Topeka, Kansas,
- Broken Bow, Nebraska,
- Goodland, Kansas, and
- Amarillo, Texas.

Management has developed the following "transition matrix." This matrix gives the probability of a truck being returned at each of the locations given the city where it was picked up:

Return City

		Enid	Topeka	Broken Bow	Goodland	Amarillo
Pick-Up City	Enid	.3	.35	.2	.1	.05
	Topeka	.25	.15	.3	.2	.1
	Broken Bow	.05	.1	.2	.3	.35
	Goodland	.4	.05	.1	.15	.3
	Amarillo	.35	.4	.1	.1	.05

For example, if a truck is picked up in Goodland, there is a 30% chance that it will be returned in Amarillo. Beginning with a truck in Broken Bow, Nebraska, simulate the rental and location of a truck for a 20-week period. Start by developing a discrete random variable generator for each city. Use the random numbers given. From the simulation experiment, determine the percentage of time a truck will be returned in each city. Discuss how this simulation might be changed to yield more accurate results.

Week	Pickup	r	Return	Week	Pickup	r	Return
1	Broken Bow	.45		11		.03	
2		.69		12		.47	
3		.33		13		.06	
4		.69		14		.55	
5		.88		15		.86	
6		.16		16		.25	
7		.7		17		.63	
8		.7		18		.18	
9		.07		19		.63	
10		.37		20		.18	

City	Number of Returns
Enid	
Topeka	
Broken Bow	
Goodland	
Amarillo	
Total	20

5-15. Gary is an enterprising businessperson. He is in the process of starting up a pizzeria. Specifically, he needs to know if 10 tables in his dining area is enough. Further, he would like to ensure that the average time a party must wait for a table is less than 2 minutes. Do a manual simulation of Gary's pizzeria to determine if the facility can meet Gary's criteria with its current setup. Assume there are 10 tables in the pizzeria, and at the beginning of the simulation 8 of the tables are occupied. The process generators for this problem follow. To do the simulation, you will also need to know the departure time of each group that is already in the pizzeria. These are shown in the table below.

Group Number	1	2	3	4	5	6	7	8
Departure time	68	66	81	85	27	18	21	59

Grp #	r_1	Inter-arrival (min)	Time of Arrival (min)	r_2	# in Group	Time Seated	Tables Avail	Wait Time for Table	r_3	# of Piz Ord	Cook Time	Time into Oven	Time Pizza Served	r_4	Eating Time	Depart Time	Total Time in Rest
9	.191			.204	3				.806					.959			
10	.655			.313	3				.478					.302			
11	.513			.982	6				.596					.593			
12	.474			.246	3				.322					.938			
13	.638			.019	2				.019					.285			
14	.439			.791	4				.154					.564			
15	.227			.346	3				.217					.389			

Interarrivals

Interarrivals are exponentially distributed with a mean of $1/\lambda = 10$ min/group. Thus, $\lambda = 6$ groups/hr. This results in the following process generator for interarrivals.

$$x = \frac{-\ln(r_1)}{6} * (60 \text{ min/hr})$$

 # in Group

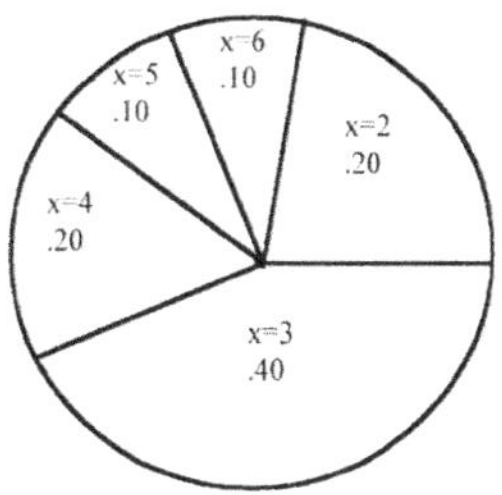

Order Size

# in Group	#of Pizzas Ordered (x)	p(x)
2	1	1.00
3	1 (or 2)	.60 (.40)
4	2	1.00
5	2 (or 3)	.20 (.80)
6	3 (or 4)	.50 (.50)

Cooking Time...

The pizza oven is of the conveyor belt type. It takes 10 minutes to cook a pizza and pizzas must be spaced 2 minutes apart.

Consumption and Conversation Time

This process is uniformly distributed (a, b) where a = 10 and b = 50. The continuous process generator for a uniform distribution is $x = r_4(50 - 10) + 10$ which generates "eat" times.

Note: also, assume a 5-minute delay between the time a group is seated until the time its pizza(s) enters the oven.

5-16. A major airport authority used Monte Carlo simulation to plan the replacement of an international terminal at one of its airports. The early program plan was for a building with dual levels, 1.37 million square feet, and 21 contact gates, with a cost estimate of $1.4 billion. The estimated project duration was 125 months. A value-planning workshop with 40 or so industry experts was held over an intense 2-week period. This value-planning workshop looked at all aspects of the project; a subset of this group performed a risk analysis of the estimated duration and estimated cost. Below is a table of activities on the critical path of the network.

Activity	Optimistic (months)	Expected (months)	Pessimistic (months)
Value Planning	0.50	0.50	0.50
Planning	2.0	6.0	8.0
Award Stage II Design	1.0	2.0	4.0
Stage II	10.0	16.0	20.0
Value Engineering	0.25	0.25	0.25
Project	3.0	6.0	8.0
Stage III Design	14.0	20.0	26.0
Demo Old Control Tower	4.0	4.0	7.0
New Ticketing	14.0	18.0	24.0
New Baggage Hall	6.0	8.0	10.0
Construct Departures and Arrivals	16.0	18.5	20.0
Fitout Baggage	6.0	8.0	10.0
Fitout Ticketing	12.0	14.0	16.0
Asbestos Abatement Main	4.0	5.0	8.0
Demolition Main	2.0	3.0	4.0
Renovate Main	9.0	12.0	16.0
Demolition Ticketing	10.0	12.0	16.0

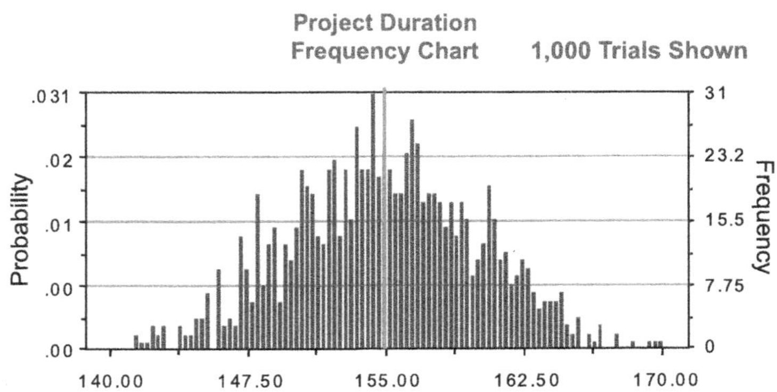

The results from an @Risk simulation are shown below:

Given that the mean project duration was 155 months with a standard deviation of 7, complete the following table and plot the results in Excel.

Duration	Probability of Completing in Less than Duration
125	
145	
155	
165	

5-17. As part of a process reengineering study, you collected the following data for interarrival times.

0.202693	0.777556	0.407303	0.016905	0.063665
0.014461	0.515572	0.718809	0.562519	0.068552
0.726741	0.827333	0.257001	0.52073	0.56361
0.763765	0.169967	0.864643	1.423221	0.357062
0.161565	0.535109	0.00114	0.571689	0.044286
0.244942	0.337265	0.144999	0.03313	0.018785
0.368699	0.434368	0.042632	0.240196	0.027319
0.163853	0.269635	0.339452	0.822752	0.231464
0.749167	0.301527	0.090666	0.464595	0.02154
0.04488	0.423597	1.197478	0.117271	0.145224
0.792627	0.710691	0.725581	0.29873	0.308896
0.070382	0.568303	0.197198	0.066977	0.543177
0.633093	0.318796	0.405549	0.423037	0.458229
0.163983	0.185042	0.082472	0.031344	0.356039
0.659358	0.570516	0.189155	1.217896	0.294747
0.743034	0.38452	0.356751	0.044138	0.83982
0.715777	0.37971	0.163473	0.722322	0.770268
0.698327	0.251889	0.188892	2.512673	0.996764
1.012138	0.256679	0.370203	0.23874	0.836308
0.023978	0.228313	0.569765	0.207689	0.083142

Using @Risk, fit the data and determine the best fit distribution.

5-18. You have just recently accepted a job with a local defense contractor. Your starting salary is $63,000. You believe that you can save up to 8% and your employer will match 5.5% of your gross salary that you invest in a S&P 500 fund. You would like to have $1 million in today's dollars in a mutual fund and retire in 20 years. Using historical inflation, bond returns, and stock market returns, develop a simulation model using @Risk to assess the probability of achieving these results. Assume that you will receive on average a 5% raise to account for inflation, promotions, and bonuses. Below is a table containing the percentage return (based on S&P 500) for the stock market and the inflation rate since 1951. Write a brief paragraph describing the results of your simulation. Discuss your investment strategy and the probability of success. Use output figures from @Risk to justify your strategy. Also, adjust your stocks and bonds allocation. Depending on risk, other sources of income, and so on, one general rules is that the percentage of bonds should correspond to your age.

Year	S&P 500 Average (%)	Inflation (%)	T. Bond Average (%)	Year	S&P 500 Average (%)	Inflation (%)	T. Bond Average (%)
1951	23.70	7.90	−0.30	1985	31.20	3.60	25.71
1952	18.20	2.20	2.27	1986	18.50	1.90	24.28
1953	−1.20	0.80	4.14	1987	5.80	3.70	−4.96
1954	52.60	0.50	3.29	1988	16.50	4.10	8.22
1955	32.60	−0.40	−1.34	1989	31.50	4.80	17.69
1956	7.40	1.50	−2.26	1990	−3.10	5.40	6.24
1957	−10.50	3.60	6.80	1991	30.20	4.20	15.00
1958	43.70	2.70	−2.10	1992	7.50	3.00	9.36
1959	2.06	0.80	−2.65	1993	10.00	3.00	14.21
1960	0.34	1.60	11.64	1994	1.30	2.60	−8.04
1961	26.60	1.00	2.06	1995	47.20	2.80	23.48
1962	−8.80	1.10	5.69	1996	22.70	2.90	1.43
1963	22.60	1.20	1.68	1997	33.10	2.30	9.94
1964	16.40	1.30	3.73	1998	28.30	1.60	14.92
1965	12.40	1.70	0.72	1999	20.90	2.20	−8.25
1966	−10.00	2.90	2.91	2000	−9.00	3.40	16.66
1967	23.80	2.90	−1.58	2001	−11.90	2.83	5.57
1968	10.80	4.20	3.27	2002	−22.00	1.59	15.12
1969	−8.20	5.40	−5.01	2003	28.40	2.27	0.38
1970	3.60	5.90	16.75	2004	10.70	2.68	4.49
1971	14.20	4.30	9.79	2005	4.80	3.39	2.87
1972	18.80	3.30	2.82	2006	15.60	3.24	1.96
1973	−14.30	6.20	3.66	2007	5.50	2.80	10.21
1974	−25.90	11.00	1.99	2008	−36.70	3.80	20.10
1975	37.00	9.10	3.61	2009	5.90	−0.40	−11.12
1976	23.80	5.80	15.98	2010	14.80	1.60	8.46
1977	−7.00	6.50	1.29	2011	2.10	3.20	16.04
1978	6.50	7.70	−0.78	2012	15.90	2.10	2.97
1979	18.50	11.30	0.67	2013	32.20	1.50	−9.10
1980	31.70	13.50	−2.99	2014	13.50	1.60	10.75
1981	−4.70	10.40	8.20	2015	1.40	0.10	1.28
1982	20.40	6.10	32.81	2016	11.70	1.20	0.69
1983	22.30	3.20	3.20	2017	21.83	2.65	2.80
1984	6.20	4.30	13.73				

5-19. Your company is bidding to design and install new multimillion-dollar radar systems at the Leeds, United Kingdom, airport. Use the following assumptions:

- This will be a 3-year contract based on the following cash flows.
- All values are in then year or actual dollars.
- You will be paid on delivery and certification of the system. (Remember, this will be in 2011 or 3 years from the contract award date of 2018, so you must account for inflation.)
- Year 0: Startup costs are estimated to be between \$8 million and \$10 million (assume this varies uniformly). These costs include refurbishing your factory, investment in capital equipment, and other normal project startup expenses. You will fund this work using existing cash reserves.
- You will borrow at 7% the expenses for years 1 and 2, which will be paid back at the end of the project.
- Year 1: These expenses are estimated to be \$8 million for mainly research and development and preliminary design. Because research and development is a highly uncertain process, assume this varies according to a normal distribution with a standard deviation of \$1.5 million.
- Year 2: Expenses at the end of the second year are of concern. During the second year, you will award subcontracts and start integration. You are concerned about the reliability of your subcontractors and their ability to adhere to their bid price. You estimate that expenses could vary uniformly between \$11 million and \$14 million.
- Year 3: The last year of the contract will be devoted to testing, installation, and customer acceptance. Typically, the airport authority has a poor track record of working within the contract guidance. Therefore, you think costs for the last year could vary between \$13 million and \$17 million with a heavy weight toward the high end, and so you model using an up ramp (fit with the triangular function and set the expected = maximum value).

Conduct a simulation using @Risk to develop your bid price. Assume an inflation value of 3% and a profit of 25%. Develop a plot of year 3 payout versus probability to present to management, and develop a one-paragraph recommendation. Remember that others are bidding on this project and it is key to keeping your Leeds facility operational.

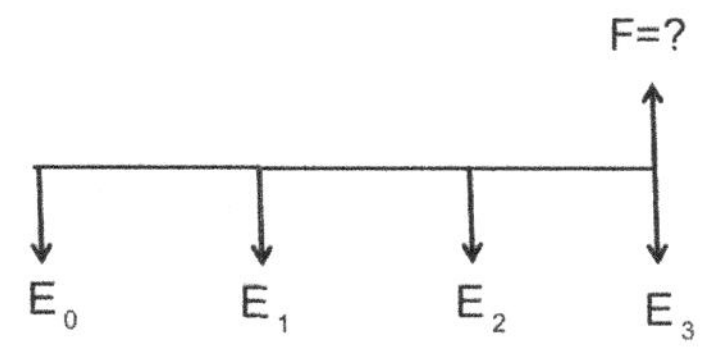

5-20. As vice president of engineering, you are responsible for many projects. Your company has recently bid on a major avionics upgrade for the US Air Force under a fixed-price contract. The proposal group in your company developed a preliminary design configuration as follows:

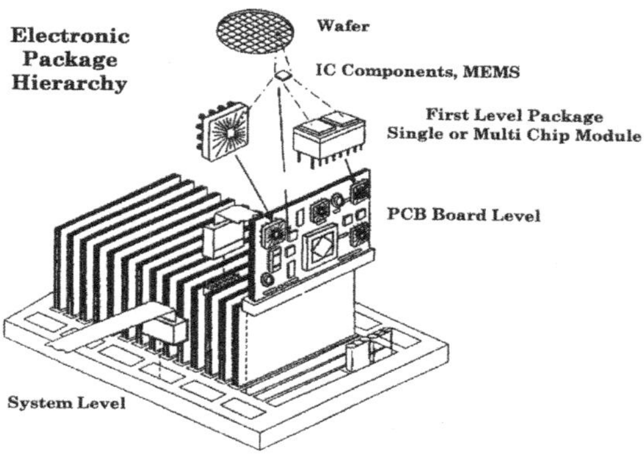

Hardware: Electronic Box
12″ x 16″ x 10″ Box
9 Circuit Cards Plus Motherboard
Wiring Harness
Fan
Read-Out Displays

Software
20,000 Source Lines of Code
Systems Engineering
Systems Requirements Study
Technical Integration
Logistics Technical Documents

Assuming a 1 January award date, the program manager has developed the following project management schedule:

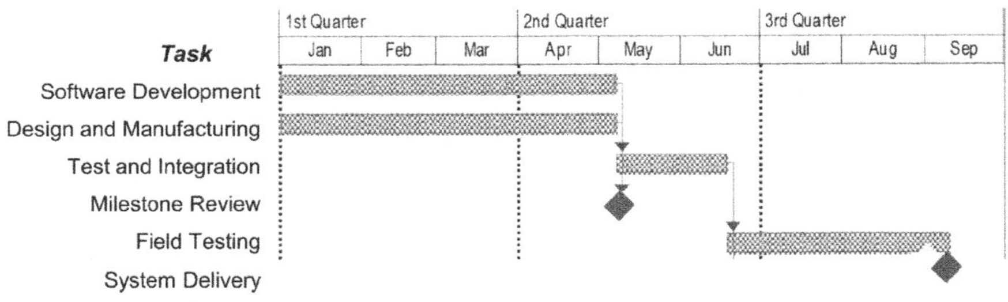

Gantt chart used for initial project planning

Based on this simplified schedule, you express your concern to both the project manager and the divisional vice president about meeting the 180 calendar days allotted for delivery of the system with the current staff. Your major concerns are software development and systems testing. With follow-on systems scheduled for bid during this and the next fiscal year, delivery of this system on time and within budget is critical. Since time drives cost, you must have an accurate estimate of the total project duration and assess the risk of various project durations.

You are trying to develop a more quantifiable and defensible methodology for estimating these projects for your company. Simulation seems to be the best tool for assessing risk for project delivery times. You would like to make this part of the normal estimating process for your division. Because of the visibility of this project, you decide to pursue simulation as an analytical tool to assess risk for the senior leadership of the company.

After careful research, you developed the following data based on historical data and expert solicitation in support of a simulation for this project:

Item	Distribution[3]	@Risk Fit	Notes
Software Develop-ment	*Uniform* from 74 to 112 days	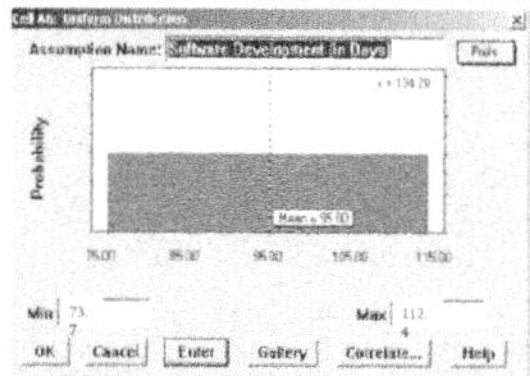	Based on the constructive Cost Model (COCOMO) for a 35-person team assuming semidetached and embedded for the limits.
Design and Manufact-uring	*Normal* with a mean of 90 and a standard deviation of 15 days	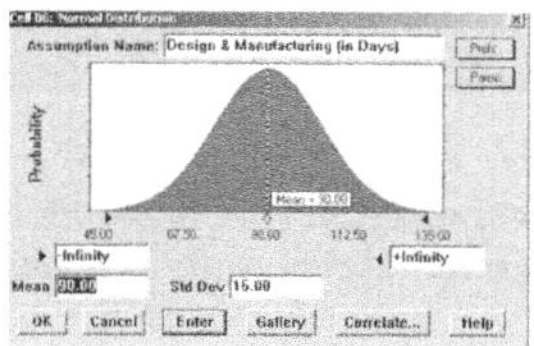	Uses many currently existing systems—small standard deviation. Usually outsourced and have little control over delivery. Plan for 25 engineers initially to be on this task. We estimate that for design and manufacturing, for every additional five engineers applied to the task that the mean and standard deviation would decrease by 10% (i.e., 90, 81, 72.9, etc.).
Systems Testing and Integration	*Triangular* function with 20 days minimum, peak at 30 days, and a maximum of 60 days	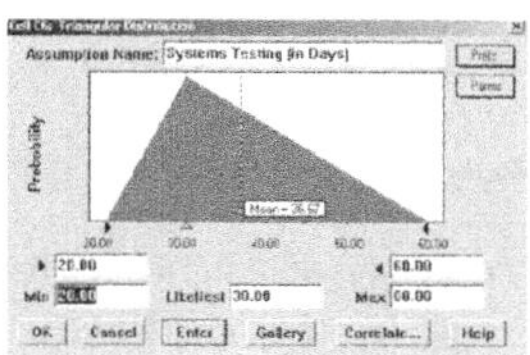	Greater potential for variation of greater than 30 days based on historical results. Plan for 10 engineers initially on this task. Additional engineers will probably reduce the maximum value. Assume 5 more engineers decrease maximum value 10%.
Field Testing	60 days fixed	NA	Specified not to exceed in contract.

[3] Calendar days were used to simplify the analysis.

Please re-create the following plot using @Risk along with another meaningful plot for management. Also, write a two-page executive summary on how best to achieve the 180-day stated objective. Assume that you cannot dedicate more than 100 engineers to this program. If you cannot reach the 180-day goal with 100 engineers, how much risk are you willing to assume?

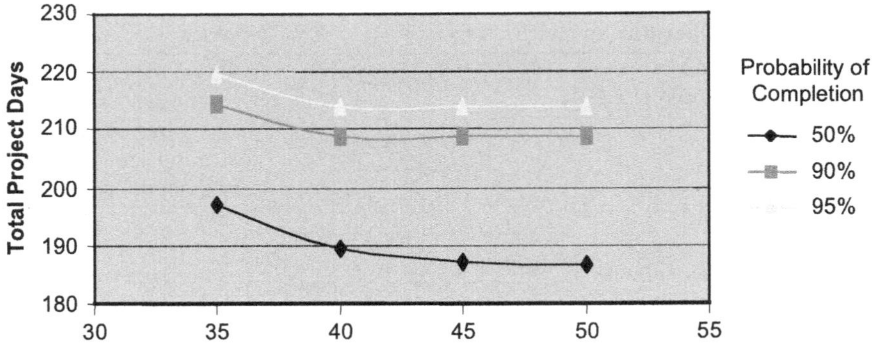

Probability of completion for 35, 40, 45, and 50 person software development teams

Note that the spreadsheet model for software development time using the Basic COCOMO model (software cost estimating will be discussed in Chapter 8) produced the following:

COCOMO Model		Software Development Team Size	Project Time Minimum	
a factor	3	35	73.7	Days
b factor	1.12	40	64.5	Days
Lines of Code	20000	45	57.3	Days
EAF Factor	1	50	51.6	Days
Effort Minimum	86.0 Person-Months			

		Software Development Team Size	Project Time Maximum	
a factor	3.6	35	112.4	Days
b factor	1.2	40	98.3	Days
Lines of Code	20000	45	87.4	Days
EAF Factor	1	50	78.6	Days
Effort Maximum	131.1 Person-Months			

5-21. You are in charge of costing and developing the schedule for a new satellite radome for a third world country to support its emerging wireless communications business. You need a rough first order estimate to apply to the United Nations Development Program for funding. As a first step, you develop the following work breakdown structure. Using the following information, develop a bid price and schedule: Project management (1.0) will be a concurrent activity for all other activities. Design and analysis (2.0) must be completed before the actual physical construction (3.0) and the software (4.0) are developed. These will be parallel activities. Integration and testing can only occur after the radome is constructed and the software developed. After integration and testing, you will turn the facility over to the phone company for operation. (Hint: this will describe the program (or project) evaluation and review technique (PERT) diagram for your simulation)

Satellite radome[4]

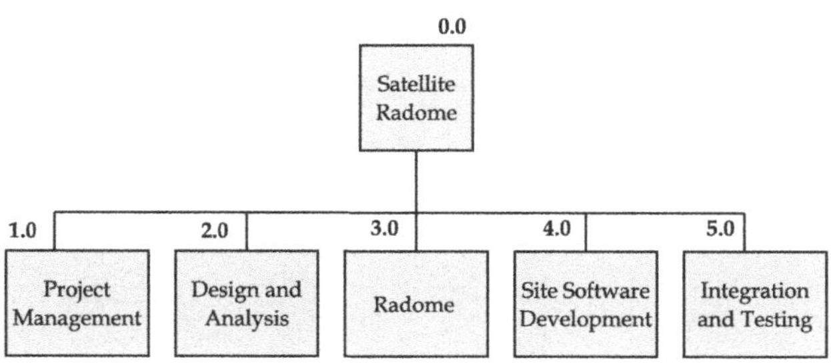

Level I WBS

[4] Photo taken 15 January 2018 from the US Air Force at http://www.afcent.af.mil/Units/379th-Air-Expeditionary-Wing/News/Display/Article/644008/met-provides-secure-communications-to-warfighter.

The local currency is euros. Everything will be built with US parts and labor, but you will be paid in euros. The current exchange rate is 1.3 dollars for every euro. The following table lists the behavior of each of the program elements. You have developed cost engineering relationships for each of these major elements.

Task	Duration	Cost ($)
Project Management	Will Last for the Whole Project	8% of Project Cost
Design and Analysis	Varies Uniformly between 5 and 7 Months	$50,000 Month
Radome	Varies Uniformly between 3 and 5 Months	Fixed-Cost Bid of $2 million
Site Software	Ramp Up between 4 and 6 Months	$80,000 Month
Integration and Testing	Varies Uniformly between 2 and 3 Months	$35,000 Month

Develop two plots of time versus probability of completion and total cost in dollars ignoring inflation. Then develop your bid price in euros. Assume you will receive 25% upfront and 75% on project delivery. Use a profit of 20% for this project.

5-22. Consider the software-centric project shown in the Gantt chart below:

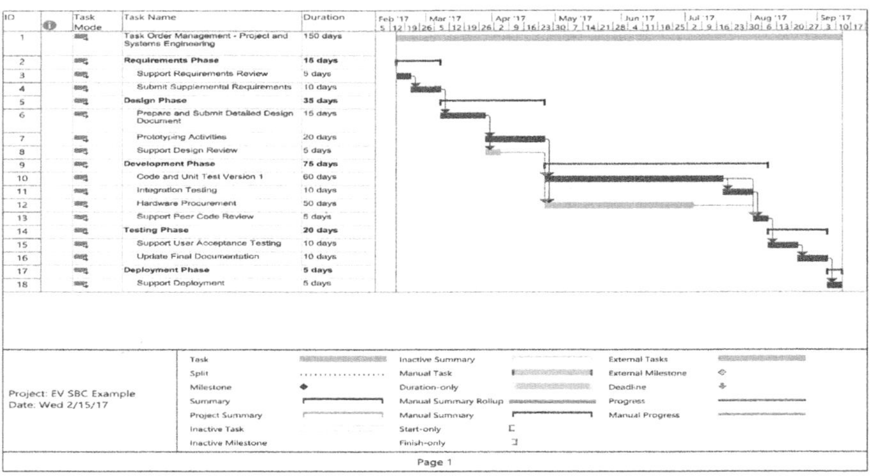

This is a firm fixed-price contract with a 150-workday requirement. Overruns mean reduced profit. Management has asked you to assess the risk of meeting the 150-workday requirement. You decide SBC would be the best means to develop a risk profile. After interviewing SMEs, combing the archives for historical records, and so on, you develop the distributions shown in the table below.

Task	Bid Value	Variability
Task Order Management—Project and Systems Engineering	150	This is fixed and tied to the total project duration and is not used in the risk assessment.
Support Requirements Review	5	This is fixed by the contract.
Submit Supplemental Requirements	10	SMEs say we usually miss this deadline by at least 50%. Thus, use an up-ramp distribution between 8 and 15 days.
Design Phase		
Prepare and Submit Detailed Design Document	15	This is fixed by the contract.
Prototyping Activities	20	SMEs say that for this contract this is probably realistic, but you could expect some variation of 20%. Thus, use a uniform distribution between 16 and 24 days.
Support Design Review	5	This is fixed by the contract.
Development Phase		
Code and Unit Test Version 1	60	Because this is the most expensive and longest activity to collect historical results for 36 like projects, which is summarized in the table below, you will need to fit a distribution to this value.
Integration Testing	10	This is fixed by the contract.
Hardware Procurement	50	You have extensive experience with the vendor and they have always met your deadlines. Thus, you have high confidence in this number and decide to treat it as a fixed value.
Support Peer Code Review	5	This is fixed by the contract.
Testing Phase		
Support User Acceptance Testing	10	The timeframe specified by the contract is seldom met because of customer requirements. You decide to use a triangular function with a min = 8, most like =10, and max =15 days.
Update Final Documentation	10	This is fixed by the contract.
Deployment Phase		
Support Deployment	5	This is fixed by the contract.

Historical Code and Unit Test Data

62	53	75	72	74
74	64	75	70	66
58	53	50	57	63
64	65	52	72	62
59	60	65	70	60
71	51	60	63	69
75	75	60	50	63

Using SBC,

a. Develop a cost-risk profile for this project, and make a recommendation to management about the probability of meeting the 150-day requirements.
b. If meeting the 150-day requirement is not realistic, what value would you recommend for this project? What are your top three recommendations to add additional resources to reduce the project duration? Why?

Summarize your results in a two-page writeup to management. (The writeup should be professional with figures and tables titled and introduced in the body of the text, presented in logical order, good problem statement, written for decision makers, and so on.) Justify the risk level. If the risk level is not warranted to meet the 150-workday requirement, make recommendations about where efficiencies can be obtained in order to reduce the risk level and meet the contract requirements. Rerun your simulation to ensure that your recommendations reduce the risk to an acceptable level. Make sure you address the most sensitive input parameters.

References

Farr, John V., Faber, Isaac, and Richards, Jed, May 2016, "Assessing Bias of Expert Solicited Parameters for Simulation Based Costing," Proceedings of the 2016 Industrial and Systems Engineering Research Conference, H. Yang, Z. Kong, and M. D. Sarder, eds., Anaheim, CA.

Galway, L., 2007, "Subjective Probability Distribution Elicitation in Cost Risk Analysis—A Review," Rand Project Air Force Research Report, accessed 2 July 2015 at http://www.rand.org/pubs/technical_reports/TR410.html.

Ganguly, Anirban, Nilchiani, Roshanak, and Farr, John, V., March 2010, "Defining a Set of Metrics to Evaluate the Potential Disruptiveness of a Technology," *Engineering Management Journal*, Volume 22, Issue 1, pp. 34–44.

Gove, R., 2007, "Development of an Integration Ontology for Systems Operational Effectiveness," Masters of Science Thesis, Stevens Institute of Technology.

Gove, R., Sauser, B., and Ramirez-Marquez. J., 2008, "Integration Maturity Metrics: Development of an Integration Readiness Level," White Paper, Stevens Institute of Technology.

Kahneman, Daniel, 2011, *Thinking, Fast and Slow*, Macmillan, Basingstoke, United Kingdom.

Mankins, John C., 6 April 1995, "Technology Readiness Levels," White Paper, accessed 10 April 2018 at http://www.artemisinnovation.com/images/TRL_White_Paper_2004-Edited.pdf.

Marks, C., Smead, K., and Alt, J., 2013, "Enhancing Subject Matter Expert Elicitation Techniques," Tech. rep. TRAC-M-TR-13-048, TRADOC Analysis Center, Monterey, CA.

NASA, 2015, "Cost Estimating Handbook," accessed 1 August 2018 at https://www.nasa.gov/pdf/263676main_2008-NASA-Cost-Handbook-FINAL_v6.pdf.

Sauser, Brian J., Forbes, Eric, Long, Michael, and McGrory, Suzanne E., 20–23 July 2009, "Defining an Integration Readiness Level for Defense Acquisition," International Symposium of the International Council on Systems Engineering, Singapore.

chapter six

Life cycle framework and techniques

6.1 Introduction to developing life cycle models

The specific purposes of using a life cycle cost (LCC) perspective in acquisition management, capital budgeting, product development, product upgrades, and so on include

- Estimating the total ownership costs (TOC) to the stakeholder, and assessing the affordability of the program,
- Reducing and capturing TOC through using LCC trade-offs in the systems engineering and product development process,
- Estimating future expenditures,
- Controlling cost using LCC contractual provisions in procurements,
- Assisting in day-to-day procurement decisions, and
- Understanding TOC implications to determine whether to proceed to the next development phase.

According to NATO (2009),

> *Data is required in order to conduct a life cycle cost analysis. In terms of time, effort, and resources consumed, collection of data is a major part of the LCC effort. Life cycle cost analysis can best be defined as the development of all costs associated with materials and labor throughout a product/service life cycle; typically from development/acquisition to disposal. Life cycle costing is a data driven process, as the amount, quality and other characteristics of the available data often define what methods and models can be applied, what analyses can be performed, and hence, the results that can be achieved.*

Figure 6.1 illustrates the relationship between data maturity and level of assumptions to be applied.

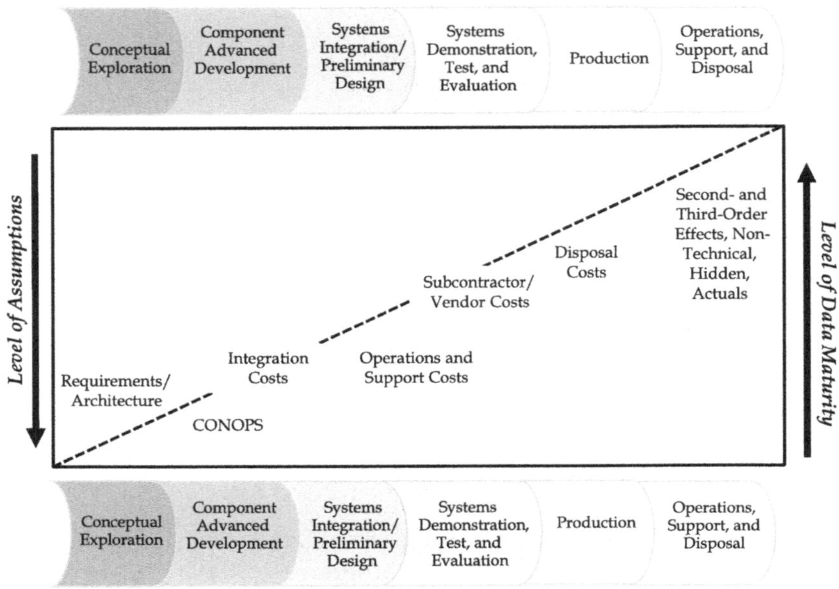

Figure 6.1 Level of assumptions and data maturity needed for LCC analysis (modified from NATO, 2009)

Life cycle costs in general are viewed from a pre- and post-production perspective, with some typical main categories as shown in Figure 6.2. This figure shows a partial list of LCC general categories that can be used to develop more detailed costs. Every system is unique, and this figure is by no means an all-encompassing list.

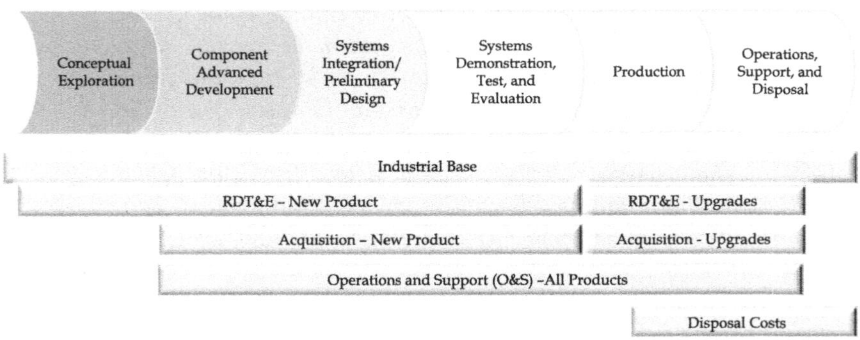

Figure 6.2 Some general LCC categories

6.2 Developing LCC models

Life cycle cost analysis (LCCA) is an economic evaluation technique that determines the total cost of owning, operating, and disposing of a system over its lifespan. Conducting LCC analysis requires us to not only understand the concept of the time value of money (including price escalation, inflation, cost of capital, depreciation, and taxation) but also to capture the true TOC.

When building a model for LCCA, there are two principal types of uncertainty that LCC model builders are advised to consider:

1. Uncertainty regarding the cost-generating activities, and
2. Uncertainty regarding the expected cost of these activities.

Both present unique challenges. In many respects, developing the categories is more challenging than estimating the costs. As espoused by the Deming Institute (2013), "The most important figures that one needs for management are unknown or unknowable but successful management must nevertheless take account of them." This was the motive for developing and presenting a categorization methodology, with Figure 6.2 representing the top-level categories.

Once the categories have been developed, the next step is to ascertain the costs and then develop a LCC model. A simple process for developing a life cycle model is shown in Figure 6.3.

The LCC categories shown in Figure 6.2 are applicable to the development of a large system. They are not applicable to all products. For example, consider a LCCA of a new facility such as a school. The main categories might include initial investment cost, operational cost, maintenance and repair cost, and finally residual or retirement value.

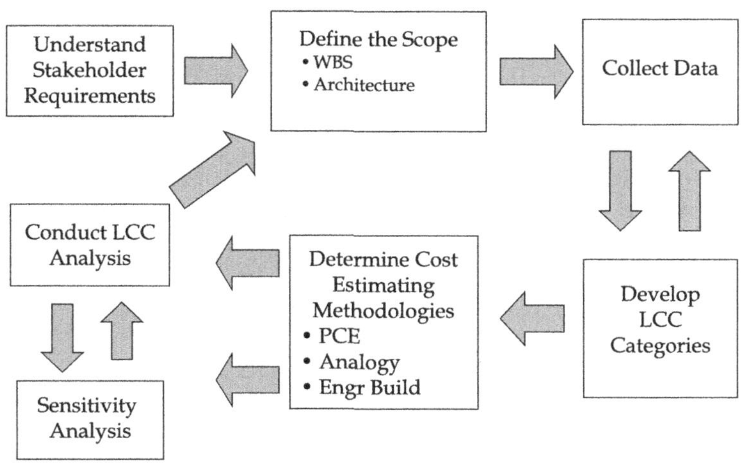

Figure 6.3 Process for developing a LCC model

6.3 Life cycle costs categories

Figure 6.1 shows some of the cost categories that should be included in developing a LCC model. The list is by no means all-inclusive. The figure is presented to demonstrate the complexity and time phasing of when to plan for costs and when the actual costs are incurred. The problem is further exacerbated by the fact that until the product goes into production, we do not know the detailed hardware, software, and interface components, and thus we have no way of developing life cycle categories and the appropriate costs. Therefore, as we progress through the various phases in the product life cycle, we must update our LCC categories. Below is a brief discussion of each of the main categories shown in Figure 6.2.

6.3.1 Industrial base and supplier/vendor relationships

Much has changed because of globalization. Many specialized, large-ticket items require an investment in the industrial base to support the infrastructure. This is especially true for defense systems such as tanks, submarines, and so on. Cultivating supplier/vendor relationships has become critical in an era when just-in-time inventory practices are the rule instead of the exception. New items might require primary contractors to invest in upgrading subcontractor capabilities to ensure quality and responsiveness. In our global environment, where offshore outsourcing and international partnerships are key, creating and sustaining the industrial base for major products while understanding and accounting for these costs throughout the product life cycle are important and significant. Table 6.1 summarizes some of these associated cost categories and elements.

6.3.2 Research, development, testing, and evaluation

Research, development, testing, and evaluation (RDT&E) is the most commonly accepted term used to encompass development activities prior to production. From a development perspective (pre-production), these are the most difficult costs to ascertain because the product architecture has not yet been developed. Maturing technology for production is often impossible to plan and cost. Post-production RDT&E costs are easier to

Table 6.1 Some industrial base LCC categories

Cost Category	Cost Element
Infrastructure Investment	Workforce Development and Retention
	Physical Infrastructure
	Minimum Sustainment Production
Other Costs	Other Costs

Table 6.2 Some RDT&E LCC categories

Cost Category	Cost Element
Systems Engineering and Project Management	Systems Engineering Costs
	Project Management
Support and Test Program	Test and Evaluation Sets and Expenses
	Training Costs
	Data Costs
	Demonstration Costs
	Software and Hardware Development
Prime and Subcontractor Development	Prime and Subcontractor Infrastructure
	Licensing Agreements
Product Development Costs	Hardware Acquisition
	Hardware Modification
	Software Acquisition
	Software Modification
	Software Licensing
	Systems Integration
	RDT&E
Contingency	Evolving Requirements
Other Costs	Other Costs

ascertain because the makeup of the system in known. Table 6.2 lists some of the cost categories and elements used in developing a LCC model.

6.3.3 *Acquisition*[1]

Once the product is designed, one can more accurately determine LCC because a detailed engineering bottom-up model is now possible. However, if you refer to Figure 1.7, you will see that during concept exploration 70% to 75% of the cost decisions are made. The problem is that you cannot develop realistic LCC because the subsystems and components of the system are not fully defined early in the life cycle. In other words, we cannot get a good handle on realistic LCC costs until late in the systems demonstration phase—once the costs are committed. At this point, we can start to populate our LCC model. Table 6.3 is a partial list of cost categories and elements that are associated with acquisition or production.

One item often overlooked is inventory-holding costs. These costs can range between 15% and 35% of the item value per year simply in lost opportunity costs, rent, pilferage, insurance, and so on. Modern inventory management practices were created simply to minimize these costs.

[1] The term "acquisition" is meant to include any aspect of producing the technology, such as buying, manufacturing, producing, and so on.

Table 6.3 Some LCC categories for acquisition expenses

Cost Category	Cost Element
Infrastructure	Physical Plant
	Storage and Spares
	Startup Costs
	Tooling
Product Development	Initial Item Management
	Initial Training
	Initial Technical Data
Inventory Holding Costs	Opportunity Costs
	Finance Costs
	Infrastructure
Production	Software Acquisition
	Software Modification
	Hardware Acquisition
	Hardware Modification
	Interface Acquisition
	Licensing
	Warranty Considerations
	Initial Spares
	Transportation
Quality Assurance	Test Program Sets and Cost
Contingency	Evolving Requirements
Other	Other

6.3.4 *Operations and support*

Too often, operations and support (O&S) costs play a secondary role in the trade space used during concept exploration. We often focus on the development cost because

- Once we have stakeholder buy-in and the product is under development, it can be hard to terminate the project,
- We simply do not know how to calculate LCC, and
- The LCCs are so overwhelming that many programs will never enter into production if the TOC influences the decision process.

Obviously, the operational life of a product drives the post-production LCC. Table 6.4 presents some rough orders of magnitude for purchase price as a function of TOC. As you can see in this table, developing good post-production costs for O&S is critical to capturing the TOC. Table 6.5

Table 6.4 Development costs as a function of TOC

Product	Purchase or Development Cost as a Percentage of TOC
Automobiles	30%–40%
Major Defense Systems	10%–20%
Commercial Buildings	10%–15%

Table 6.5 Some LCC categories for operations and support expenses

Cost Category	Cost Element
Personnel Costs	Labor Maintenance Costs
	Other Support Personnel
	Operational Crews and Management
Consumable Goods	Energy
	Consumable Components
	Some Technology (e.g., personal computers)
Maintenance Costs	Overhaul Support
	Spares Consumption Costs
	Training and Management Costs
	Spares Replenishment Costs
	Pilferage and Damage of Spares
	Infrastructure Support
	Warranty and Vendor Maintenance
Inventory Holding Costs	Opportunity Costs
	Finance Costs
	Infrastructure
Continuing System Improvement	Software Modifications
	Hardware Modifications
	Integration/Interfaces
Contractor Support	Contract Management
Sustainment Support	Systems Engineering and Project Management
	Product Improvement Programs/Parts
	Documentation
	Value Engineering
	Software Maintenance/Licensing
	Packaging and Transportation
	Support Equipment Upgrades/Replacement
Indirect Support	Operational Support Personnel
	Operational Personnel Training
Infrastructure	Training Facilities
	Support Personnel Facilities and Costs
General Training and Education	Basic and Initial Skill Training
	Education Infrastructure
Acquisition	Acquisition
Other	Other Costs

Table 6.6 Some LCC categories for disposal expenses

Cost Category	Cost Element
Environmental	Cleanup
	Regulatory
Post-Production Support	Personnel
	Spares
	Facilities
Retirement	Sell-Off Costs
	Storage Costs
	Early Retirements
Other	Documentation

lists some LCC categories for the O&S phase. In building a LCC model, ascertaining these costs is critical because of their relative contribution to TOC.

6.3.5 *Disposal or retirement*

Unless special conditions apply, planning for disposal costs is relatively straightforward. The problems arise in special cases such as when asbestos, nuclear energy, funding retirement plans, some drugs, and so on are involved. For example, as of February 2005, Wyeth Pharmaceuticals had placed over $21.5 billion in reserves for associated lawsuits for the diet drug Fen-Phen (New York Times, 2005). According to the same article one Houston lawyer represented 8,000 cases and one case received an award of $1 billion. Table 6.6 lists some of the cost categories and elements for disposal or retirement of a product.

6.3.6 *Summary of life cycle categories*

Example 6.1 shows that, when conducting analysis using figures of merit such as net present value (NPV), you must account for all hidden costs. Total ownership costs or LCC must be used when conducting correct analysis.

6.4 *Billable rates*

Determining billable rates is governed by complex rules and regulations that vary depending on the customer. Depending on whether you are trying use cost recovery, market-based billing, or value, your ability to charge for labor can vary greatly.

EXAMPLE 6.1

Your government facility uses an overhead crane to move major pieces of equipment. The current crane has been very reliable since the mid-1980s, but management would like to replace it with a new crane that has modern electronic controls. Your boss has asked you to cost out refurbishing the existing crane or replacing it with a new one. Based on your conversations with several vendors, you came up with the following information:

Refurbish the existing crane:

- Steel repairs and reinforcement: $525,000
- Strip and recoat all exposed surfaces: $480,000

Replace the crane:

- Cost of new crane: $3,689,000 (includes installation)

Solution

You know that you must look at this from a total owner cost perspective and must make some assumptions about life cycle, operations and maintenance costs, inflation, and so on. You are mandated by law to use an interest rate of 1.8% when analyzing new construction and capital replacement projects in your state.

1. List all assumptions related to this project from a TOC perspective.

The most important assumption is whether the existing crane will last as long as a new crane. Will you have to refurbish or replace the existing one before the new one needs replacing? We are going to assume that both cranes have a life of 30 more years before they require major repairs or replacement. When challenged by management, you should probably state these assumptions to provide validity to your analysis:

- The existing crane, which is very well built, can easily last 30 more years.
- There will be a hidden cost of $50,000 for removing the existing crane. There is no secondary market for the crane because of its poor shape, old motors, and so on, beyond scrap metal.

(continued)

- The annual maintenance cost of the existing crane is about $1,000 per year.
- The power requirements to operate both cranes are the same—about $300/year.
- Assume that, on average, the old crane has to be repaired every 5 years at a cost of $3,000.
- Assume that the new crane will operate for 10 years with no repairs needed and will then require the same maintenance intervals of 5 years with repair costs of $5,000 (more complicated control system).
- The new crane is easier to maintain, and so the annual maintenance costs are about $500/year.
- Assume the following inflation rates: energy=5%, general rate=3%.

2. Conduct the appropriate analysis to present to management to help in deciding which option to pursue.

Below are the first 10 years of a 30-year spreadsheet. All the items listed are costs, so we simply left everything positive. Note how the appropriate adjustment factors were applied and the NPV was calculated based on 1.8%.

B24 =B23+NPV(C4,C23:AF23)

	A	B	C	D	E	F	G	H	I	J	K	L
1	**Interest and Inflation Rates**											
2	Energy Inflation Rate		5.00%									
3	General Inflation Rate		3.00%									
4	NY State MARR		1.80%									
5												
6	Year	0	1	2	3	4	5	6	7	8	9	10
7												
8	***Existing Crane***											
9	Steel Repairs and Reinforcement	$525,000										
10	Strip and Recoat All Surface	$480,000										
11	Annual Maintenance		$1,030	$1,061	$1,093	$1,126	$1,159	$1,194	$1,230	$1,267	$1,305	$1,344
12	Power		$315	$331	$347	$365	$383	$402	$422	$443	$465	$489
13	Major Repairs						$3,478					$4,032
14	Net Cash Flows	$1,005,000	$1,345	$1,392	$1,440	$1,490	$5,020	$1,596	$1,652	$1,710	$1,770	$5,864
15	***Net Present Value***	***$1,074,173***										
16												
17	***New Crane***											
18	Cost of New Crane	$3,689,000										
19	Remove Existing Crane	$50,000										
20	Annual Maintenance		$515	$530	$546	$563	$580	$597	$615	$633	$652	$672
21	Power		$315	$331	$347	$365	$383	$402	$422	$443	$465	$489
22	Major Repairs											$6,720
23	Net Cash Flows	$3,739,000	$830	$861	$894	$927	$963	$999	$1,037	$1,077	$1,118	$7,880
24	***Net Present Value***	***$3,796,754***										

Note how the NPV function is used in Excel in that you cannot include year 0 costs when using the built-in function.

3. Make a recommendation to management.

Over a 30-year life, the new crane will cost roughly $2.7 million more in today's dollars. All things equal (safety, reliability, etc.), the logical choice is to refurbish the existing crane.

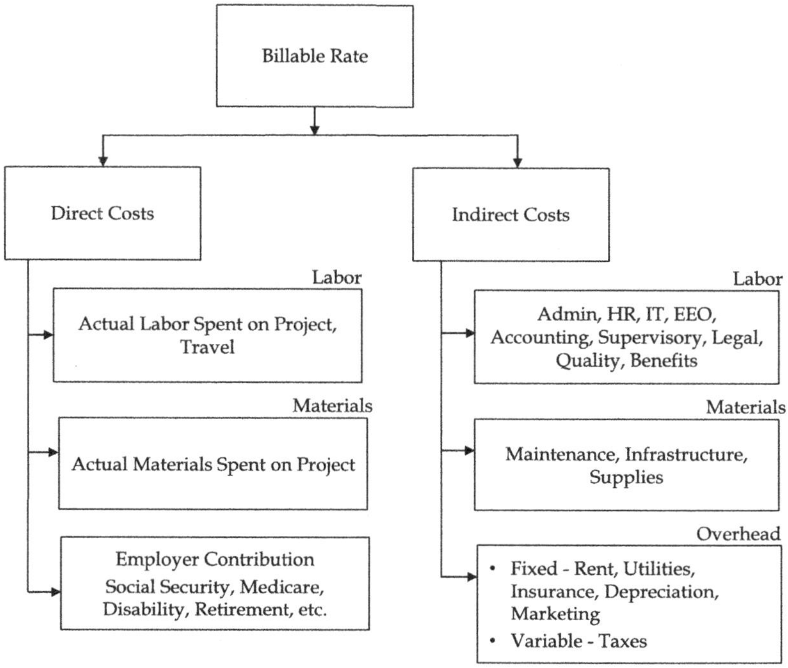

Figure 6.4 Components of billable rates

As shown in Figure 6.4, billable rates consist of direct and indirect costs. Labor costs are the key to good cost estimating. Obviously, the most complex component is overhead. "Overhead" refers to the indirect costs associated with running a business—in other words, the costs that are not directly factored into producing goods or services. Figure 6.4 breaks costs into direct and indirect costs. Direct costs can be defined as any cost that can be attributed to a project or organization that supports the manufacturing of a tangible product or the delivery of a service. Indirect costs cannot be directly allocated to any project or billable service and usually benefit more than one project or product.

Figure 6.5 shows how net pay is affected by payroll deductions. In essence, net pay translates to gross pay, which translates to billable rates. Though not really needed for good estimating, it is important for hiring and staffing projects.

6.5 Costing labor

Determining labor costs requires more than simply multiplying the wages by the number of hours worked. There are three ways to view labor costs: (1) gross or hourly pay, (2) net pay, and (3) billable rate.

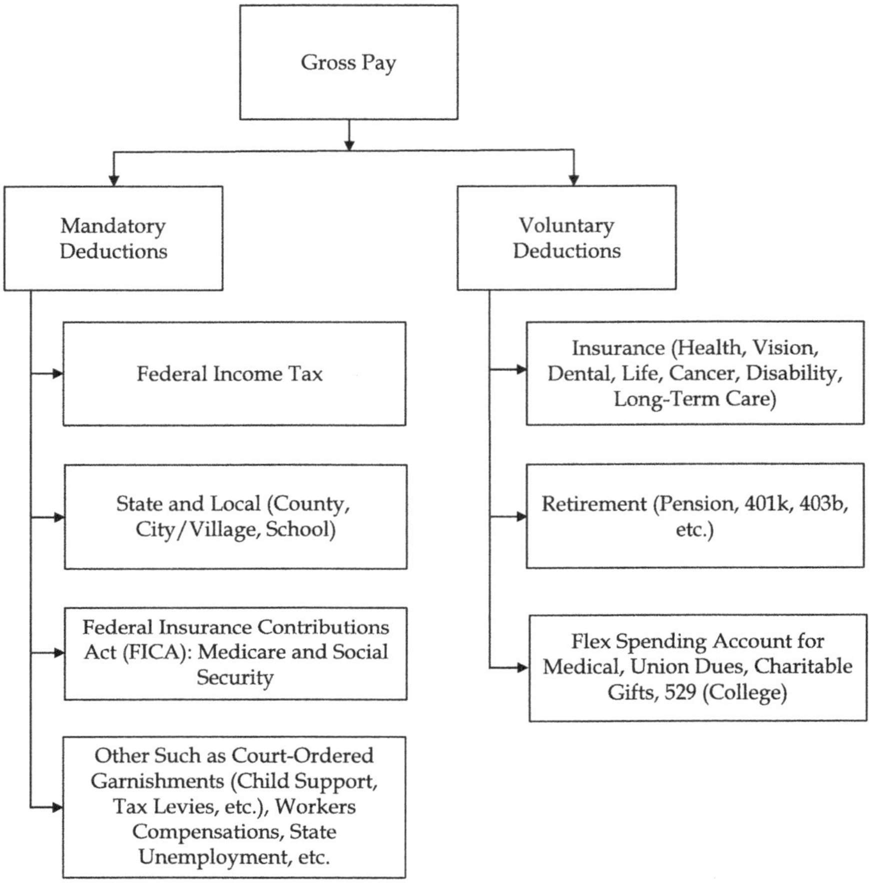

Figure 6.5 Payroll deductions

Gross pay is simply what is advertised for the job: "Software engineers can make $100,000 per year." Beyond being viewed as a qualitative measure of the pay, gross pay means very little from an estimating perspective. It is simply the starting point from which deductions are removed from your pay. Some deductions that might be subtracted from your gross pay (as of 2017) include the following:

- Social Security tax is a 6.2% deduction that funds the employee's Social Security benefits. The tax is limited to the first $127,000 of wages earned and is mandated under the Federal Insurance Contribution Act.
- Medicare tax is a 1.45% deduction. For workers making over $200,000, an additional 0.9% is withheld.

- Federal Unemployment tax is currently at 6% of the first $7,000 earned.
- State Unemployment tax varies by state.
- Workers Compensation Insurance also varies by state and falls somewhere between .88% and 3.5%.
- State taxes can vary greatly by locale. For example, New York State withholds from regular wages at a state tax of 9.62% and New York City residents pay additional 4.25%. Note that all of these are above the school, town/city, and county taxes paid in most states.
- Employee contributions to medical, dental, and vision insurances.
- Retirement plan contributions, such as pensions, 401k/403b, employee stock ownership plans, and so on.
- Meals and uniforms might be deducted in some instances.
- Third-party payroll processors might charge a fee.
- Other court-mandated deductions, such as union dues, child support, and so on.

As you can see, trying to estimate net pay is complex and requires knowledge of local, state, and federal laws. Fortunately, for cost estimating we are more interested in converting gross salary to billable rates.

In general, gross rates are adjusted to include overhead, general and administrative (G&A) costs, profits, and contingencies. Overhead in general refers to the costs that support the direct cost of revenue-generating projects of the company. Factors such as vacation, sick time, personal days, employer's Social Security and Medicare contributions, pension/retirement plans, and so on are all included in overhead costs. You can think of it this way: if the company had no jobs or projects, it would have no overhead expenses. Also included might be bonding and insurance on jobs, permitting, and legal costs. G&A costs are those that support the overall management and operation of the business. A company would still have G&A expenses even if there was no billable work. This might include office rent, depreciation, utilities, training, insurance, manager salaries, and so on.

6.6 Summary

In this chapter, we presented an overview of LCC categories and methods for determining the associated costs. Unfortunately, some items do not fit in a neat box in a life cycle model. Systems engineering, project management, and quality are some of these elements. For example, much has been written about the cost of quality (Campanella, 1999, Harrington, 1987).

Reece and Farr (2016) developed the following broad categories when calculating the TOC for cloud versus distributed storage and computing: location, facility, power, space, heating/cooling/water, equipment maintenance, communication lines, support personnel, software licensing, fire protection, security, upgradability/phased growth of computing/data center issues, cloud storage, and administrative supplies. Under these broad categories, about 150 detailed categories were identified where cost comparison should be performed! This is a pretty well-understood system, yet the number of categories illustrates the complexity of even simple problems when trying to conduct LCCA. Conducting meaningful LCCA is complex and requires both domain and cost analysis knowledge, even for well-understood problems.

When developing a LCC model or conducting an analysis, certain characteristics should be incorporated for the model to be useful. These include the following:

- The model should be useful as a management and analysis tool and should be responsive to design changes and various operational scenarios (e.g., reliability, maintainability, supportability, concept of operations).
- All significant cost drivers needed to quantify TOC issues should be incorporated.
- The model should be sensitive to key performance parameters from the trade space studies.
- Accurate input data should be readily available at the appropriate level of detail.
- The model should be flexible and scalable.
- Inputs and outputs should be expressed in terms that are familiar to the stakeholders.

The main purpose of this text is to convey how LCC analysis must be used to fully understand how to determine and interpret the TOC of a system. As discussed, we are often fixated on the upfront costs, which can lead to bad investment decisions.

QUESTIONS

6-1. What types of projects have high costs for
  a. The conceptual and design phases?
  b. The advanced development and detailed design phases?
  c. The production phase?
  d. Operating and maintaining the system?
  e. The divestment phase?

6-2. As we get further into the life cycle of a product, our ability to capture the TOC improves. In essence, we go from trying to predict to capturing. Transitioning from analogous and parametric prediction methods, to detailed engineering builds, and finally to good cost accounting requires a thorough understanding of the architecture or product and data collection. At what point in the product life cycle should you start developing a detailed engineering LCC model? Why?

6-3. The tsunami warning system managed by the United States Department of Commerce, National Oceanic and Atmospheric Administration, is designed to detect tsunamis and provide prompt notification to all nations bordering the Pacific Ocean. You have been tasked with soliciting bids to install at two sites over a 2-year period sensors that use pressure detectors to measure changes in water depth as a tsunami wave passes overhead. The sensors then transfer the information to a surface buoy, which relays it to the monitoring stations by satellite. The low bidder provided you the following detailed costs:

Year	0	1	2	3	4	5
Development and Procurement						
Salary	$74,000	$76,220	$0	$0	$0	$0
Travel	$15,000	$11,000	$ 5,150	$0	$0	$0
Hardware Acquisition	$30,000	$30,000	$0	$0	$0	$0
Software	$10,000	$10,300	$0	$0	$0	$0
System Integration	$15,000	$10,300	$0	$0	$0	$0
System Testing	$20,000	$10,000	$20,600	$0	$0	$0
Hardware Modification	$ 5,000	$ 3,000	$ 3,000	$0	$0	$0
Software Modification	$11,000	$ 3,000	$ 3,090	$0	$0	$0
Documentation	$ 3,000	$ 3,090	$0	$0	$0	$0
Training	$11,000	$11,330	$0	$0	$0	$0
Warranty Considerations	$ 600	$ 642	$ 687	$735	$786	$842
Operations and Support						
Salary	$0	$0	$36,000	$37,080	$38,192	$39,338
Repairs and Maintenance	$0	$8,000	$ 8,240	$ 8,987	$ 9,257	$ 9,535
Travel	$0	$0	$ 7,030	$ 7,241	$ 7,458	$ 7,682
Costs						
Cost by Year in Actual Dollars	$194,600	$176,882	$83,797	$54,043	$55,694	$57,396
Total Project Cost in Actual Dollars	$622,412					
Cost in 2008 Dollars @ 3%	$194,600	$171,730	$78,987	$49,457	$49,483	$49,510
Total Project Cost in 2008 Dollars	$ 593,767					

Are there any cost categories missing from the LCC presented by the contractor? Your request for proposal called for a commercial off-the-shelf solution that was proven and did not require development. However, you believe that the contractor is in essence having you pay for further adaption and development of their sensors for this application. Based on the large integration, testing, modification, and O&S costs, do you believe that more details are warranted?

6-4. Document Storage Plus has hired you to conduct a LCCA of its operations. Currently, trucks pick up documents from a host of customers, and they are then transported back to a distribution center, placed on pallets, wrapped, and trucked to warehouses typically near the customers. There are many distribution centers. For example, one might handle the New York City area, and another might handle North and South Dakota. You propose to develop several large regional storage facilities and rail the documents to minimize costs. Given that customers seldom require access to those documents (less than 5% are ever delivered back to the customer), what are the advantages and disadvantages from an LCC perspective of these two distribution systems?

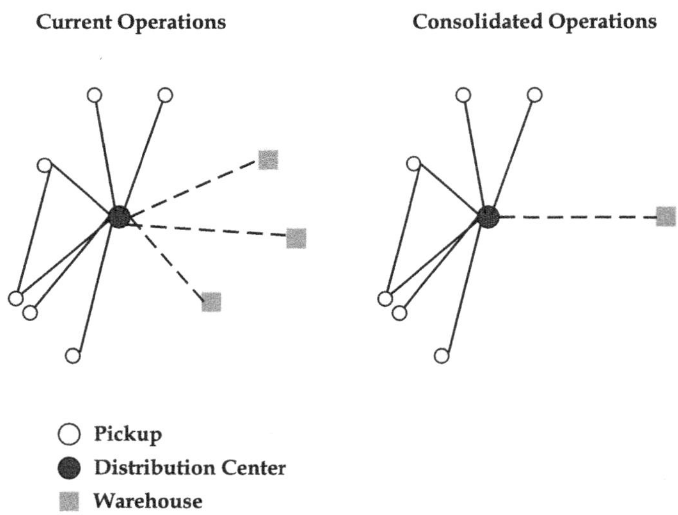

6-5. Consider the general life cycle model shown below. Develop some of the subcategories for each of the main categories for a new CAD system. Do you think this generalization is a good way to view the problem and structure cost categories?

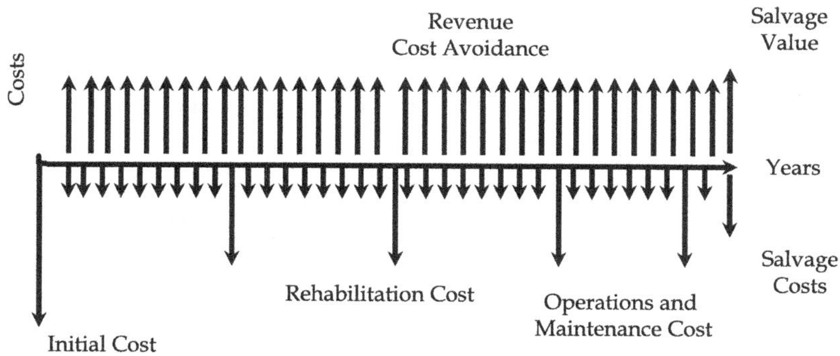

6-6. You must pay a full-time employee for 2,080 hours (52 weeks × 40 hours) a year. However, only 60% to 65% of the billable hours is available for client projects because of vacation, sick time, holidays, staff meetings, disability, administrative activities such as training, new business development, and so on. What are the advantages of using 1099 labor, part time, versus hiring full-time employees? Develop a list of pros and cons for each type of employee.

PROBLEMS

6-1. Several HVAC contractors have bid to replace the air conditioners in a major capital renovation project. You have narrowed the choices to two final competitors.

Air Conditioner System A	Air Conditioner System B
Initial Cost: $200,000	Initial Cost: $240,000
Annual Maintenance Costs: $12,000	Annual Maintenance Costs: $8,000
System Life: 10 years	System Life: 12 years

You plan on using an annual effective interest rate of 5%. Using LCC, what is the most cost-effective system? (Hint: note the different systems' lifespan; therefore, you cannot use NPV.)

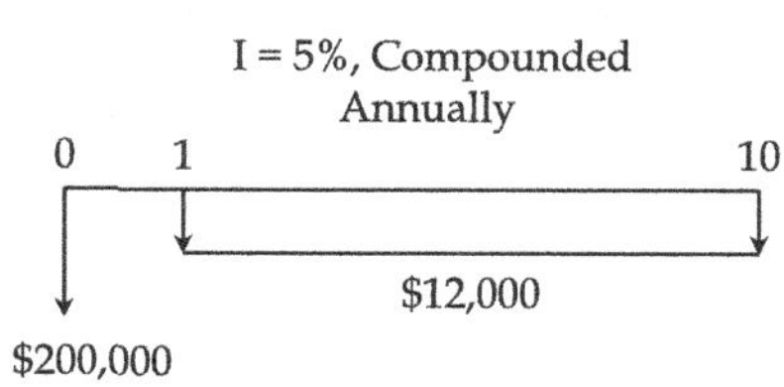

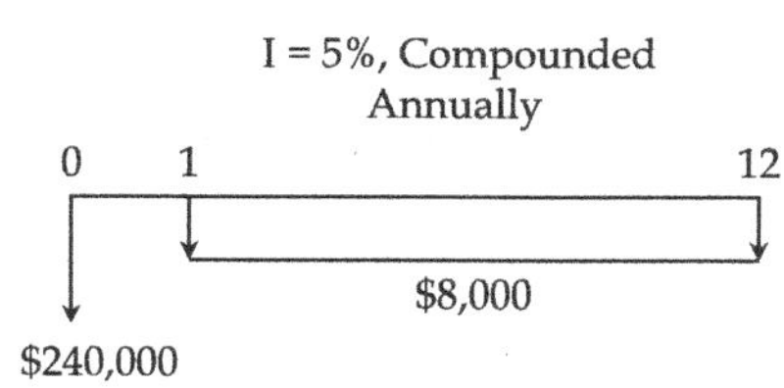

6-2. Using NPV analysis, determine which of these three hot water heaters is the most economical. (Hint: make a plot of NPV vs. year.)

		Electric Water Heater	Natural Gas Water Heater	Solar Water Heater
Energy Produced		4300 kWh	200 Therms	4300 kWh
Cost per Unit		$0.10	$1.00	
Energy Inflation Rate		10%	10%	NA
Inflation Rate		3%	3%	3%
Purchase Cost		$300.00	$1,500	$5,000
	Year			
Cost to Operate	1	$ 430.00	$ 200.00	
	2	$ 473.00	$ 220.00	
	3	$ 520.30	$ 242.00	
	4	$ 572.33	$ 266.20	$220.00
	5	$ 629.56	$ 292.82	
	6	$ 692.52	$ 322.10	
	7	$ 761.77	$ 354.31	
	8	$ 837.95	$ 389.74	$275.00
	9	$ 921.74	$ 428.72	
	10	$1,013.92	$ 471.59	
	11	$1,115.31	$ 518.75	
	12	$1,226.84	$ 570.62	$300.00
	13	$1,349.52	$ 627.69	
	14	$1,484.48	$ 690.45	
	15	$1,632.92	$ 759.50	
	16	$1,796.22	$ 835.45	$350.00
	17	$1,975.84	$ 918.99	
	18	$2,173.42	$1,010.89	
	19	$2,390.76	$1,111.98	
	20	$2,629.84	$1,223.18	

Does the life expectancy of the hot water heater affect your answer?

6-3. Your company is looking at buying a fleet of cars for service technicians. Below is a table of characteristics of the three options you are investigating.

Car	Purchase Price	Gas Mileage (MPG)	Annual Maintenance Cost	Salvage (end of year 4)
Midsize	$17,000	25	$1,200	$ 9,000
Hybrid	$27,500	45	$ 500	$10,000
Economy	$16,000	31	$1,000	$ 5,000

On average, each service technician travels 40,000 miles per year. Using varying gas prices from $2.00 to $4.00 per gallon in $0.25 increments, conduct a sensitivity analysis and develop a meaningful plot for management to make an informed decision about the fleet makeup. What is your recommendation to management? (Hint: assume an inflation rate of 3% and convert everything to an annual cost, or assume a life of 4 years and calculate NPV.)

6-4. You are evaluating whether to replace the electric motors at a wastewater treatment plant. Using an inflation rate of 3.5% and salvage value of 15% of the initial costs, determine which of these pumps is the most economical from a life cycle perspective. Note the different life expectancies for the various options.

Input	Option A	Option B	Option C	Do Not Replace
Initial Cost	$ 5,000	$2,250	$21,500	0
Annual Energy Cost	$11,000	$6,700	$ 5,500	$11,000
Annual Maintenance	$ 500	$ 500	$ 1,000	$ 500
Annual Inspection/ Certification Cost	$ 2,500	$2,500	$ 2,500	$ 2,500
Life Expectancy	8	6	12	5

6-5. The following table contains cost data needed to compare a hybrid versus a conventional SUV. Assume that you will average 15,000 miles per year. What is the breakeven point in terms of fuel costs when the hybrid becomes the more economical option?

Cost Categories	Cost Categories Decomposed	2016 Toyota Highlander Hybrid Limited Platinum	2016 Toyota Highlander Limited Platinum
Purchase Price	Purchase Price (year 0)	$51,347	$46,229
	Taxes and Fees	$ 3,342	$ 3,053
Ownership Costs	Fuel Consumption	29/27 MPG	20/26 MPG
	Insurance (annual costs)	$ 1,808	$ 1,732
	Maintenance and Repair (total for 5 years)	$ 5,548	$ 5,548
	Financing (cash purchase)	NA	NA
Resale Value	Projected Resale Price (year 5)	$26,452	$24,542

Using a MARR of 3% and a 50/50 mix of highway and in-town driving, conduct a sensitivity analysis of fuel price versus equivalent annual costs. Plot these results in a scatter plot of gas prices (vary from $2.00 to $4.00) to help make your decision.

6-6. Below is a bid for a microwave transmission tower and support equipment to be installed in a remote region of southwest Asia. This system consists of an antenna and on-site computers with operators. You expect this system to have a life of 10 years.

Item	Hours	Cost
Management	44,093	$ 7,731,153
Systems Engineering	20,094	$ 3,619,772
Software	42,915	$ 7,183,903
Hardware	3,006	$ 500,355
Installation	38,183	$ 6,937,304
Integration and Testing	41,492	$ 7,062,878
Training	9,635	$ 1,695,175
Travel	0	$ 2,142,724
Infrastructure	0	$ 1,174,913
Proposal Development	682	$ 491,848
Total Bid	**200,100**	**$38,540,024**

Use the various life cycle categories listed in Tables 6.5 and 6.6 to identify which cost categories are relevant as you develop the total LCC.

6-7. You need to replace the HVAC system in your home. Your local contractor provides you with two quotes: (1) $6,000 for a new condensing unit, air handler, heater, thermostat, and installation for a 15 seasonal energy efficiency ratio (SEER) Amana unit, and (2) $6,981 for an 18 SEER Amana unit. Both are five-ton units. Research the savings of a 15 versus an 18 SEER unit in north Florida and determine which is the best option. (Assume all other things equal: reliability, life of product, maintenance, etc.)

6-8. Your team lead has asked you for a "fully costed" project estimate. You asked your team members, who subsequently polled each individual responsible for the work breakdown structure tasks and passed those estimates back up to you. The totals resulted in the direct costs shown in the table below. The current contract assigns overhead charges of 25% of labor costs and G&A charges at 10% of all direct costs. You also want to add a 5% contingency to all costs.

		Resources ($100k)	
Activity ID	Activity Name	Materials	Labor
A	Requirements Assessment	0	1
B	Subsystem Design	10	4
C	Subsystem Production	24	10
D	Integration Software Development	2	14
E	Testing and Delivery	15	9
	Total	**51**	**38**

a. Calculate the fully costed project.
b. What budgeting strategy did you use to come up with these figures, and what is one advantage of using such a strategy?
c. Is your budget category based or activity based? What is one advantage of reflecting budget items in this way?

6-9. Major infrastructure renovations in a dense urban environment can be very disruptive to local businesses. Because of reduced foot and car traffic from the construction, many small businesses simply cannot survive. With major sewer, electrical, steam, network, cable, power, drainage, and roads/pavement work, these contracts can be very complex, lengthy, and expensive. Often cities award two contracts starting at opposite ends of the roadway in order to expedite construction. The mayor's office has asked you to justify the cost of two versus one construction contract. The major unknown is the economic value of lost businesses. You collect the following information:

Miscellaneous	
Interest Rate (compounded daily)	5% APR
Two Construction Contracts	
Award Fee (millions)	$16
Length of Contract (months)	14
Cost per Month (millions)	$2.80
Businesses Lost per Month	0.5
One Construction Contract	
Award Fee (millions)	$5
Length of Contract (months)	30
Cost per Month (millions)	$1.50
Businesses Lost per Month	1

Make a recommendation to the mayor's office. (Hint: you do not have any data to justify the economic value of a lost business. Thus, you decide to develop a plot using NPV analysis of total contract cost [Y axis] versus cost per lost business [X axis] to present to the mayor. This information can at least give you some analytics to support a recommendation. Vary the value of a lost business between $0 and $1 million. Then make a recommendation accordingly.)

6-10. You are looking at starting you own software development business. You know that in your area software programmers are paid $75,000 per year, and software engineers command around $100,000 per year. As owner of the company, you want to be paid

$200,000 per year. Develop the billable rates for your employees and yourself. Use the Internet, and include the details of your calculations in order to justify the rates to potential customers, including the government.

6-11. You are a one-person company, and your expenses from last year are shown in the table below:

Month	Newspaper	Water	Elect	Lawn/Plants	Phone	Cable/WiFi	Office Supplies	Office Insurance	Rent	Liability Insurance	Professional Publications	Health Insurance
Jan	$17.97	$183.64	$260.00		$150.33	$99.93	$710.60		$1,575.00	$500.00	$140.14	$1,250
Feb	$17.97	$171.11	$260.00		$150.33	$102.09	$1,918.00		$1,575.00		$102.09	$1,250
Mar	$17.97		$260.00		$150.33	$102.09			$1,575.00		$179.30	$1,250
April	$17.97	$177.85	$260.00		$150.33	$179.30	$888.00		$1,575.00		$140.14	$1,250
May	$17.97		$260.00	$86.50	$150.33	$140.14	$140.64		$1,575.00	$500.00	$140.14	$1,250
Jun	$17.97	$216.91	$260.00	$257.34	$150.33	$140.14	$26.45		$1,575.00		$139.04	$1,250
July	$17.97		$260.00	$216.25	$150.33	$139.04	$138.25	$1,324.00	$2,000.00		$86.50	$1,250
Aug	$17.97	$177.85	$260.00	$173.00	$150.33	$140.36			$2,000.00		$257.34	$1,250
Sept	$17.97		$260.00	$173.00	$150.33	$119.31	$1,675.57		$2,000.00	$500.00	$216.25	$1,250
Oct	$17.97	$177.85	$260.00	$173.00	$150.33	$140.05	$1,284.02		$2,000.00		$173.00	$1,250
Nov	$17.97		$260.00	$173.00	$150.33	$140.05	$134.55		$2,000.00		$140.14	$1,250
Dec	$17.97	$194.29	$260.00		$150.33	$140.05			$2,000.00	$500.00	$140.14	$1,250
Total	$215.64	$1,299.50	$3,120.00	$1,252.09	$1,803.96	$1,582.55	$6,916.08	$1,324.00	$21,450.00	$2,000.00	$1,854.22	$15,000
Professional Societies		$300.00										
Accountant		$1,200.00										
Ads/Marketing		$2,900.00										
Total Expenses		**$62,218.04**										

Assume 80% billable time, 40-hour workweeks (1 week each off for Christmas and Thanksgiving, 2 weeks off for summer vacation, and 1 additional week off), and 8-hour billable days. Ignore G&A costs for this example.

Answer the following questions:

a. You want to gross $150,000. What will be your billable rate?

b. Compare to the General Service Administration (GSA) rate (see any GSA schedule on the Internet), in the spreadsheet to determine what you need to charge to pay your bills.

c. If you could charge as an executive consultant estimate your gross pay.

Bibliography

Andrews, Richard, 2003, "An Overview of Acquisition Logistics," Fort Belvoir, VA, Defense Acquisition University, accessed 2 April 2007 at https://acc.dau.mil/CommunityBrowser.aspx?id=32720.

Department of Defense, 1 April 1983, "Life Cycle Cost in Navy Acquisition," MIL-HDBK-259 (NAVY), Washington, DC.

Department of Defense, October 2001, "Indirect-Cost Management Guide—Navigating the Sea of Overhead," 3rd Edition, Defense Systems Management College Press (web version only).

Department of Defense, August 2006, "Risk Management Guide for DoD Acquisition," 6th Edition, Version 1.

Dhillon, B. S., 1989, "Life Cycle Costing: Techniques, Models and Applications," OPA (Amsterdam) B.V., Gordon and Breach Science Publishers, Philadelphia, Pennsylvania.

Emblemsvåg, Jan, 2003, "Life-Cycle Costing: Using Activity-Based Costing and Monte Carlo Methods to Manage Future Costs and Risks," John Wiley and Sons, Hoboken, NJ.

References

Campanella, Jack, 1999, "Principles of Quality Costs: Principles, Implementation, and Use," American Society of Quality.

Deming Institute, 29 August 2013, W. Edwards Deming Institute Blog, accessed 8 May 2018 at https://blog.deming.org/2013/08/unknown-and-unknowable-data/Department of Defense, 1 April 1983, "Life Cycle Cost in Navy Acquisition," MIL-HDBK-259 (NAVY), Washington, DC.

Harrington, James, 1987, "Poor-Quality Cost," American Society of Quality.

NATO, Sepember 2009, "Code of Practice for Life Cycle Costing," RTO Publication SAS-069.

New York Times, February 2005, accessed 1 August 2018 at https://www.nytimes.com/2005/02/17/business/fenphen-case-lawyers-say-theyll-reject-wyeth-offer.html.

Reece, Justin, and Farr, John V., June 2016, "Cost Considerations for Cloud Versus Distributed Computing Services," Center for Nation Reconstruction and Capacity Development, White Paper 2016-3, accessed 1 August 2018 at http://www.usma.edu/cnrcd/CNRCD_Library/Reece%20Senior%20Thesis.pdf.

section two

Estimation of complex systems

chapter seven

Costing of complex systems

7.1 Introduction

Engineering has evolved greatly as new methods, processes, and tools have been evolved for understanding and building large, complex systems. The challenges that many engineers face involve connecting the subsystems properly into a system of systems (SoS) product that allows for the proper interactions without huge propagations of error. Figure 7.1 shows how systems theory can be used to understand the framework for a complex system. Unfortunately, as system complexity increases, our cost estimating techniques are not very mature or reliable.

Developing cost estimates based upon parametric cost estimates (PCEs), bottom-up, hybrid, and analogies techniques for systems that have already been developed can be tricky. Analysts who do this find that the sum of the most likely costs of the components of a system in development does not equal the final cost of the entire system (especially software), since costs do not scale linearly because of complexity, integration challenges, and the like. The same certainly holds true when scaling systems up to SoS and enterprises. Any cost engineer will tell you that we can cost hardware and software separately with some confidence. However, when we integrate, we simply do not know how to estimate these costs, much less scale to the larger software-centric systems. This is the challenge of costing complex systems.

7.2 Issues surrounding complex systems

Figure 7.2 contains a plot of costs committed and costs incurred over a typical systems life cycle. The figure clearly shows the importance of upfront systems engineering and managing requirements.

From a life cycle cost (LCC) perspective, what is even more critical is that, early in the life cycle, we are committed to schedules, costs, and ultimately funding when we simply do not have the techniques to accurately estimate costs. The top-down tools we used to estimate costs early in the product development cycle are gross rules of thumb at best. When combined with requirements creep, unstable funding, and so on, cost estimates be off by up to 100%!

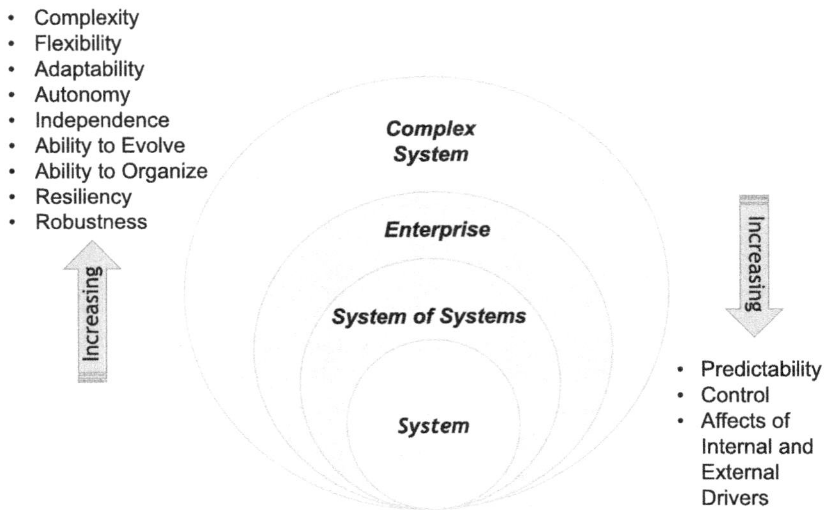

Figure 7.1 Complex systems theory taxonomy (from Farr et al., 2014)

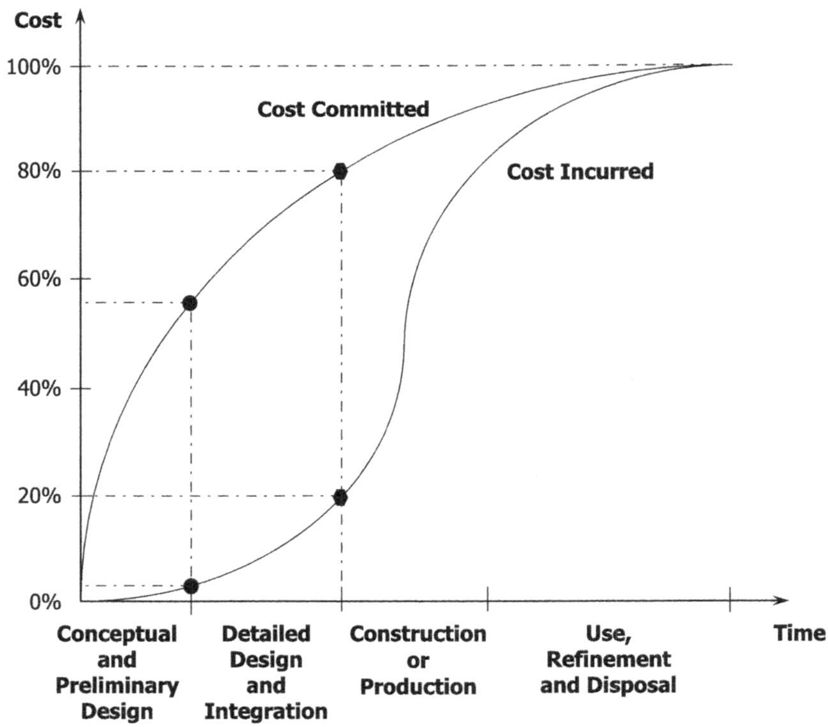

Figure 7.2 Costs committed versus costs incurred

As shown in the product life cycle model presented in Chapter 1, the techniques for estimating systems costs vary depending on where we are in the life cycle. Taking our seven-phase model of conceptual exploration, component advanced development, systems integration and preliminary design, systems demonstration and test and evaluation, production, and operations support and disposal, different techniques might be used to estimate costs. For example, early in conceptual exploration, the only technique that might be satisfactory is a parametric cost estimation technique such as the Constructive Systems Engineering Cost Model (COSYSMO), which will be explained in detail later in this chapter. As we move further into the product development cycle (say, at the end of preliminary design), estimating will be conducted using a bottom-up approach or engineering build of the system. Finally, as we enter into production, we will modify our engineering bottom-up model to more accurately reflect the final design elements of hardware, software, and interfaces and integration. Table 7.1 was modified from a commercial vendor and demonstrates that, very early in the product development cycle, we simply do not know enough about the system to accurately develop costs. Unfortunately, this is when budgets are allocated, bids developed, and so on. In order for LCC to become more accurate, we must use software and other formal engineering tools sooner in the design.

Table 7.1 Cost and schedule estimates as a function of technical baseline work products (modified from Barker, 2008)

Baseline Created	Technical Work Products from Which Estimates Are Developed	Methodologies Used to Develop Cost Estimates
Customer	Customer Requirements • Capabilities • Characteristics Concept of Operations	**Top-Down** • Based on Number/Complexity of Requirements • Based on Number/Complexity of Scenarios • Based on Number/Complexity of External Interfaces **Analogous** • Estimates Based on Complexity of Technical Work Products Compared to Similar Complexity of Similar Projects ***Estimates Are Based on Experience and Historical Data with a ±75% Accuracy***

(*continued*)

Table 7.1 (continued)

Baseline Created	Technical Work Products from Which Estimates Are Developed	Methodologies Used to Develop Cost Estimates
System	System Requirements Preliminary Architecture	**Top-Down** • Based on Number/Complexity of Requirements • Based on Number/Complexity of Scenarios • Based on Technology Maturity • Based on Architecture Complexity **Analogous** • Estimate Based on Complexity of Technical Work Products Against Known Projects **Bottom-Up** • Estimates Based on Architecture ***Estimates Are Based on Experience and Formal Design and SE Tools with a ±50% Accuracy***
Component (hardware, software, process)	Component Requirements • Hardware and Software Systems Architecture • Document All Hardware, Software, Processes, and Interfaces Test Architecture	**Bottom-Up** • Estimates Based on Architecture, Technologies Selected, Testing Plan, etc. ***Estimates Are Based on Formal Design (WBS, COCOMO, COSYSMO, Function Point, etc.) and SE Tools with a ±10% Accuracy***
Design, Test, and Production	System into Production Hardware, Software, and Processes Design and Test Strategy Service Agreements	**Bottoms Up** • Estimates Based on Detailed Design, Test Schedules, Implementation Details, and Other Technical Work Products • Delivered Solution Architecture ***Estimates Are Detailed Bottom-Up Based on All Technical Work Products***

7.3 Systems engineering and management costs

7.3.1 Hardware costs

If we use a hierarchal approach (a system of systems or enterprise is composed of systems, systems are composed of subsystems, and subsystems are composed of components), any of these levels will be the building block of a bottom-up estimate. In its simple form, hardware can be separated into the physical components (that comprise the building blocks) plus the labor for estimating purposes. We can think of this as levels of our work breakdown structure (WBS). Note that part of developing LCC for any component of systems is developing your WBS using phases of product life cycle and assigning hardware, software, and so on for every phase.

As a first cut, we could use these categories as a way to classify costs. Unfortunately, depending on where you are in the product life cycle, you will need to adjust costs to account for technology maturity, which might include technical readiness levels, system readiness levels, learning curve issues, and so on.

As you transition from a top-down cost estimating relationship such as COSYSMO, you could use rough relations to estimate these costs over the product life cycle and refine them as the design becomes more final. The model presented must evolve as you move further down the life cycle.

7.3.2 *Software*

Since software dominates most complex systems, we devoted Chapter 8 solely to traditional software costing methods, including the COnstructive COst Model (COCOMO) and function point analysis. The COCOMO family of models presented is probably the most widely used software estimation models in the world. Both of these techniques can be classed as analogy methods.

7.3.3 *Interfaces and integration at the component and system level*

No overarching methodology exists for costing the integration of hardware and software and developing the interfaces. Interfaces and integration challenges are the key reason why the cost of systems scale non-linearly. We know from the Department of Defense, NASA, and other developers of large SoS problems that we do not know how to estimate their costs with any degree of accuracy, especially when little historical data exists.

7.3.4 *Systems engineering and project management costs*

One area that has received significant attention because it is often underfunded and has been connected to major cost overruns is systems engineering and project management (SE/PM). Figure 7.3 shows some of the SE/PM functions that comprise this category. Stern and colleagues (2006) reported that the average SE/PM costs for major aircraft programs had increased from 8% of the total development costs in the 1960s to about 16% in the 1990s. The SE/PM components are significant to controlling costs, schedule, and quality during product design. However, what are the SE/PM concerns post-production? These are also significant for upgrades and supportability issues.

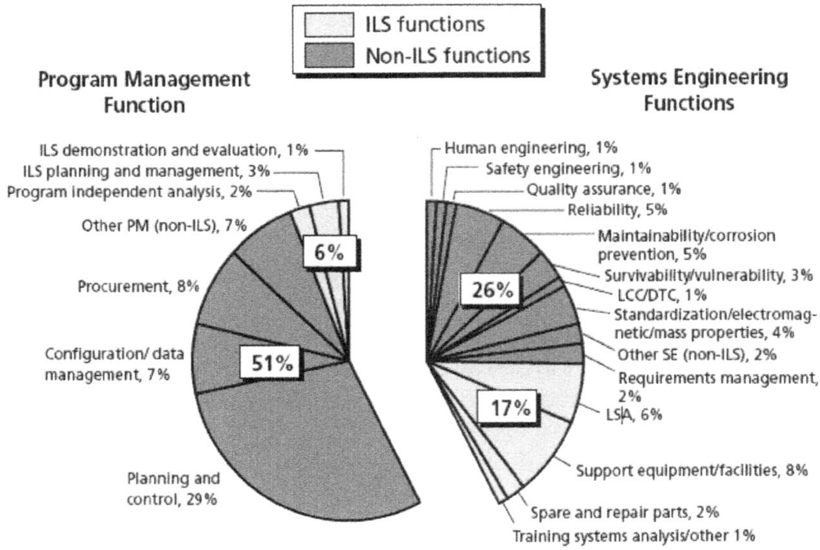

Figure 7.3 SE/PM as a function of integrated logistics support for a typical Air Force program (Stern et al., 2006)

According to Stern and colleagues (2006) of Rand, there is roughly a 50/50 split of systems engineering and project management costs for most large defense programs. As shown in Figure 7.4, these costs can be significant. Unfortunately, COSYSMO only provides a technique for estimating systems engineering costs during the development phase. We are

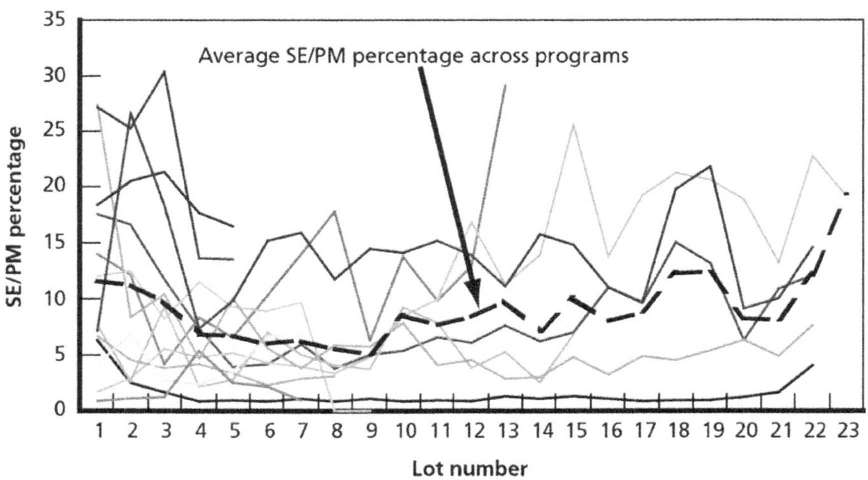

Figure 7.4 Average systems engineering and project management costs for 22 major Air Force programs (Stern et al., 2006)

in the process of trying to identify quantitative means for estimating project management costs (see Young et al., 2011). For simulation-based costing to evolve, this will be needed.

The COCOMO family of models (these will be discussed in detail in Chapter 8) only estimates costs for the work conducted between the beginning of the concept exploration phase and the end of the integration and test phase. Costs and schedules of other phases (like the requirements phase) must be estimated separately. Any significant variation from ideal development (e.g., poor management, programming skill level) should be conducted using the intermediate or detailed version of the model.

According to Valerdi (2005, 2006), the COSYSMO model can help people reason about the cost considerations of systems engineering on projects. Similar to its predecessor, COCOMO II, it was developed at the University of Southern California as a research project with the help of BAE Systems, General Dynamics, Lockheed Martin, Northrop Grumman, Raytheon, and SAIC. COSYSMO follows a parametric modeling approach used to estimate the quantity of systems engineering labor, in terms of person months, required for the conceptualization, design, test, and deployment of large-scale software and hardware projects.

Each parameter in the COSYSMO algorithm is part of the cost estimating relationship that was defined by systems engineering and costing experts. COSYSMO is typically expressed as

$$PM_{NS} = A\left(\sum_{k}\left(\omega_{e,k}\Phi_{e,k} + \omega_{n,k}\Phi_{n,k} + \omega_{d,k}\Phi_{d,k}\right)\right)^{E}\prod_{j=1}^{14}EM_j \tag{7.1}$$

Where

PM_{NS} = effort in person-months (nominal schedule)

A = calibration constant derived from historical project data

E = represents diseconomies of scale

k = {REQ, IF, ALG, SCN}

w_k = weight for "easy," "nominal," or "difficult" size driver

Φ_k = quantity of "k" size driver

EM = effort multiplier for the *j*th cost driver; the geometric product results in an overall effort adjustment factor to the nominal effort

The size of the system is the weighted sum of the system requirements (REQ), system interfaces (IF), algorithms (ALG), and operational scenarios (SCN) parameters and represents the additive part of the model, while the EM factor is the product of the 14 effort multipliers and represents the multiplicative part of the model, which is shown in Table 7.2.

Intermediate COCOMO is designed to estimate the software effort derived from an analysis of software requirements and the design,

Table 7.2 Summary of LCC estimating methods (Young et al., 2010)

Method	Description	Advantages	Disadvantages
Actual Costs	Uses costs experienced during prototyping, hardware engineering development models, and early production items to project future costs for the same system	• Could provide detailed estimate • Reliance on actual development data	• Development data may not reflect cost correctly • Higher uncertainty • Often mistakenly uses contract prices to substitute for actual cost • Various levels of detail involvement • Requires existing actual production data
Analogy/ Comparative Method	Extrapolates available data from similar completed projects and adjust estimates for the proposed project	• Reliance on historical data • Less complex than other methods • Save time	• Subjective/bias may be involved • Limited to mature technologies • Reliance on single data point • Hard to identify appropriate analog • Software and hardware often do not scale linearly
Cost Accounting	Formulated based on the expenditures of reliability, maintainability, and decomposed component cost characteristics	• Reliance on detailed data collection	• Accounting ethics (i.e., cook the book) • Post-production phase strongly preferred • Requires large and complex data collections • Labor intensive
Detailed Engineering Builds/ Bottom-Up	Estimates directly at the decomposed component level, leading to a total combined estimate	• Most detailed at the component level through work breakdown structures • System oriented • Highly accurate • High visibility of cost drivers	• Resource intensive (time and labor) • May overlook system integration costs • Reliance on stable systems architectures and technical knowledge

(*continued*)

Table 7.2 (*continued*)

Method	Description	Advantages	Disadvantages
Expert Judgment/ Delphi Method	Produced by human experts' knowledge and experience via iterative processes and feedbacks	• Available when there are insufficient data, parametric cost relationships, or unstable system architectures	• Subjective/bias • Detailed cost influence/driver may not be identified • Programs' complexities can make estimates less reliable • Human experience and knowledge required
Parametric/ Cost Estimating Relationship	Uses mathematical expressions and historical data to create cost relationships models via regression analysis	• Statistical predictors provide information on expected value and confidence of prediction • Less reliance on systems architectures • Less subjective	• Heavy reliance on historical data • Attributes within data may be too complex to understand • Resource intensive (time and labor) • Difficult to collect data and generate correct cost relationships • Limited by data and independent variables
Top-Down	Based on the overall project characteristics and derived by decomposing into lower-level components and life cycle phases	• Fast and easy deployment • Minimal project detail required • System oriented	• Less accurate than others • Tends to overlook lower-level component details or major cost drivers • Limited detail for justification

implementation, and testing of software. COSYSMO estimates the system engineering effort associated with the development of the software system concept, overall software system design, implementation, and testing—not the actual software development costs. Intermediate COCOMO estimates of the software effort will surely account for the additional effort required by any additional testing of the software system. The COSYSMO effort will account for additional test development and management since the systems engineers are required to perform additional validation and verification of the system (Valerdi, 2006). Case Study 7.1 at the end of this chapter demonstrates the use of COCOMO and COSYSMO to estimate the cost of a software-centric system.

Obviously, there are some shortcomings to this type of approach that are inherent in any top-down model developed early in the life cycle, including the following:

- The model is developed based on historical data; unless you have significant experience in that domain, the model should not be used.
- Engineers estimate from the bottom up of an engineering build.
- Requirements should not be used for estimating because there is no direct correlation between requirements and effort; COSYSMO does recognize this implicitly by distinguishing between pure and equivalent requirements.

7.4 From requirements to architectures

From a set of system requirements, a concept of operations and/or a functional description is developed where the system-level requirements or "whats" are translated to "hows" using tools such as functional block diagrams. This functional hierarchy process and these interdependencies are shown in Figure 7.5. The functional description provides the basis for either a physical architecture or a work breakdown structure (see Figure 7.6) or a cost breakdown structure. Note that the WBS shown in Figure 7.6 was

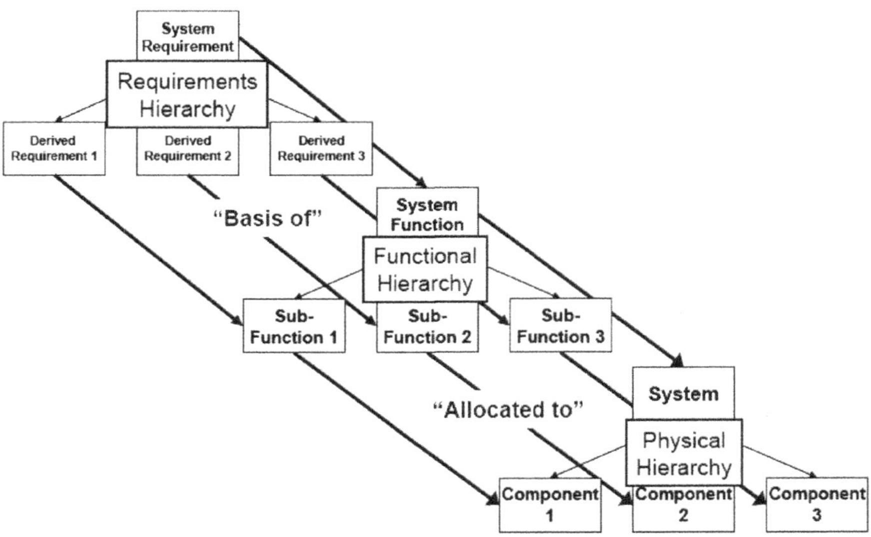

Figure 7.5 Role of functional and physical views of a system (Stevens Institute of Technology, 2009)

SOFTWARE-INTENSIVE SYSTEM WBS				
1	2	3	4	5
SOFTWARE INTENSIVE SYSTEM				
	PRIME MISSION PRODUCT			
		APPLICATIONS S/W		
			BUILD 1	
				CSCI 1...n
				CSCI TO CSCI INTEG. AND CHKOUT
			BUILD 2...n	
				CSCI 1...n
				CSCI TO CSCI INTEG. AND CHKOUT
			APPLICATIONS S/W INTEG., ASSEMBLY, TEST, & CHKOUT	
		SYSTEM S/W		
			BUILD 1	
				CSCI 1...n
				CSCI TO CSCI INTEG. AND CHKOUT
			BUILD 2...n	
				CSCI 1...n
				CSCI TO CSCI INTEG. AND CHKOUT
			SYSTEM S/W INTEG. ASSEMBLY, TEST AND CHECKOUT	
		INTEG., ASSEMBLY, TEST AND CHECKOUT		
		HW/SW INTEGRATION		
	SYSTEMS ENGINEERING/PROGRAM MANAGEMENT			
	SYSTEM TEST AND EVALUATION			
	TRAINING			
	DATA			
	PECULIAR SUPPORT EQUIPMENT			
	COMMON SUPPORT EQUIPMENT			
	INITIAL SPARES AND REPAIR PARTS			

Figure 7.6 Example WBS (Department of Defense, 2011)

developed using physical categories and is only applicable for one phase during our product life cycle. This process is applicable for both new systems and systems that are being reengineered.

7.5 Summary

Costing systems is complex and consists of a variety of techniques, including analogies, PCEs, and detailed bottom-up modeling. Table 7.2 summarizes the pros and cons of these various methods. There is probably no one-size-fits-all solution and certainly no substitute for experience. Readiness levels to be used as scaling factors show promise with additional research to provide insight into developing immature technologies.

In the evaluation and reengineering of existing systems, the functional analysis serves as a basis for developing a work breakdown structure or a cost breakdown structure leading to the collection of costs by functional area (Blanchard, 2009). If you can develop architectures and a WBS, you have a well-understood system suitable for realistic cost estimates.

Until we can quantify integration costs, which, in essence, also means developing a costing algorithm for testing, we will not be able to conduct meaningful bottom-up analysis for complex systems. Early-stage estimates will continue to be based on PCE or analogies.

Case Study 7.1 Development costs of the AgriClean environmental monitor

(modified from Sweeton, 2009)

The marketing and research and development departments have identified a need for a wireless sensor that detects and reports very small changes in temperature, humidity, and acidity. The sensor was specially designed to withstand natural elements and to be waterproof. The marketing department recognized an opportunity to develop this product into a system of sensors that could monitor waste levels in agricultural ponds and waste runoff points. They developed their prototype with the expectation that it would be transitioned into the company's medium-scale product line monitoring system called Watchman, which monitors and measures waste levels for various environmental hazards.

COSYSMO was used to estimate the systems engineering costs of the AgriClean. The table below shows the categories and ratings assigned to all drivers. For example, it was estimated that for the software needed to implement the technology transition strategies, there would be five easy, one medium, and zero hard requirements. As specified by the formula, the number of easy requirements was multiplied by the weight for an easy requirement. The same was done for the medium and hard requirements. The three are added together. This is repeated for the other size drivers as shown in Table 7.3. The products were added together, resulting in a total size driver of 75.1.

Twelve of 14 cost drivers were considered. The software written was primarily internal, so level of service requirements would not be needed. Because the software written did not originate from a previously existing system, migration complexity was not a factor either. For each cost driver, a subjective determination was made of the level of understanding of that factor based on characterizations provided in the COSYSMO model. Each rating from VL (very low) to VH (very high) is given a weight, also provided by the model. Table 7.4 shows drivers, the level of understanding,

Table 7.3 COSYSMO size drivers for AgriClean

Size Drivers	Easy	Medium	Hard	Total Drivers
# System Requirements	5	1	0	
Weights	0.5	1	5	3.5
# System Interfaces	2	2	1	
Weights	1.7	4.3	9.8	21.8
# Algorithms	7	4	0	
Weights	3.4	7.5	18.2	49.8
# Operational Scenarios	0	0	0	
Weights	9.8	22.8	47.4	0
Size Driver				75.1

and the associated weight. The weights multiplied result in a total cost driver of 0.230155.

The work in person-months is found by multiplying the size driver by the cost driver. The historical constant A, as well as the diseconomies of scale exponent E, were not considered; therefore, they were set to 1. Then, the cost was found by multiplying by the average monthly rate of $16,000.

$$\text{Person-Months} = \text{Size Driver} * \text{Cost Driver}$$

$$17.28466 = 75.1 * 0.230155$$

With a total cost of 17.28 *16,000 = $276,555.

Table 7.4 COSYSMO cost drivers for AgriClean

Cost Drivers	Level of Understanding	Weight
Requirements Understanding	H	0.77
Architecture Understanding	VH	0.65
Technology Risk	VL	0.7
Documentation	VL	0.82
# and Diversity of Installation/Platforms	N	1
# Recursive Levels in the Design	N	1
Stakeholder Team Cohesion	L	1.22
Personnel/Team Capability	H	0.81
Personnel Experience/Continuity	VH	0.67
Process Capability	L	1.21
Multisite Coordination	N	1
Tool Support	N	1
		0.230155

Table 7.5 Non-software development and fixed-cost-related strategies

Recommended Technology Transition Strategy	Cost	Costing Technique
Strategic Planning	$ 16,000	Historical Data
Support for Separate Budget	$ 4,100	Historical Data
Technology Transition Agreement	$ 24,500	Historical Data
Cross-Pollinate Personnel	$ 13,000	Historical Data
Buy Materials Earlier	0	No Cost
Metrics Support, Commercial Off-the-Shelf Products	$ 15,000	Historical Data
Support for Gated Reviews (lab manager salary + cost of process)	$224,000	Historical Data; COSYSMO
Relationship Manager's salary	$ 82,000	Historical Data
TOTAL	**$378,600**	

Table 7.5 shows each recommendation and its cost, taking into account work done in each of the 2 years, the original targeted duration. It used the appropriate pay rates for the personnel involved.

The total cost of developing AgriClean is the sum of the cost of the software development–related strategies, which include the development and system engineering activities and the cost of the management-related and fixed-cost strategies.

For the integration of AgriClean into the Watchman project, the cost is:

COCOMO	$ 377,640
COSYSMO	$ 276,555
Management-Related/Fixed-Cost Strategies	$ 378,600
Total Cost	$1,032,795

QUESTIONS

7-1. Suppose we define System Cost = Hardware + Software + Interface/Integration + Support. Which of these is the most difficult to ascertain early in the product life cycle? Which of these is the most difficult to determine the recurring costs?

7-2. From Figure 1.7, during the concept advanced development phase, 85% of the costs are typically committed. Between requirements creep, cost variances of up to 75%, technical readiness levels of some components of less than three, and so on, budgeting during concept advanced development is at best a rough order of magnitude estimate. How can we develop more accurate estimates early during the product life cycle when funds are typically programmed?

7-3. Below are bids from two actual software-centric systems. These systems consist of a sensor network, data processing, some data mining, and ultimately storage.

Project 1		Project 2	
Management	$ 7,731,153	Management Leadership	$ 4,483,465
Systems Engineering	$ 3,619,772	Overhead	$15,943,536
Software	$ 7,183,903	Data Cleaning and Processing	$ 1,993,215
Hardware	$ 500,355	Software Development	$ 4,569,854
Integration	$ 6,937,304	Simulation	$11,369,329
Testing	$ 7,062,878	Software Overhead	$ 3,086,863
Training	$ 1,695,175	Contract Data Requirements List	$ 4,415,556
Travel	$ 2,142,724	Systems Engineering	$ 3,070,027
Documentation	$ 1,666,761	Integration and Test	$ 9,900,734
		Installation	$ 8,475,736
		Travel	$ 2,828,208
Total	$38,540,025	Total	$70,136,523

Can you develop any rules of thumb for project management, systems engineering, integration and testing, and the like based on this limited data set?

PROBLEMS

7-1. Many companies use analogies early in the development process to estimate total project costs. For example, one large IT services provider uses the following rules of thumb when calculating total project costs:

Percentage of Total Project Cost	Cost Category
8%	Project Management
15%	Systems Engineering
40%	Development
26%	Testing and Evaluation
11%	Installation

Your engineers and estimating department have calculated the following development costs for an IT solution:

- Software: $7,183,903
- Hardware: $500,355
- Integration: $6,937,304

Using the above information, develop the total project costs. How does this compare with the total costs for Project 1 of Question 7-3? What are the main differences?

7-2. Many contractors simply use percentages when developing bids. Below are some values used by one contractor for estimating capital projects for water distribution systems.

Subtotal Items Applied to the Total of All Installation, Equipment, Facilities, etc.	Percentage
Markup for Bonds and Insurance	3
Mobilization and Demobilization Costs	5
Contractors Overhead	8
Contractors Profit	4
Contingency	25
Applied to the Total of All Installation and Physical Plants Added to the Subtotal Items	
Allowance for Engineering and Design	8
Permitting, Legal, and Administrative	12
Engineering Services during Construction	8

Preliminary estimates for a new water distribution upgrade in Nassau County, Florida, total $98 million. What will be the estimated total project costs to the county using the previous percentages?

References

Barker, Bruce, April 2008, personal note.

Blanchard, Benjamin S., 2009, "Cost Management," in *Handbook of Systems Engineering*, 2nd Edition, eds. Andrew P. Sage and William B. Rouse, John Wiley and Sons, Hoboken, NJ.

Department of Defense, 3 October 2011, "Work Breakdown Structure for Defense Material Items," accessed 1 August 2018 at https://www.ncca.navy.mil/references/MIL_STD_881C_OCT032011.pdf.

Farr, John V., Enos, James R., and McCarthy, Daniel J., 2014, "An Application of System Dynamics to Quantify Investments in Stbility and Reconstruction Operations by the Department of Defense," in *Case Studies in System of Systems, Enterprises, and Complex Systems Engineering*, Taylor & Francis/CRC Press, Boca Raton, FL.

Stern, David E., Boito, Michael, and Younossi, Obaid, 2006, *Systems Engineering and Program Management—Trends and Costs for Aircraft and Guided Weapons Programs*, RAND Corporation, Santa Monica, CA.

Stevens Institute of Technology, 2009, "SYS 650 System Architecture and Design," Course Notes.

Sweeton, Julie E., April 2009, "Comparison of Life Cycle Costs between Two Technology Transition Approaches," EM 620 Project.

Valerdi, Ricardo, August 2005, "The Constructive Systems Engineering Costing Model (COSYSMO)," PhD Dissertation, University of Southern California.

Valerdi, Ricardo, September 2006, "Academic COSYSMO User Manual—A Practical Guide for Industry and Government," Version 1.1, MIT Lean Aerospace Initiative.

Young, Leone Z., Farr, John V., and Valerdi, Ricardo, March 2010, "The Role of Influential Complexities in Systems Engineering Cost Estimating Processes," 8th Conference on Systems Engineering Research, Hoboken, NJ.

Young, Leone Z., Wade, Jon, Farr, John V., Valerdi, Ricardo, and Kwak, Young Hoon, 7–10 June 2011, "An Approach to Estimate the Life Cycle Cost and Effort of Project Management for Systems Centric Projects," International Society of Parametric Analysts and the Society of Cost Estimating and Analysis Joint Annual Conference and Training Workshop, Albuquerque, NM.

chapter eight

Software-intensive systems

8.1 Introduction

Almost every aspect of our modern society is controlled by software. You need look no further than the defense industry to see how dramatic and pervasive software has become. Consider the following military examples:

- The F4 fighter had no digital computer or software (early 1970s).
- The F16A fighter had 50 digital processors and 135,000 lines of code or KLOC (late 1970s).
- The F16D fighter had 300 digital processors and 236 KLOC (late 1980s).
- The B-2 bomber has over 200 digital processors and 5,000 KLOC (late 1990s).
- The Future Combat Systems program would have had over 16,000 to 50,000 KLOC (late 2000s) if had not been canceled because of complexity and integration challenges.

Software requirements (the percentage of functionality provided by software) have grown from less than 10% in the 1980s to as much as 80% for complex systems in our current world; think of the ATM, pay at the pump gas system, cars, planes, and the like.

Software is also redefining the consumer's world. Microprocessors embedded in today's automobiles require software to run, permitting major improvements in their performance, safety, reliability, maintainability, and fuel economy. The average car in 1990 had 1 million lines of code; in 2018, the average car had over 100 million lines of code with software and electronics contributing to over one-third of the cost of a car. Nearly all complex systems, including airplanes, automobiles, phones, TVs, and the like, have seen a dramatic growth in software similar to that shown in Figure 8.1. New devices in the consumer electronics sector have dramatically changed how we play music, conduct personal computing, and manage our daily activities. As software becomes more deeply embedded in most goods and services, creating reliable and robust software is becoming an even more important challenge. Despite the pervasive use of software, and partly because of its relative immaturity, especially with regards to integrating complex hardware and

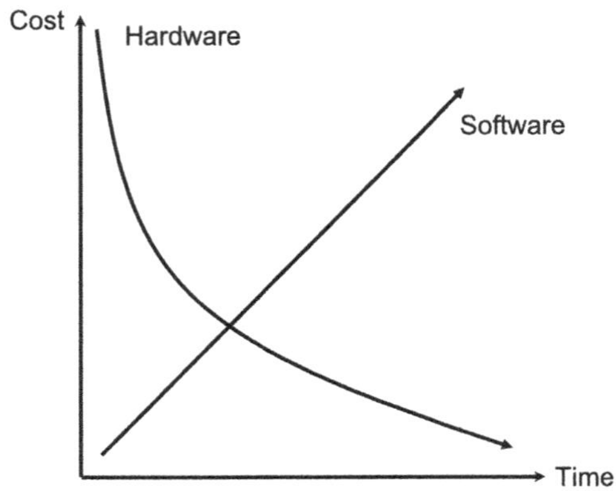

Figure 8.1 Growth of hardware and software costs for most complex systems

software applications, understanding the economics of software presents an extraordinary challenge.

Engineers typically know how to estimate hardware; we can simply count up the components. However, software and integration and interfaces continue to be the challenge in costing complex systems. Therefore, we wrote this chapter to expose readers to the myriad of methods to estimate software costs. As you will see, historical analysis dominates software cost estimation.

Because of complexity, software development and maintenance and upgrades are error-prone, time-consuming, and complex activities. Experience has revealed that many software programs' efforts falter because the management of these projects falls into several common traps:

- Inadequate or poorly written requirements or scope or requirements creep,
- Little understanding or visibility of the total system, code written in stovepipes,
- Failure to accept already developed and "good enough" solutions,
- Poor configuration management and control at key points during development,
- Homegrown software with little standardization or open standards, lack of modular design inhibits reuse (i.e., spaghetti code) and documentation,
- No attention to total ownership cost considerations,
- The churn of management and software developers,
- Not properly using software engineers and architects,
- "Not invented or written here" mentality, and
- Lack of historical data and skills in estimating effort and schedule.

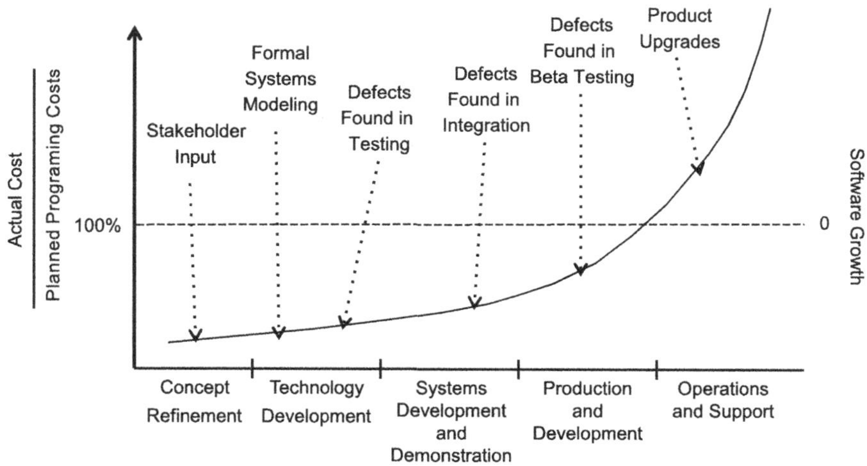

Figure 8.2 Software cost growth during the life cycle

Figure 8.2 shows the role of software cost growth during the life cycle. Note the rapid growth of costs as the software becomes integrated with the products.

8.2 Software estimating techniques

8.2.1 Overview

Figure 8.3 shows various techniques for estimating software development time, resources, and costs. Note the varying degrees of complexity in the literature, from simple algorithms to complex Bayesian techniques. Most techniques are a combination of experience and historical results. In Chapter 1, we classed these types of life cycle cost (LCC) techniques as either parametric cost estimating, analogy, or some combination of these.

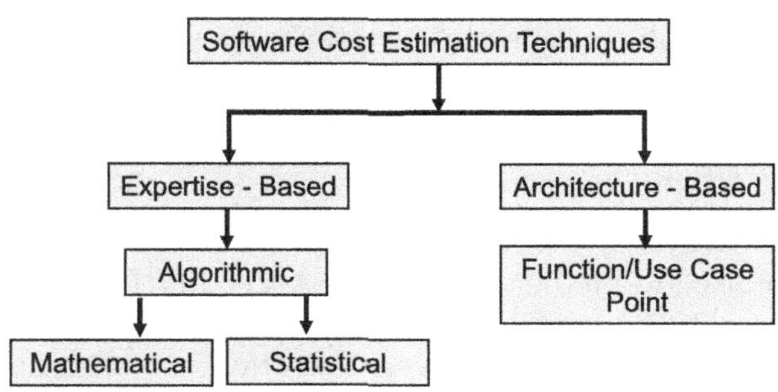

Figure 8.3 Software estimation techniques (modified from IBM, 2007)

Probably the most important step in developing a software (or any) cost estimate is developing some type of functional representation to capture all elements in the life cycle, including the following (modified from Department of Defense, 2005):

- A product-oriented family tree composed of hardware, software, services, data, and facilities. The family tree results from systems engineering efforts during the acquisition of a defense materiel item.
- A work breakdown structure that displays and defines the product, or products, to be developed and/or produced. It relates the elements of work to be accomplished to each other and to the end product. A work breakdown structure can be expressed down to any level of interest. However, the top three levels are as far as any program or contract need go unless the items identified are high cost or high risk. Then, and only then, is it important to take the work breakdown structure to a lower level of definition.

8.2.2 *Expertise-based and hybrid models*

Most models are a mix of expertise based and hybrid because of the subjective nature of many of the inputs and algorithms. Expertise is nothing more than subjective human estimating combined with some simple heuristics. One large defense contractor uses the expertise and algorithm method to estimate software costs:

1. Estimate the number of function points based on requirements, like projects, and so on;
2. Use the Intermediate Constructive Cost Model (COCOMO) to estimate the resources required; and then
3. Multiply the software development time by 175% to estimate costs.

This is one example of an experience-based algorithm combined with a mathematical model to produce a hybrid technique. Most companies use "rules of thumb" with hybrid techniques to estimate software development costs such as Example 8.1.

8.2.3 *Algorithmic models*

The original COCOMO is an algorithm-based model developed by Boehm (1981) and is used to predict the effort and schedule for a software product development. The model is based on inputs relating to the size of the software and a number of cost drivers that affect productivity. COCOMO drew on a study of about 60 projects at TRW Inc., with software ranging in size from 2,000 to 100,000 lines of code. Most companies use a modified

EXAMPLE 8.1

One major services company uses application development costs to estimate total project costs and other cost categories. They take the application development costs and multiply it by 2.5 to determine the total project cost. They then use the following percentage to determine the cost elements:

- Project Management = 8% × Total Project Cost
- Systems Engineering = 15% × Total Project Cost
- Application Development = 40% × Total Project Cost
- Testing = 26% × Total Project Cost
- Service Delivery/Fielding = 11% × Total Project Cost

Obviously, these types of expertise-based models are based on extensive experience and are critical to developing rough order of magnitude estimates early in the product life cycle.

version of one of the COCOMO family of models to estimate software development times and efforts.

The original COCOMO consists of a hierarchy of three increasingly detailed versions (modified from NASA, 2015):

1. Basic COCOMO, sometimes referred to as COCOMO 81, computes software development effort (and cost) as a function of program size and is good for quick, early, rough order of magnitude estimates of software costs.
2. Intermediate COCOMO computes software development effort as a function of program size and a set of "cost drivers" that includes a subjective assessment of product, hardware, personnel, and project attributes.
3. Detailed COCOMO incorporates all characteristics of the intermediate version with an assessment of the cost drivers' impact on each step (analysis, design, etc.) of the software engineering process.

The COCOMO family of models uses either thousands of delivered source instructions (KDSI) or thousands of lines of code (KLOC) as a unit of measure for software size. Note that code created by applications should not be included. Also, when using COCOMO, you should not include commented delivered source instructions or lines of code (DSI or LOC) when making estimates, but you should include declarations. The literature often interchanges these terms when discussing COCOMO.

The main difference between DSI and LOC is that a single LOC may contain several DSIs. Only Basic COCOMO uses DSI, whereas the other versions use source lines of code (SLOC).

There are some problems with using DSI/LOC to estimate software, mainly the ease of programming of higher-level languages. Imagine two applications that provide the same exact functionality with one written in C and the other application written with a higher-level language, such as Java. The same functionality would require many more lines of C code than Java. According to Lum and colleagues (2003), "Typically either physical lines or logical lines are used when counting SLOC. Comments and blanks should never be included in any count of lines of code. The physical SLOC measure is very simple to count because each line is terminated by the enter key or a hardline break. A logical statement is a single software instruction, having a defined beginning and ending independent of any relationship to the physical lines on which it is recorded or printed."

The most commonly used measure of source code program length is the number of LOCs (Fenton and Pfleeger, 1997). The abbreviation NCLOC is used to represent a non-commented source line of code. NCLOC is also sometimes referred to as effective lines of code (ELOC). NCLOC is therefore a measure of the uncommented length. Thus, by measuring NCLOC and commented lines of code (CLOC) separately, we can define

$$\text{Total length (LOC)} = \text{NCLOC} + \text{CLOC} \tag{8.1}$$

The commented length is also a valid measure, depending on whether or not line documentation is considered to be a part of programming effort. LOC/KLOC is the primary input in determining software development costs.

8.2.3.1 Original or basic COCOMO model

Basic COCOMO, which is also referred to as COCOMO 81, is a static model that uses a non-linear single-valued input equation to compute software development effort (and cost) as a function of software program size. The main input into the model is estimated KDSI. The model takes the form

$$E = aS^b \tag{8.2}$$

where

E = effort in person-months

S = size of the software development in KDSI

a, b = values dependent on the development mode

where

organic	$a = 2.4$	$b = 1.05$
semidetached	$a = 3.0$	$b = 1.12$
embedded	$a = 3.6$	$b = 1.20$

The Basic Model also presents an equation for estimating the development schedule (time to develop or TDEV) of the project in months:

$$TDEV = cE^{d} \tag{8.3}$$

Factors c and d also depend on the development mode, where

organic	$c = 2.5$	$d = 0.38$
semidetached	$c = 2.5$	$d = 0.35$
embedded	$c = 2.5$	$d = 0.32$

Boehm and his colleagues at the University of Southern California have defined three development modes. Note that the following detailed descriptions of these modes were modified from NASA (2015):

- Organic mode: In the organic mode, the project is developed in a familiar, stable environment, and the product is similar to previously developed products. The product is relatively small and requires little innovation. Most people connected with the project have extensive experience in working with related systems within the organization and therefore can usefully contribute to the project in its early stages without generating a great deal of project communication overhead. An organic mode project is relatively relaxed about the way the software meets its requirements and interface specifications. If a situation arises where an exact correspondence of the software product to the original requirements would cause extensive rework, the project team can generally negotiate a modification of the specifications that can be developed more easily.
- Semidetached mode: In this mode, the project's characteristics are intermediate between organic and embedded. "Intermediate" may mean either of two things:
  - An intermediate level of project characteristics, or
  - A mixture of the organic and embedded mode characteristics.

Therefore, in a semidetached mode project, it is possible that

- The team members all have an intermediate level of experience with related systems,
- The team has a wide mixture of experienced and inexperienced people,

- The team members have experience related to some aspects of the system under development but not others, and
- The size of a semidetached mode product generally extends up to 300 KDSI.
- Embedded mode: In this development mode, the project is characterized by tight, inflexible constraints and interface requirements. The product must operate within a strongly coupled complex of hardware, software, regulations, and operational procedures. The embedded mode project does not generally have the option of negotiating easier software changes and fixes by modifying the requirements and interface specifications. The project therefore needs more effort to accommodate changes and fixes. The embedded mode project is generally charting its way through unknown territory to a greater extent than the organic mode project. This produces a project that uses a much smaller team of analysts in the early stages, as a large number of people would get swamped in communication overhead.

Example 8.2 shows how the Basic COCOMO model is used to determine software costs.

Once the product design has been completed for an embedded mode project, the best strategy is to bring on a very large team of programmers to perform detailed design, coding, and unit testing in parallel. Otherwise the project would take much longer to complete. This strategy, as we will see, leads to the higher peaks in the personnel curves of embedded mode projects and to the greater amount of effort consumed compared to an organic mode project working to the same total development schedule.

EXAMPLE 8.2

A large transportation and distribution company is planning to develop a new computer program to keep track of high-value inventory items. An initial study has determined that the size of the program will be roughly 32,000 DSI. You have an experienced team of developers who know the domain. Using the Basic COCOMO and the organic mode, we determined that

Effort: $2.4*(32)^{1.05} = 91$ Staff-Months (SM)
Productivity: 32,000 DSI/91 SM = 352 DSI/SM
$\approx$ 18 DSI/Day $\approx$ 2 DSI/Hour
Development Time = $2.5*(91)^{.38} = 13.8$ Months $\approx$ 14 Months

Table 8.1 Phase distribution of effort: organic mode

Phase	Small (2 KDSI)	Intermediate (8 KDSI)	Medium (32 KDSI)	Large (128 KDSI)
Plans and Requirements	*6%*	*6%*	*6%*	*6%*
Product Design	16	16	16	16
Detailed Design	26	25	24	23
Code and Unit Test	42	40	38	36
Integration and Test	16	19	22	25
Total	100	100	100	100

8.2.3.2 Phase distribution of effort and schedule for organic mode

After estimating the effort and schedule of a project, we need to determine how to distribute them among different phases of the project. This distribution varies as a function of the size of the product and not the number of developers. Larger software projects require relatively more time and effort to perform integration and test activities and are able to compress the programming portion of the program by having more people programming components in parallel (NASA, 2015). Smaller software projects have a more uniform, flat distribution of labor throughout the development cycle and have relatively more resources devoted to the phases other than integration and testing.

Using Tables 8.1 and 8.2, we can calculate the distribution of the effort and schedule among different phases of an organic mode software product. Example 8.3 demonstrates how these tables are used. Then, using the same type of calculations as in Example 8.2, we can calculate the number of personnel required for each phase. Tables 8.3 and 8.4 present the percentage distribution of the basic software effort and schedule within the

Table 8.2 Phase distribution of schedule: organic mode

Phase	Small (2 KDSI)	Intermediate (8 KDSI)	Medium (32 KDSI)	Large (128 KDSI)
Plans and Requirements	*10%*	*11%*	*12%*	*13%*
Product Design	19	19	19	19
Detailed Design and Code and Unit Test	63	59	55	51
Integration and Test	18	22	26	30
Total	100	100	100	100

EXAMPLE 8.3

Using Table 8.2, we can calculate the number of programmers and development time by phase for Example 8.1:

Programming Effort = 91 SM
Development Time = 13.8 SM

Phase	%	Effort	%	Schedule
Product Design	16	14.6	19	2.6
Detailed Design	24	21.8		
Code and Unit Test	38	34.6	55	7.6
Integration and Test	22	20.0	26	3.6
Total	100%	91 SM	100	13.8

Table 8.3 Phase distribution of effort: semidetached mode

Phase	Small (2 KDSI)	Intermediate (8 KDSI)	Medium (32 KDSI)	Large (128 KDSI)
Plans and Requirements	*7%*	*7%*	*7%*	*7%*
Product Design	17	17	17	17
Detailed Design	27	26	25	24
Code and Unit Test	37	35	33	31
Integration and Test	19	22	25	28
Total	100	100	100	100

development phases of a semidetached mode product. Tables 8.5 and 8.6 present the percentage distribution of the basic software effort and schedule within the development phases of an embedded mode product. Note that if you include the plans and requirements, the total will exceed 100%. These tables separate these two entities.

Table 8.4 Phase distribution of schedule: semidetached mode

Phase	Small (2 KDSI)	Intermediate (8 KDSI)	Medium (32 KDSI)	Large (128 KDSI)
Plans and Requirements	*16%*	*18%*	*20%*	*22%*
Product Design	24	25	26	27
Detailed Design and Code and Unit Test	56	52	48	44
Integration and Test	20	23	26	29
Total	100	100	100	100

Table 8.5 Phase distribution of effort: embedded mode

Phase	Small (2 KDSI)	Intermediate (8 KDSI)	Medium (32 KDSI)	Large (128 KDSI)
Plans and Requirements	*8%*	*8%*	*8%*	*8%*
Product Design	18	18	18	18
Detailed Design	28	27	26	25
Code and Unit Test	32	30	28	26
Integration and Test	22	25	28	31
Total	100	100	100	100

By comparing Tables 8.1 through 8.6, we can see some differences between the effort and schedule distribution of the products developed in different modes. The main differences are as follows (NASA, 2015):

- The embedded mode project consumes considerably more effort in the integration and test phase. This results from the need to follow and verify software requirements and interface specifications more carefully in the embedded and semidetached mode.
- The embedded mode project consumes proportionally less effort in the code and unit test phase. This results from the proportionally higher effort required for the other development phases.
- The embedded mode project consumes considerably more schedule in both the plans and requirement phase and the product design phase. This is because of the project's need for more thorough, validated requirements and design specifications, and the greater need to perform these phases with a relatively small number of people.
- The embedded mode project consumes considerably less schedule in the programming phase. This results from the strategy of employing a great many people programming in parallel in order to reduce the project's overall schedule.

Table 8.6 Phase distribution of schedule: embedded mode

Phase	Small (2 KDSI)	Intermediate (8 KDSI)	Medium (32 KDSI)	Large (128 KDSI)
Plans and Requirements	*24%*	*28%*	*32%*	*36%*
Product Design	30	32	34	36
Detailed Design and Code and Unit Test	48	44	40	36
Integration and Test	22	24	26	28
Total	100	100	100	100

The Basic COCOMO model was developed when 100,000 lines of code was a large program and few computer-aided software engineering (CASE) tools existed. We typically write codes in excess of 1,000 KLOC for most applications that require significantly more integration. Therefore, these algorithms were not developed for large codes, languages such as Java, C++, and so on, and the extensive amount of CASE tools that exist today. Conversely, our applications today must be interoperable between many other platforms and require significant integration effort. Many companies have developed their own internal effort adjustment factors (EAFs) for use with COCOMO. One large defense contractor uses Intermediate COCOMO with an EAF of 1.75 because most of their work is defense related with significant reporting and security requirements and entails a significant amount of integration with a large focus on commercial off-the-shelf (COTS) solutions.

8.2.3.3 Intermediate COCOMO

Intermediate COCOMO, which is also referred to as COCOMO II, takes Basic COCOMO as a starting point and then accounts for personnel, product, computer, and project attributes. The Intermediate COCOMO model computes effort as a function of program size and a set of cost drivers (Pressman, 1997). The Intermediate COCOMO equation is

$$E = a * KLOC^{b} * EAF \tag{8.4}$$

The a and b factors for the Intermediate COCOMO model vary slightly from the Basic COCOMO and are presented below:

organic	a = 3.2	b = 1.05
semidetached	a = 3.0	b = 1.12
embedded	a = 2.8	b = 1.20

The EAF is calculated using 15 cost drivers (Boehm et al., 2000). The cost drivers are grouped into four categories: product, computer, personnel, and project. Each cost driver is rated on a six-point ordinal scale ranging from low to high importance. Based on the rating, an effort multiplier is determined using Table 8.7 (Boehm, 1981). The product of all effort multipliers is the EAF. Thus, the correct mathematical expression is

$$E = aS^{b}\prod_{1}^{15} EAF \tag{8.5}$$

There are two reasons why the Intermediate model produces better results than the Basic model. First, it has more fidelity because it considers

Table 8.7 Software development effort multipliers[1]

Cost Driver	Very Low	Low	Nominal	High	Very High	Extra High
ACAP Analyst Capability	1.46	1.19	1.00	0.86	0.71	--
AEXP Application Experience	1.29	1.13	1.00	0.91	0.82	--
CPLX Product Complexity	0.70	0.85	1.00	1.15	1.30	1.65
DATA Database Size	--	0.94	1.00	1.08	1.16	--
LEXP Language Experience	1.14	1.07	1.00	0.95	--	--
MODP Modem Programming Practices	1.24	1.10	1.00	0.91	0.82	--
PCAP Programmer Capability	1.42	1.17	1.00	0.86	0.70	--
RELY Required Software Reliability	0.75	0.88	1.00	1.15	1.40	--
SCED Required Development Schedule	1.23	1.08	1.00	1.04	1.10	--
STOR Main Storage Constraint	--	--	1.00	1.06	1.21	1.56
TIME Execution Time Constraint	--	--	1.00	1.11	1.30	1.66
TOOL Use of Software Tools	1.24	1.10	1.00	0.91	0.83	--
TURN Computer Turnaround Time	--	0.87	1.00	1.07	1.15	--
VEXP Virtual Machine Experience	1.21	1.10	1.00	0.90	--	--
VIRT Virtual Machine Volatility	--	0.87	1.00	1.15	1.30	--

the effect of more cost drivers. Second, it is advertised as applicable for use for systems that can be divided into "components." DSI value and cost drivers can be chosen for individual components instead of for the system as a whole. COCOMO can estimate the staffing, cost, and duration of each of the components, allowing you to experiment with different development strategies to find the plan that best suits your needs and resources. Note that if all the cost drivers are rated as nominal, which means that that factor has no effect on overall EAF, the Intermediate model reduces to the Basic model.

[1] Appendix 8-A contains a description of these factors.

The COCOMO II manual (Center for Software Engineering, 2000) provides quantifiable values to develop the effort multipliers.

8.2.3.4 Advanced COCOMO

The Advanced or Detailed model differs from the Intermediate model in that it uses different EAFs for each phase of a project. This phase dependency yields better estimates than the Intermediate model. The four phases used in the Detailed COCOMO model are requirements planning and product design (RPD), detailed design (DD), code and unit test (CUT), and integration and test (I&T). The RPD phase includes the beginning stages of product design and planning. The DD phase is calculated for a more detailed level of design. The CUT includes coding activities as well as individual unit testing, and I&T is the final stage of testing. All components are put together, and the system is tested as a whole.

8.2.4 Function points

In the late 1970s, IBM felt the need to develop a language-independent approach to estimating software development effort. They tasked employee Allan Albrecht with developing this approach. The result was the function point (FP) technique. Function points measure size in terms of the amount of functionality in a system. Function point analysis is a structured technique of classifying the components of a system. It is a method used to break systems down into smaller components so that they can be best understood and analyzed. Function point analysis provides a structured technique for problem solving. FPs are computed by first calculating an unadjusted function point count. Counts are made for the following categories (Fenton and Pfleeger, 1997):

- External inputs: those items provided by the user that describe distinct application-oriented data (such as file names and menu selections);
- External outputs: those items provided to the user that generate distinct application-oriented data (such as reports and messages rather than the individual components of these);
- External inquiries: interactive inputs requiring a response;
- External files: machine-readable interfaces to other systems; and
- Internal files: logical master files in the system.

Once this data has been collected, a complexity rating is associated with each count according to Table 8.8.

Each count is multiplied by its corresponding complexity weight, and the results are summed to provide the unadjusted function points (UFP). The adjusted FP count is calculated by multiplying the UFP by

Table 8.8 Function point complexity weighting factors

Item	Simple	Average	Complex
External Inputs	3	4	6
External Outputs	4	5	7
External Inquiries	3	4	6
External Files	7	10	15
Internal Files	5	7	10

a technical complexity factor (TCF). Components of the TCF are listed in Table 8.9.

Each component is rated or graded from 0 to 5, where 0 means the component has no influence on the system and 5 means the component is essential (Pressman, 1997). The TCF can then be calculated as

$$TCF = 0.65 + SUM(Fi) / 100 \tag{8.6}$$

Where F(i) is the components of TCF and varies between 0 and 5 based on complexity. The final function point calculation is

$$FP = UFP \times TCF \tag{8.7}$$

Then you use Table 8.10 to convert function points to SLOC.

Table 8.9 Components of the technical complexity factor

Factor	Description
F1	Reliable Backup and Recovery
F2	Data Communications
F3	Distributed Functions
F4	Performance
F5	Heavily Used Configuration
F6	Online Data Entry
F7	Operational Ease
F8	Online Update
F9	Complex Interface
F10	Complex Processing
F11	Reusability
F12	Installation Ease
F13	Multiple Sites
F14	Facilitate Change

Table 8.10 Number SLOC per function point as a function of programming language (Quantitative Software Management, n.d.)

Language	Average	Median	Low	High
ABAP (SAP)	28	18	16	60
ASP	51	54	15	69
Assembler	119	98	25	320
Brio +	14	14	13	16
C	97	99	39	333
C++	50	53	25	80
C#	54	59	29	70
COBOL	61	55	23	297
Cognos Impromptu Scripts +	47	42	30	100
Cross System Products (CSP) +	20	18	10	38
Cool:Gen/IEF	32	24	10	82
Datastage	71	65	31	157
Excel	209	191	131	315
Focus	43	45	45	45
FoxPro	36	35	34	38
HTML	34	40	14	48
J2EE	46	49	15	67
Java	53	53	14	134
JavaScript	47	53	31	63
JCL	62	48	25	221
LINC II	29	30	22	38
Lotus Notes	23	21	19	40
Natural	40	34	34	53
.NET	57	60	53	60
Oracle	37	40	17	60
PACBASE	35	32	22	60
Perl	24	15	15	60
PL/1	64	80	16	80
PL/SQL	37	35	13	60
Powerbuilder	26	28	7	40
REXX	77	80	50	80
Sabretalk	70	66	45	109
SAS	38	37	22	55
Siebel	59	60	51	60
SLOGAN	75	75	74	75
SQL	21	21	13	37
VB.NET	52	60	26	60
Visual Basic	42	44	20	60

FP is widely accepted as the standard methodology for using an architecture to predict software size. It is also widely used to quantify scope and requirements creep. FP growth can be traced through the life cycle to better understand cost growth. Example 8.4 contains a simple FP example.

With the advent of object-oriented programming, more complex software estimating techniques are being used. With the evolution of SLOC, to FP, to now use case points (UCP, based on Unified Modeling Language),

EXAMPLE 8.4

Consider the context diagram for an ATM system:

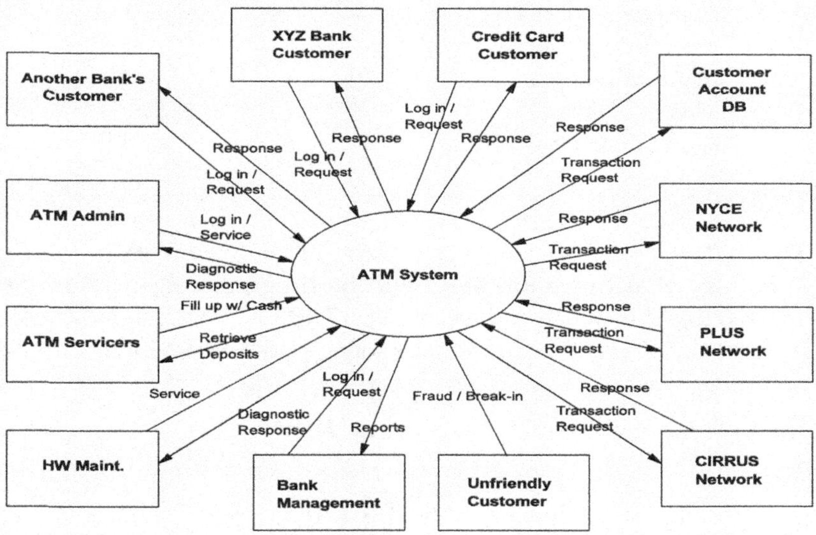

Function points have been estimated for the architecture along with complexity ratings and are shown below:

Item	Simple	Average	Complex	Total
External Inputs	3*3	4*4	2*6	37
External Outputs	2*4	2*5	3*7	49
External Inquiries	2*3	2*4	4*6	38
External Files	2*7	3*10	1*15	59
Internal Files	3*5	4*7	1*10	53

$$UFP = 37 + 49 + 38 + 59 + 53 = 236$$

(continued)

We now need to compute the technical complexity factors for this problem, shown in the following table:

Reliable Backup and Recovery	5
Data Communications	5
Distributed Functions	5
Performance	3
Heavily Used Configuration	5
Online Data Entry	1
Operational Ease	5
Online Update	4
Complex Interface	5
Complex Processing	1
Reusability	5
Installation Ease	2
Multiple Sites	5
Facilitate Change	3
Total	49

Thus FP = 236 * (0.65+ 49/100) = 236 * 1.14 = 269 function points. Assuming that this application will be programmed in Java (see Table 8.10), this will produce about 17 KLOC. We can now use Intermediate COCOMO to estimate software development resources and time.

An ATM is a highly embedded system. Thus

$$E = 3.6S^{1.2} \prod_{1}^{15} EAF$$

For our problem, S = 17KLOC. Also, the system must be highly reliable and is very complex because of encryption. Thus E = 3.6(17)1.2 * 1.65 * 1.4 = 249 staff-months. The development time or TDEV = $2.5(249)^{.32}$ = 14.6, say 15 months. Using $12,000/staff-month, this project would cost about $3 million and take roughly 15 months.

we are better able to translate architecture to functions to cost. Table 8.11 summarizes some of the characteristics of SLOC, FP, and UCP. All have merit, and their use is driven mainly by historical records, knowledge management, and the maturity of the systems engineering processes of an organization.

Table 8.11 Characteristics of SLOC, FP, and UCP analysis (modified from Schofield et al., 2013)

Characteristic	Source Lines of Code	Function Points	Use Case Points
Useful at the Project Level for Estimating or Planning	With Historical SLOC Data	With Historical FP Data	With Historical UCP Data
ISO/Standards Based	No	ISO 20926	No
Translation from Requirements	Difficult	Achievable	Achievable
Easy to Calculate	Very Easy	Easy	Moderate Difficulty
Capture Complexity	Some	Moderate	Extensive
Transferrable to Like Project, Good Knowledge Management	Yes	Moderate	Not Really Meant for Other Projects

8.2.5 Costing software with agile development

In recent years the software development community has almost entirely moved away from traditional project management practices and toward a technique commonly referred to as agile development. This change came after a series of high-profile software development failures. The approach focuses on short, well-scoped development cycles referred to as sprints (i.e., a set period of time during which specific work has to be completed) by a small team known as a scrum team. Each cycle is part of a larger project but gives much greater flexibility and feedback from users along the way. The nature of this type of development is that easy application of traditional costing techniques becomes difficult due to the intentional lack of long-term planning.

There are several communities of practice for costing software developed using an agile approach. However, there is no one standard approach. As the field of costing evolves, it is the author's hope that an industry standard will emerge. However, for the reader who must provide cost estimates in the present, there are some heuristics that have proven useful.

The good news is that most of the techniques in this book are easily modified to cost software built using agile. While agile is highly iterative, there are some cost centers used for estimation:

- Sprint cycle length,
- Number of sprint cycles,
- Number of epics (i.e., can be defined as a large amount of work that has one common objective), and
- Number of team members on the scrum team.

The traditional "lines of code" would be a poor metric for costing in this setting. Ceschi and colleagues (2005) claimed that none of the companies they had surveyed currently used COCOMO and that 40% used FP estimation on their agile projects. Other issues, such as the number of sprint cycles (horizon is often well known) and size and composition of the team (number and type of engineers, architects, developers, etc.) must be considered.

Another consideration is that agile teams often benefit from product design teams who work with users to provide initial "markups." The general steps that complement agile projects include the following:

- User interviews (requirements generation),
- Product design and markup,
- Chopping product and mark down,
- Product development, and
- Product delivery

Agile can be used in whole or with any subset of these product development stages; however, in practice, product, design, and user interviews are often part of a product manager's role. Typically, a well-scoped project with requirements is initiated by an agile development team.

8.3 Summary

Relative to systems, estimating software is mature, and most companies have their own parametric cost estimating applicable to their domains, processes, and so on. The biggest challenge is converting requirements to LOCs. Numerous models and techniques are published in the open literature to convert a LOC-type measure to some type of development effort. However,

- Software cost estimation models are not a substitute for a detailed estimate by a subject-matter expert,
- Software estimation models highly depend on the user's experience in the application domain, analysis ability, and understanding of the requirements, and
- Like any cost estimate, developing the elements to be cost is equally critical to the process and accuracy of making the estimate.

In our software-centric world, costing software has become increasingly complex. However, as software languages and CASE systems have evolved, using number of lines of code as a predictor has become less reliable. For example, what might take many thousands of lines of C++ code might only take a few hundred lines of code in Java.

Software is redefining the consumer's world. In actuality, most companies understand their products and capabilities and do a good job at estimating software development costs by using mainly homegrown relationships based on years of experience. Beyond requirements creep, the main contributor to cost overruns is integration and the associated scalability challenges that are mainly software driven.

QUESTIONS

8-1. Assuming that software is a significant component of your system, can you give some reasons why software maintenance might account for 70% or more of the TOC? Can you give some other examples where 70% doesn't seem reasonable?

8-2. Conduct online research regarding how major investments in software affect the profit and loss of a business. For example, how do depreciating investments in software affect a financial statement? Should leasing versus buying be considered? Refer back to the categories used in after-tax cash flow analysis described in Chapter 3.

8-3. Software costs can be determined using a wide variety of means. In reality, most estimates involve a combination of software estimating techniques. List the advantages and disadvantages for each of the following techniques. Also, discuss when the technique should be the primary method for estimating software costs. Software estimating techniques include:

- Top-down,
- Bottom-up or engineering build,
- Analogy/expert judgment, and
- Algorithms.

8-4. Why is LOC a bad measurement of productivity?

PROBLEMS

8-1. Develop a general spreadsheet model using Basic COCOMO. You estimate that your software will have 75,000 DSI to achieve the desired functionality. Develop plots of effort and development time for organic, semidetached, and embedded modes. Vary the size of the software from 60,000 to 80,000 DSI.

8-2. Develop a general spreadsheet implementation of Intermediate COCOMO. Also, input labor and overhead rates to get a total cost model. This model should be general enough to be used as a first-cut software-estimating tool. The HLOOKUP and/or VLOOKUP functions in Excel could be a great help in developing a general model.

8-3. You are developing a new IT application that must operate within a strongly coupled complex environment of hardware, software, and operational procedures. You are in charge of estimating the development and installation costs of the data collection system. Given the following information, estimate the total cost of the system. The system will consist of the following components:

- Three COTS servers @ $9,000 each,
- Two COTS sensors @ $13,000 each, and
- 10,000 lines of code to integrate the systems.

Your company uses the following rules of thumb when estimating costs:

- The total project cost is estimated to be 2.5 times the application development costs, which is allocated based on project management (8%), systems engineering (15%), development (40%), test (28%), and services delivery (11%),
- Engineering costs are billed at $18,000 month, and
- Software COTS integration is best estimated using the embedded mode and the organic COCOMO with an effort adjustment factor of 1.33.

8-4. You are costing a custom business-to-business implementation of an Internet-based inventory tracking system for a new customer. The owners would like for you to design and implement this system at 10 stores along with a central backup at the corporate headquarters. You are bidding the job for 3 years along with training and hardware and software support. You will be paid on delivery and acceptance of the system. Your customer wants an annual hardware and software maintenance estimate for consideration. Following are the key assumptions:

- Ignore the time value of money—few will consider the time value of money for this type of problem.
- Hardware support: 5%, 10%, and 15%[2] per year (years 1, 2, and 3) of the total hardware bid cost.
- Software support: 2.5% of the total software cost for years 1, 2, and 3.
- Training: 2% of total cost (not including profit), one-time cost at year 0.
- Hardware bid cost: actual cost plus 20%.
- Software: values from Basic COCOMO, organic mode, increased by a factor of 25% for all new code.

[2] Assume these percentages include inflation, increased labor costs, and so on. These are used to convert to actual dollars.

- Normally you use 8% and 15% for program management and systems engineering costs, respectively, of the total development costs (i.e., exclusive of training and support costs).
- Software programming, integration, and test and evaluation costs are $12,000 per professional staff month (PSM; this includes all overhead costs).

Consider the three basic components of any system: hardware, software, and interfaces. Your senior manager gave you the following basic rules for estimating costs.

Hardware:

Item	Number	Cost/Number
Central Server	1	$4,000
Single Point of Sale Registers	12	$ 900
Printer	12	$ 600
Bar Code Scanner	12	$ 100
Surge/Battery	12	$ 125
Infrastructure (wiring, setup, etc.)	11	$ 1500
Miscellaneous (cables, startup supplies, etc.)	12	$ 300
Monthly Connect Charges	11*36	$ 40

Allows for two extras sets (hardware, printer, scanners, and surge/battery) as spares.

Software:
Past experience has shown that you can reuse about 50 KLOC, and you will need to write about 1.2 KDSI for these types of applications. Since you have a lot of experience in these types of applications, use the organic mode to estimate the development time of the new code. Assume the existing code has a value of $25,000.

Integration:
Your best methodology for estimating the integration development time is experience. Using the following projects, extrapolate interface development and testing and evaluation for your business-to-business systems. A simple linear regression model using Excel is probably suitable given the number of data points.

Number of Single Point of Sale Systems	Integration Coding Time	Testing and Evaluation Time
4	.2 PSM	1 PSM
15	1.3 PSM	2 PSM
10	.8 PSM	1 PSM
17	1.1 PSM	1.5 PSM

Use 25% profit based on the total project cost to develop an estimate of this project using Excel. Your model should be very robust so that you can conduct meaningful sensitivity analysis.

8-5. You are planning on developing an online interface for mining open source newspaper content to look at media bias. The application will consist of Structured Query Language (SQL) and HTML. After collecting requirements and conducting the analysis and design of the software application, you plan to use Intermediate COCOMO. You believe that the SQL portion of the code will consist of 700 FP, and the HTML portion of the code will be about 300 FP.
   a. Estimate the number of lines of code needed for this application based on TCF=1.
   b. You believe that this product will have very high product complexity, and you will be dealing with some very large databases. Using $8,000 per professional staff month, what is the total cost and development time for this project?

8-6. Under the control of the FCC, the Emergency Alert System (EAS) was initiated in 1963 during the Kennedy administration to allow the president to address the entire nation. The EAS is an interagency effort with the FCC, FEMA, and NOAA to permit the system to be used for state and local emergencies. Planned estimates for the next software upgrade will have 55,000 DSI to achieve the desired functionality. Assume that an engineer for the developers bills her time out at $12,000 per month. Develop estimates of effort, development time, productivity, average staffing, and costs using COCOMO. Given the complexity associated with interoperability, assume an intermediate level of project characteristic describes this upgrade.
   a. Estimate the effort in person-months.
   b. Estimate the development time.
   c. Determine the software productivity.
   d. Determine the average staffing.
   e. Estimate the software development costs.

8-7. The Admiral software package is being proposed as a replacement for an existing system called Buddy. A COTS-based supplier is also being considered. Programmers estimated that translating the 180,000 lines of Buddy's Visual Basic 6.0 code would require about 328,000 lines of JAVA code that would make up Admiral. The development team worked with the engineers and subject-matter experts to score the Admiral development project according to the category parameters. Using COCOMO II, estimate the development time and resources needed to complete this project. Assume $12,000/PSM for all programs and management. This will be a complex integration challenge.

Cost Driver	Very Low	Low	Nominal	High	Very High	Extra High	Score
ACAP Analyst Capability	1.46	1.19	1.00	0.86	0.71	--	0.86
AEXP Application Experience	1.29	1.13	1.00	0.91	0.82	--	0.82
CPLX Product Complexity	0.70	0.85	1.00	1.15	1.30	1.65	1.15
DATA Database Size	NA	0.94	1.00	1.08	1.16	--	0.94
LEXP Language Experience	1.14	1.07	1.00	0.95	NA	--	0.95
MODP Modem Programming Practices	1.24	1.10	1.00	0.91	0.82	--	0.82
PCAP Programmer Capability	1.42	1.17	1.00	0.86	0.70	--	1.00
RELY Required Software Reliability	0.75	0.88	1.00	1.15	1.40	--	1.15
SCED Required Development Schedule	1.23	1.08	1.00	1.04	1.10	--	1.00
STOR Main Storage Constraint	NA	NA	1.00	1.06	1.21	1.56	NA
TIME Execution Time Constraint	NA	NA	1.00	1.11	1.30	1.66	NA
TOOL Use of Software Tools	1.24	1.10	1.00	0.91	0.83	--	0.83
TURN Computer Turnaround Time	NA	0.87	1.00	1.07	1.15	--	0.87
VEXP Virtual Machine Experience	1.21	1.10	1.00	0.90	NA	--	NA
VIRT Virtual Machine Volatility	NA	0.87	1.00	1.15	1.30	--	1.0

8-8. You need to develop some type of algorithm to cost systems integration of your inventory management system. You have some standard library packages, but each must be integrated and tailored to the individual customers. Each case is unique depending on the type of legacy system, capabilities, customer requirements, documentation, and so on. You go back and analyze the last eight installations and capture the lines of code and the development and install effort in staff-months. You also assign a complexity factor similar to the ones used by Intermediate COCOMO. The results are summarized in the table below:

LOC	Complexity	Actual
30,000	1	12.5
50,000	2	27.4
43,000	3	29.2
8,000	5	8.1
11,000	6	15.2
4,250	4	3.2
9,000	6	14.0
14,000	4	12.2

a. Compare Intermediate COCOMO as a predictor for these results. Use the product complexity EAF to capture the difficulty of the installation. Is this model a good technique for predicting the development and install effort?
b. Develop a linear regression model using LOC and complexity as input. Is regression a good modeling technique?
c. Overlay the actual data, regression model, and the Intermediate COCOMO predictions on a plot versus lines of code. What is the best method based on this data set?

Appendix 8-A Effort multiplier for the intermediate COCOMO model

(from Boehm et al., 2000)

Required Software Reliability (RELY): This is the measure of the extent to which the software must perform its intended function over a period of time. If the effect of a software failure is only slight inconvenience, then RELY is very low. If a failure would risk human life, then RELY is very high.

Database Size (DATA): This measure attempts to capture the effect large data requirements have on product development. The rating is determined by calculating D/P, the ratio of bytes in the database to SLOC in the program. In other words, DATA captures the effort needed to assemble the data.

Product Complexity (CPLX): Complexity is divided into five areas: control operations, computational operations, device-dependent operations, data management operations, and user interface management operations.

Developed for Reusability (RUSE): This cost driver accounts for the additional effort needed to construct components intended for reuse on current or future projects.

Documentation Match to Life Cycle Needs (DOCU): Several software cost models have a cost driver for the level of required documentation. In COCOMO II, the rating scale for the DOCU cost driver is evaluated in terms of the suitability of the project's documentation to its life cycle needs.

Execution Time Constraint (TIME): This is a measure of the execution time constraint imposed on a software system. The rating is expressed in terms of the percentage of available execution time expected to be used by the system or subsystem consuming the execution time resource.

Main Storage Constraint (STOR): This rating represents the degree of main storage constraint imposed on a software system or subsystem.

Platform Volatility (PVOL): "Platform" is used here to mean the complexity of hardware and software (operating system, database management system, etc.) the software product calls on to perform its tasks. If the software to be developed is an operating system, then the platform is the computer hardware. If a database management system is to be developed, then the platform is the hardware and the operating system. If a network text browser is to be developed, then the platform is the network, the computer hardware, the operating system, and the distributed information repositories. The platform includes any compilers or assemblers supporting the development of the software system. This rating ranges from low, where there is a major change every 12 months, to very high, where there is a major change every 2 weeks.

Analyst Capability (ACAP): Analysts are personnel who work on requirements, high-level design, and detailed design. The major attributes that should be considered in this rating are analysis and design ability, efficiency and thoroughness, and the ability to communicate and cooperate.

Programmer Capability (PCAP): Current trends continue to emphasize the importance of highly capable analysts. However, the increasing role of complex COTS packages, and the significant productivity leverage associated with programmers' ability to deal with these COTS packages, indicates a trend toward higher importance of programmer capability as well. Evaluation should be based on the capability of the programmers as a team rather than as individuals. Major factors that should be considered in the rating are ability, efficiency and thoroughness, and the ability to communicate and cooperate.

Personnel Continuity (PCON): The rating scale for PCON is in terms of the project's annual personnel turnover: from 3%, very high continuity, to 48%, very low continuity.

Applications Experience (APEX): The rating for this cost driver (formerly labeled AEXP) is dependent on the level of applications experience of the project team developing the software system or subsystem. The ratings are defined in terms of the project team's equivalent level of experience with this type of application.

Language and Tool Experience (LTEX): This is a measure of the level of programming language and software tool experience of the project team developing the software system or subsystem. Software development includes the use of tools that perform requirements and design representation and analysis, configuration management, document extraction, library management, program style and formatting, consistency checking, planning and control, and so on. In addition to experience in the project's programming language, experience on the project's supporting tool set also affects development effort.

Platform Experience (PLEX): The post-architecture model broadens the productivity influence of platform experience, PLEX (formerly labeled PEXP), by recognizing the importance of understanding the use of more powerful platforms, including more graphic user interface, database, networking, and distributed middleware capabilities.

Use of Software Tools (TOOL): Software tools have improved significantly since the 1970s projects used to calibrate the 1981 version of COCOMO. The tool rating ranges from simple edit and code, very low, to integrated life cycle management tools, very high. A nominal TOOL rating in COCOMO 81 is equivalent to a very low TOOL rating in COCOMO II. An emerging extension of COCOMO II is in the process of elaborating the TOOL rating scale and breaking out the effects of TOOL capability, maturity, and integration.

Multisite Development (SITE): Given the increasing frequency of multisite developments, and indications that multisite development effects are significant, the SITE cost driver has been added in COCOMO II. Determining the cost driver rating involves the assessment and judgment-based averaging of two factors: site collocation (from fully collocated to international distribution) and communication support (from surface mail and some phone access to full interactive multimedia).

Required Development Schedule (SCED): This rating measures the schedule constraint imposed on the project team developing the software. The ratings are defined in terms of the percentage of schedule stretch-out or acceleration with respect to a nominal schedule for a project requiring a given amount of effort.

Bibliography

Jorgenson, Dale W., and Wessner, Charles W., eds., 2006, "Measuring and Sustaining the New Economy, Software, Growth, and the Future of the US Economy: Report of a Symposium Committee on Software, Growth, and the Future of the US Economy, Committee on Measuring and Sustaining the New Economy," National Research Council, accessed 1 August 2018 at http://www.nap.edu/catalog/11587.html.

Park, Robert E., January 1995, "A Manager's Checklist for Validating Software Cost and Schedule Estimates," Special Report, CMU/SEI-95-SR-004, Software Engineering Institute, Carnegie Mellon University.

Venugopalan, Nandini, "An Overview of Function Point Analysis," accessed 4 February 2008 at http://www.devshed.com/c/a/Practices/An-Overview-of-Function-Point-Analysis/.

References

Boehm, B., 1981, *Software Engineering Economics*, Prentice-Hall, Upper Saddle River, NJ.

Boehm, B., Clark, B., and Abtset C., 2000, "Software Cost Estimation with COCOMO II," Prentice-Hall, Upper Saddle River, NJ.

Center for Software Engineering, 2000, "COCOMO II Model Definition Manual," University of Southern California, Version 2.1, accessed 21 August 2017 at http://csse.usc.edu/csse/research/COCOMOII/cocomo2000.0/CII_modelman2000.0.pdf.

Ceschi, M., Sillitti, A., Succi, G., and De Panfilis, S., 2005, "Project Management in Plan-Based and Agile Companies," IEEE Software Journal, Volume 22, 21-25.

Department of Defense, 20 July 2005, "Work Breakdown Structures for Defense Materiel Items," MIL-HDBK-881A, DoD Handbook.

Fenton, N. E., and Pfleeger, S. L., 1997, *Software Metrics: A Rigorous and Practical Approach*, International Thomson Computer Press, Boston, MA.

IBM, 15 June 2007, "Software Estimation, Enterprise-Wide. Part I: Reasons and Means," accessed 18 November 2008 at http://www.ibm.com/developerworks/rational/library/jun07/temnenco/index.html.

Lum, Karen, Bramble, Michael, Hihn, Jairus, Hackney, John, Khorrami, Mori, and Monson, Erik, 30 May 2003, "Handbook for Software Cost Estimation" JPL D-26303, Rev. 0, Jet Propulsion Laboratory, Pasadena, CA accessed 1 August 2018 at http://everyspec.com/NASA/NASA-JPL/download.php?spec=JPL_D-26303_REV-0.017027.PDF.

NASA, 2015, "Cost Estimating Handbook," accessed 18 August 2018 at https://www.nasa.gov/offices/ocfo/nasa-cost-estimating-handbook-ceh.

Pressman, R. S., 1997, *Software Engineering A Practitioner's Approach*, McGraw-Hill, New York City, NY.

Quantitative Software Management, n.d., "QSM Resources—Function Point Languages Table," accessed 16 December 2016 at http://www.qsm.com/resources/function-point-languages-table.

Schofield, Joe, Armentrout, Alan W., and Trujillo, Regina M., May/June 2013, "Function Points, Use Case Points, Story Points: Observations from a Case Study," accessed 6 May 2018 at http://citeseerx.ist.psu.edu/viewdoc/download?doi=10.1.1.438.495&rep=rep1&type=pdf, taken from CrossTalk.

chapter nine

Cost estimating techniques

9.1 Estimating life cycle costs throughout the product development cycle

In Figure 1.9 we presented three techniques (i.e., parametric cost estimating, analogy, and bottom-up build) for determining costs throughout a typical product development cycle. We presented the applicability of each of these techniques throughout the steps in the life cycle.

9.2 Analogy

When a new system is being developed, consisting of immature technology, significant research and development, and little historical data, such as a complex system of systems, not capturing all the activities in the model can lead to an inexact life cycle cost (LCC). As shown in Figure 1.9, when we enter into the operations and support phases, estimating recurring costs is based mainly on rules of thumb, heuristics, existing data, and so on. Developing analogies early in the life cycle is often the only way to estimate a rough order of magnitude (100% or more). Even if the activities are fairly mature systems that have a significant level of commercial off-the-shelf readiness, aspects such as intensive software, subcontractor or outsourced components, and immature technology can produce imprecise models. Unfortunately, increases in software and hardware often do not scale linearly. Also, you must have substantial experience to develop any type of prediction in order to determine the appropriate independent and dependent variables.

Most, if not all, new programs are costed based on modified or improved versions of existing components. For example, no one can cost a mobile phone better than Apple and Samsung. Using historical records and experience, one can determine the cost to integrate a new camera, for example. In most cases, the new product behaves from a costing perspective like a legacy system. All elements, including software and integration, will scale.

According to NASA (2015), the basis for comparison can be in terms of capabilities, size, weight, reliability, material composition, complexity,

Table 9.1 Strengths and weaknesses of using analogies for cost estimating (Courtesy of NASA, 2015)

Strengths	Weaknesses	Applications
Based on actual historical data	In some cases, relies on single historical data point	• Early in the design process
Quick	Can be difficult to identify appropriate analog	• When less data are available
Readily understood	Requires "normalization" to ensure accuracy	• In rough order-of-magnitude estimate • Cross-checking
Accurate for minor deviations from the analog	Relies on extrapolation and/or expert judgment for "adjustment factors"	• Architectural studies • Long-range planning

and so on. NASA developed the following rules to be used for developing an analogous cost estimate:

- Using a known item's value, apply quantified adjustments to that item that measure the differences when compared to the new item.
- This requires good actual data and someone to quantify the differences.
- Recent historical data should be similar not only in performance characteristics but also similar from the standpoint of technology.
- There are some important questions to ask when assessing the relative differences between the old and the new item:
  - How different is the new compared to the old?
  - What portion of the old is just like the new?
  - How many components will be exactly the same?
  - What is the ratio of complexity between the two systems?

Table 9.1 contains the strengths and weakness of using analogies for cost estimating. Example 9.1 demonstrates how analogies can be used to determine cost.

Expert opinion could be considered an analogous estimating technique because it involves consultation with experts who use their experience and understanding of the system to arrive at an estimate of its cost. From a transparency and defensibility perspective, this is probably not an ideal way to estimate costs. Also, it is the least regarded means to determine costs. However, it can serve some purposes when there is simply no data or when new technology and processes are being used. As the estimating technique of last resort, it has a place and time when it is the best means to determine cost.

9.3 *Parametric cost estimating*

"The origins of parametric cost estimating (PCE) date back to World War II. The war caused a demand for military aircraft in numbers and models that far exceeded anything the aircraft industry had

EXAMPLE 9.1

One large IT solutions delivery company has developed some simple rules for early estimation of costs. They use one of three analogous techniques: total project cost, application development cost, or project complexity. Analogy using total project cost is shown in Table 9.2. In essence, the customer provides a total cost, and project costs are allocated as a percentage to the respective cost elements. Work can then be estimated based on these cost estimates. There are many variations; for example, this provider might use the software or application development costs from models like the Constructive Cost Model (COCOMO) and then determine the total project cost.

Table 9.2 Analogy used by a large IT solution delivery company for estimating early project development costs

Total Project Cost		% Total Project Cost		Cost Element Estimate
Total Project Cost	x	8%	=	Project Management
Total Project Cost	x	15%	=	Systems Engineering
Total Project Cost	x	40%	=	Application Development
Total Project Cost	x	26%	=	Test
Total Project Cost	x	11%	=	Service Delivery

manufactured before. While there had been some rudimentary work from time to time to develop parametric techniques for predicting cost, there was no widespread use of any cost estimating technique beyond a laborious buildup of labor-hours and materials" (Department of Defense, 1995). PCE methods for planning and strategizing system cost issues find applicability in both bottom-up and top-down estimating.

PCE uses cost estimating relationships (CERs) and associated mathematical algorithms, logic, and processes to establish cost estimates. For example, detailed cost estimates for manufacturing and testing of an end item (for instance, a hardware assembly) can be developed using very precise industrial engineering standards and analysis. Performed in this manner, the cost estimating process is laborious and time-consuming.

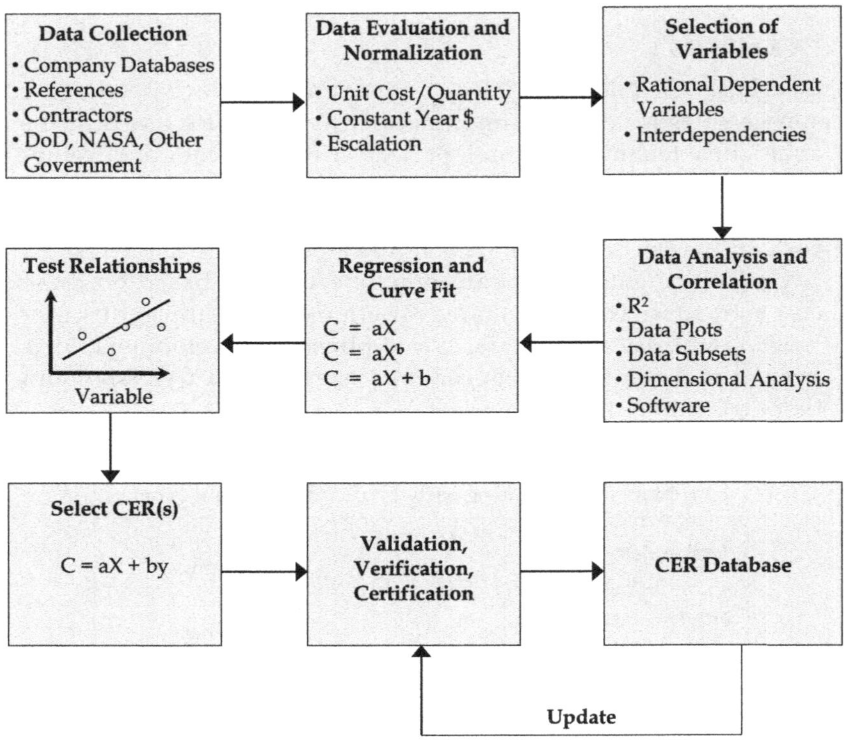

Figure 9.1 Process for determining parametric cost estimates (modified from Department of Defense, 1995)

However, if history has demonstrated that testing (as the dependent variance) has normally been valued at about 25% of the manufacturing value (the independent variable), then a detailed testing estimate need not be performed and can simply be computed at the 25% (CER) level (Department of Defense, 1995). Figure 9.1 shows a process that can be used for developing CER for PCE. Like any mathematical model, it should only be used for the range described by the "relationship" data.

The following definitions were taken from NASA (2015) and the Department of Defense (1995):

- Parametric cost estimate: Estimate derived from statistical correlation of historic system costs with performance and/or physical attributes of the system.
- Parametric cost model: A mathematical representation of parametric cost estimating relationships that provides a logical and predictable correlation between the physical or functional characteristics of a system and the resultant cost of the system. A parametric cost model

is an estimating system comprising CERs and other parametric estimating functions (e.g., cost-quantity relationships, inflation factors, staff skills, schedules). Parametric cost models yield product or service costs at designated levels and may provide departmentalized breakdown of generic cost elements. A parametric cost model provides a logical and repeatable relationship between input variables and resultant costs.

- Cost estimating relationship: An algorithm relating the cost of an element to physical or functional characteristics of that cost element or a separate cost element, or relating the cost of one element to the cost of another element. A CER can be a functional relationship between one variable and another and may represent a statistical relationship between some well-defined program element and some specific cost, for example. Many costs can be related to other costs or non-cost variables in some fashion, but not all such relationships can be turned into a CER.
- Cost estimating relationships: A mathematical expression that describes, for predictive purposes, the cost of an item or activity as a function of one or more independent variables.

Table 9.3 shows some typical relationships that could be used to develop CERs (Department of Defense, 1995). Examples 9.2, 9.3, and 9.4 are some simple CERs.

Table 9.3 Some example CER relationships

Product	Independent Variable
Construction	Floor space, roof surface area, wall surface area
Gears	Net weight, percentage of scrap, inches of teeth cut, harness, envelope
Trucks	Empty weight, gross weight, horsepower, number of driving axles, loaded cruising speed
Passenger Car	Curb weight, wheel base, passenger space, horsepower
Turbine Engine	Dry weight, maximum thrust, cruise thrust, specific fuel consumption, bypass ratio, inlet temperature
Reciprocating Engine	Dry weight, piston displacement, compression ratio, horsepower
Sheet Metal	Net weight, percentage of scrap, number of holes drilled, number of rivets placed, inches of welding, volume of envelope
Aircraft	Empty weight, speed, useful load, wing area, power, landing speed
Diesel Locomotive	Horsepower, weight, cruising speed, maximum load on standard grade at standard speed

EXAMPLE 9.2

Historical information on construction waste disposal costs has been input into a computer with statistical software. The statistical package generated the following CER equation:

$$C = 200 + 275D_{10} + 325D_{10} + 0.75M$$

Where

C = cost to dispose of construction waste in dollars
D_{10}, D_{30} = number of 10 cubic yards or 30 cubic yards roll-off dumpsters
M = number of miles between waste location and waste disposal facility

EXAMPLE 9.3

Many systems have consumables that we often ignore when developing life cycle costs. Take for example the CER for radar consumables shown in Figure 9.2. The radar's consumables in this case include all the internal cable assemblies, low-cost power supplies, and other various low-cost components that fail. The resulting CER for the radar is $Y = 2.07X + 19.04$.

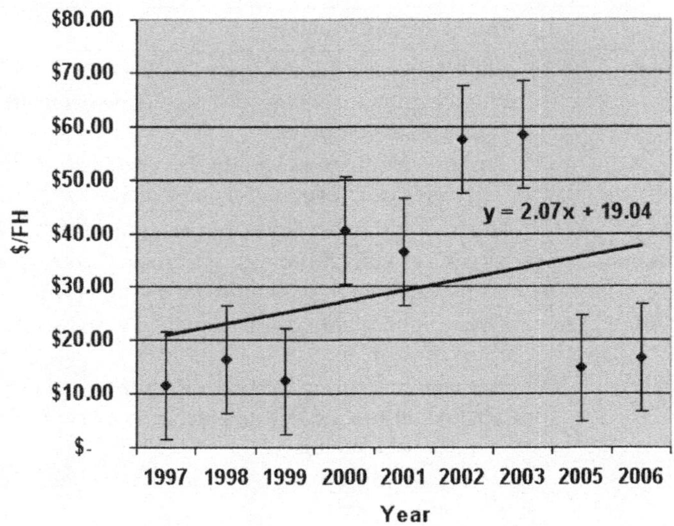

Figure 9.2 CER for radar consumables

Note that these values are in actual or then-year dollars. This relationship only covers the operations and support costs for consumable components and piece parts. It does not include consumables such as fuel, lubricants, corrosion treatment solutions, and so on. Even for a low-maintenance item such as radar, there can be significant hidden costs.

To develop CERs, we need the following pieces of information (modified from Department of Defense, 1995):

- Reliable historical cost, schedule, and technical data on a set of data points,
- Work breakdown structure, work breakdown structure dictionary, product tree, and/physical architecture,
- Analysis to determine significant cost drivers, and
- Knowledge of basic statistics and software.

EXAMPLE 9.4

A bridge is to be constructed to span a distance of 3,000 feet across a shallow river, as shown in the figure below (Griffis and Farr, 1999). A preliminary study concluded that a series of equal-span simple trusses is most suitable. It is estimated that the two end piers, A and B, will cost \$4 million each. The remaining piers, which are to be located in water, can be constructed anywhere in the crossing at a cost of \$8 million per pier. Each bridge truss of length L (feet) costs 100 times L^2.

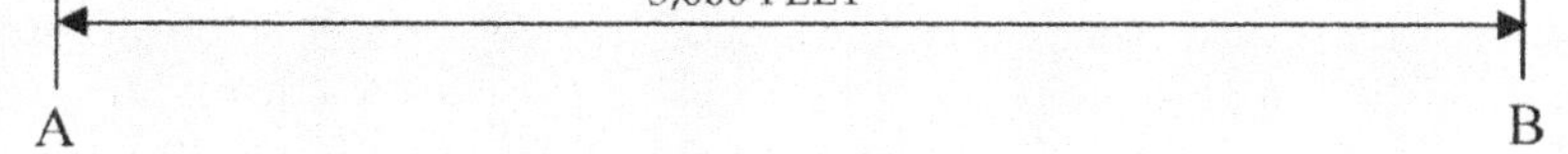

Thus, the number of interior piers, n, is given by

$$n = \frac{3,000}{L} - 1$$

(continued)

and the expected cost of the bridge, C, becomes

$$C = 2(4{,}000{,}000) + n(8{,}000{,}000) + (n+1)100L^2$$
$$= 8{,}000{,}000 + \left[\frac{3{,}000}{L} - 1\right] * 8{,}000{,}000 + \left[\frac{3{,}000}{L}\right] * 250L^2$$

Figure 9.3 shows a plot of interior truss length as a function of cost. For a planning report, this would be a very useful CER.

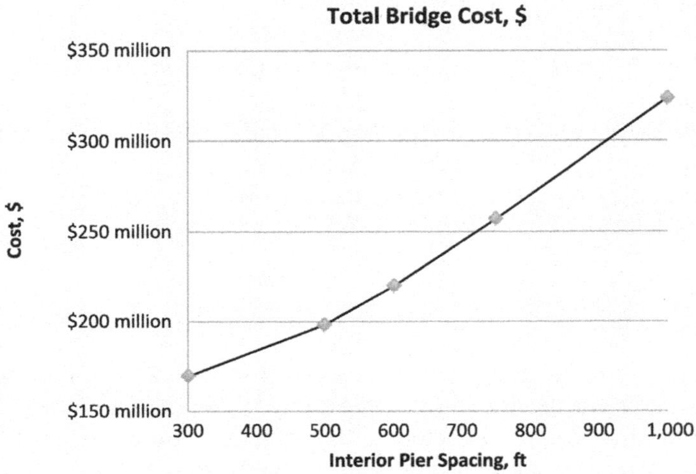

Figure 9.3 CER for cost as a function of interior truss length

Note that data for CERs can be obtained from a multitude of sources, including basic accounting records, cost reports, subject-matter experts, historical databases, other organizations, open literature, technical databases, contracts, and cost proposals.

Parametric models are not necessarily more accurate than analogy models, yet they are the only means available when a substantial database doesn't exist. Parametric models are based on relationships between costs and some product- and process-related parameters. Probably the most well-known models are the COCOMO family of models, which are shown in Figure 9.4. These models are all built from empirical data that take the form of a non-linear equation multiplied by numerous factors capturing the complexity, size, maturity, and so on, of the systems.

The parametric model is a function of some characteristics. These can be physical or performance based, for example, weight, speed,

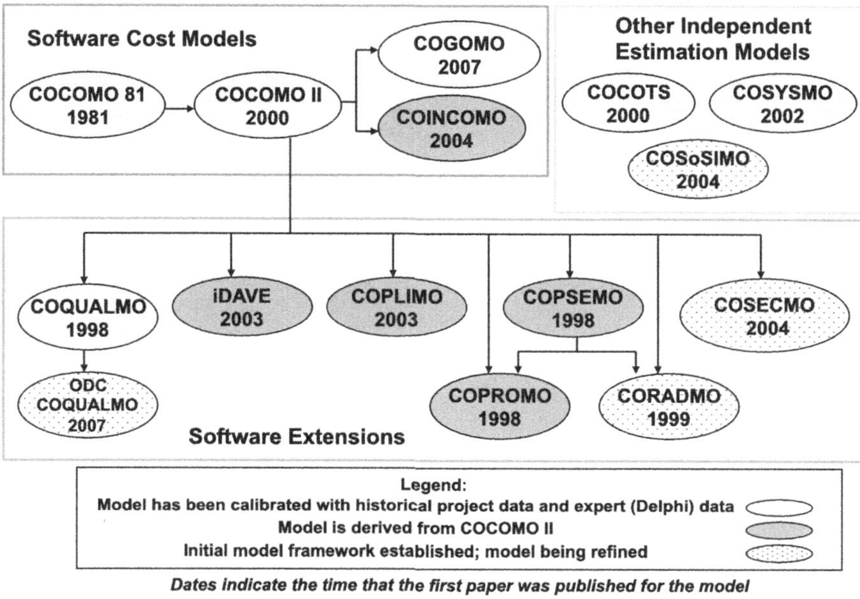

Figure 9.4 COCOMO-based family of models (University of Southern California, 2008)

power, lines of code, and so on. Normally this consists of statistically fitting of some function to historical data to develop an equation for predicting costs. This process is shown in Figure 9.5. Table 9.4 contains the strengths and weaknesses of using parametric cost models.

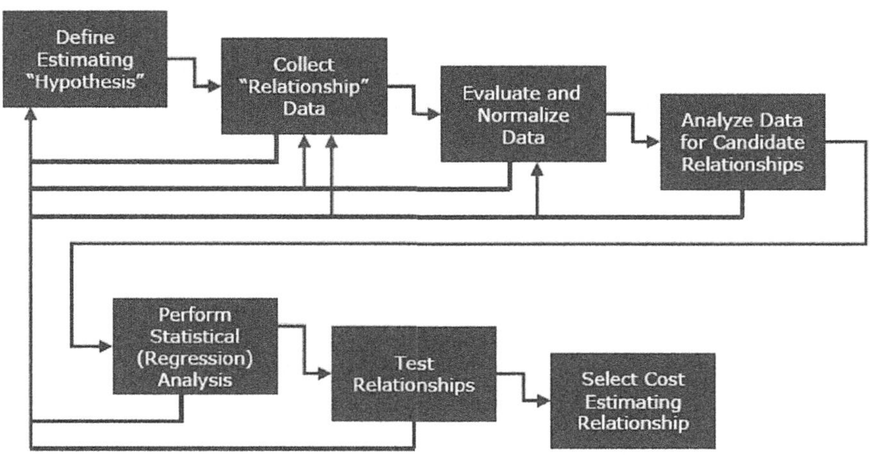

Figure 9.5 Parametric cost modeling process (Courtesy of NASA, 2015)

Table 9.4 Strengths and weaknesses of using parametric cost models (Courtesy of NASA, 2015)

Strengths	Weaknesses	Applications
Based on actual historical data	In some cases, relies on single historical data point	• Early in the design process • When less data are available • In rough order-of-magnitude estimate • Cross-checking • Architectural studies • Long-range planning
Quick	Can be difficult to identify appropriate analog	
Readily understood	Requires "normalization" to ensure accuracy	
Accurate for minor deviations from the analog	Relies on extrapolation and/or expert judgment for "adjustment factors"	

9.3.1 *The role of statistics*

Modern software, including Excel, SPSS, SAS, and others, makes sophisticated analysis of data very simple. For example, consider the regression analysis in Example 9.5. When you plot an x-y scatter plot in Excel by simply right-clicking your mouse on one of the data points, you can add a trend line with an R^2 to the plot. We will not review simple, multiple, and curvilinear regression and other predictive modeling techniques. Even complicated techniques such as neural networks, genetic algorithms, and so on, can now be easily used to develop CERs provided the modeler understands the limitations. The biggest challenge is understanding the goodness of fit for the data and the validity of the equations, even for statistically significant models.

EXAMPLE 9.5

From the data given in Problem 5-18, develop a CER for inflation and plot the results using Excel.

Figure 9.6 presents a 66-year plot of inflation. The regression line is easily generated by Excel. What is of interest if how the predictions would change if we only used the information from 1951 to 1980.

As an interesting aside for this version of the book, we updated the plot including data after 2008. Including data from 2008 to 2016 made the slope of the line decrease. If we only included data from 1951 until 2008, the CER would have predicted an increasing inflation rate!

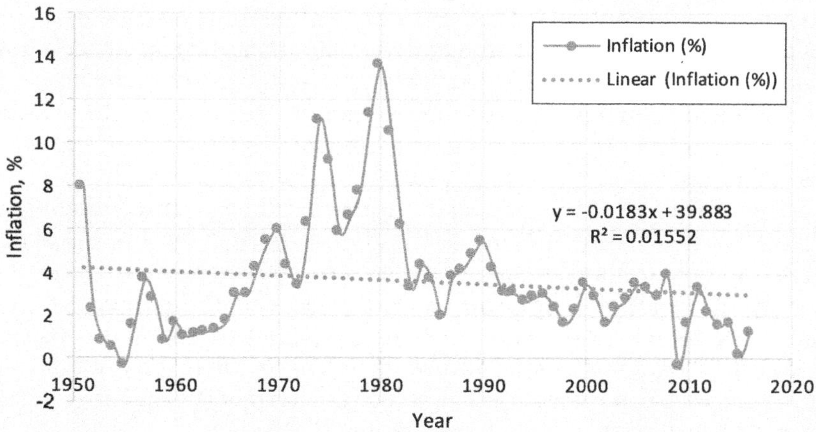

Figure 9.6 CER for inflation from 1951 to 2016

9.3.2 *Some CERs of interest*

Learning curves, and expressions such as "experience curve," "cost improvement curve," "progress curve," "progress function," "startup curve," "improvement curve," and "efficiency curve" are often used interchangeably and are based on the concept that resources required to produce each additional unit decline as the total number of units produced increases (NASA, 2015). The term "learning curve" is used when referring to an individual's or organization's performance as knowledge or insight is gained. The learning curve concept is used primarily for repetitive and labor-intensive tasks and finds most of its utility in manufacturing. In most practical applications, learning curve analysis is used to predict the cost of making the nth unit given the time and cost of making the first unit. This CER's data is gathered under the design and implementation phase of the life cycle costing process. It uses historical data, by definition, to calculate a cumulative average cost for producing an item. Consistency in improvement converts to a constant percentage reduction in time required over successively doubled quantities of units produced. This constant percentage by which the costs of the doubled quantities decrease is called the rate of learning (Federal Aviation Administration, n.d.).

Other terms that are used interchangeably with learning curve are "experience curve" and "cost improvement curve." The difference between these terms is in what measurement they define. The learning curve describes the reduction in the number of hours a job requires as workers learn their jobs. The experience curve, by contrast, applies not only to labor-intensive situations but also to process-oriented ones (NetMBA, n.d.). It covers value-added costs (including administration, marketing, and distribution) that are also reduced by a constant and predictable percentage. The cost improvement curve applies these reductions across multiple processes to reduce recurring costs over and above labor charges.

The learning curve principle is predicated on three commonly accepted observations. The first observation states that the amount of time needed to perform a task decreases the more times the task is repeated. Second, the amount of improvement decreases as the number of units manufactured increases. And third, the rate of improvement is consistent enough to be able to predict future data. The mathematical model that represents these premises is attributed to T. P. Wright (1936). The generally accepted learning curve graph is shown in Figure 9.7.

The learning curve graph represents the concept that the more units of a product are manufactured, the less time it takes to make an individual unit. Workers on the first line improve their skills and reduce the time to perform those skills at an exponentially decreasing rate.

Most learning curves range from around 70% to 100%. A learning curve of below 70% probably indicates that the worker is in a job that he or she is not capable of doing. A 100% learning curve indicates that no more knowledge is to be gained. This will occur over time if the product being made is never improved on or reworked.

9.3.2.1 *Wright's method*

Wright's (1936) model was the first CER and was originally defined as the unit curve theory; it calculates the cost to produce the nth unit given a cost to produce the first unit and the value for a decreasing slope that

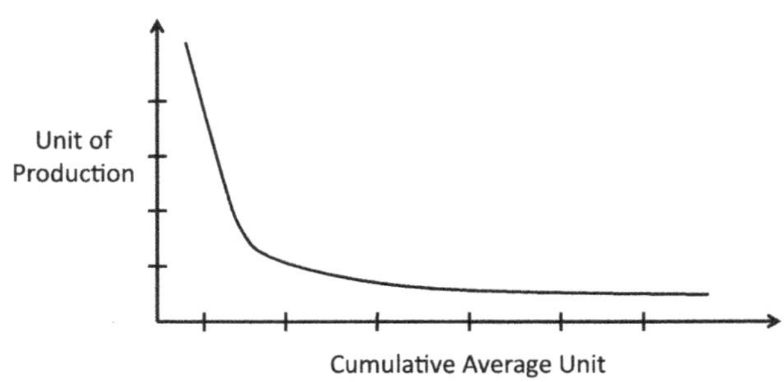

Figure 9.7 Generally accepted learning curve graph

contains the learning rate. The learning curve unit function is defined as follows:

$$Y_n = aN^b \tag{9.1}$$

where

Y_n = the time (or cost) per unit to produce the nth unit
N = the number of units produced
a = time (or cost) required to produce the first unit
b = slope of the function, which is negative since time (or cost) decreases as production goes up (log of the learning rate/log of 2)

Taking the log of both sides reduces this equation to a straight line ($Y = mX + b$). The analyst is then able to determine future predictions for N + additional units. The formula looks like this:

$$\log Y_n = b * \log N + \log a \tag{9.2}$$

If we used an example with a learning rate of 85% and time to produce the first unit as 60 hours, we could determine the time to produce the seventh unit with the following calculations based on Equation 8.1:

$$b = \log(.85)/\log(2) = -.07058/.301 = -.234$$
$$Y_7 = 60 * 7^{-.234}$$
$$Y_7 = 38.05 \text{ hours to produce unit \#7}$$

The equation for cumulative total hours (or cost) is then computed by multiplying both sides of the cumulative average equation by nth unit number.

$$NY = aN^{1+b} \tag{9.3}$$

Therefore, the equation for cumulative total labor hours from the above example is

$$NY = 60N^{1-.234} = 60N^{.766}$$

Cumulative total labor hours = $60*(7)^{.766} = 60*4.4396 = 266.38$ total hours for seven units (Martin, n.d.).

Table 9.5 shows total cumulative hours and average labor hours when the lot size is successively doubled. Note that the above calculations fall where you would expect them to for seven units (38.05 hours/266.38 hours).

Table 9.5 Effects of learning when doubling the lot size

Total (cumulative)	Average Time/Unit
60	60
102	51
173.4	43.4
294.8	36.8

9.3.3 *Summary and conclusions*

CERs/PCEs are easy to use and common throughout the cost estimating industry. Most of the software analysis tools use these simple mathematical relationships and bottom-up estimating techniques. In applying good judgment in the use of CERs, we need to remain mindful of their strengths and weaknesses (Department of Defense, 1995).

Strengths

- One of the principle strengths of CERs is that they are quick and easy to use. Given a CER equation and the required input data, one can generally turn out an estimate quickly.
- A CER can be used with limited system information. Consequently, CERs are especially useful in the research, development, testing, and evaluation phase of a program.
- A CER is an excellent (statistically sound) predictor if derived from a sound database and can be relied on to produce quality estimates.

Weaknesses

- CERs are sometimes too simplistic to forecast costs. Generally, if one has detailed information, the detail may be reliably used for estimates. If available, another estimating approach may be selected rather than a CER.
- Problems with the database may mean that a particular CER should not be used. While the analyst developing a CER should validate that CER, it is the responsibility of any user to validate the CER by reviewing the source documentation. Read what the CER is supposed to estimate, what data were used to build the CER, how old the data are, how they were normalized, and so on. Never use a cost model without reviewing its source documentation.

Analysts must be diligent when using CERs/PCEs and remember that the underlying conditions from which the relationship was developed can change, especially due to technology and complexity.

9.4 Detailed engineering builds

Obviously, the most desired method for calculating cost is direct estimation at the component level. This is often referred to as a detailed engineering build or buildup of the system and is the most accurate means of costing a system. This detailed, bottom-up application of labor and material costs is obviously the most accurate way to determine LCC. However, you must know the architecture of the system before you can cost the system. This method is not only data- and time-consuming but also expensive to produce. Therefore, we cannot accurately develop a LCC model until well into the systems demonstration phase.

Detailed engineering builds at the highest levels consist of three categories: hardware, software, and integration. Hardware costs of mature technology are well known. Also, most developers have well-developed parametric models for costing software. The most difficult elements to cost early in the product life cycle are the integration aspects. The process shown in Figure 9.8 should be used for conducting a detailed engineering build analysis. The pros and cons of this type of analysis is presented in Table 9.6. Unfortunately, you must have the details of the product/services before you can conduct an engineering build type analysis. The biggest challenge is usually eliminating a category of costs versus developing the wrong cost estimates for a known characteristic of the system

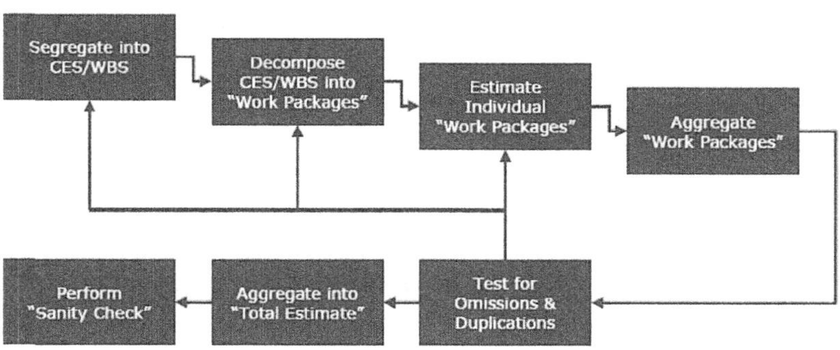

Figure 9.8 Process for conducting an engineering build cost estimate (Courtesy of NASA, 2015)

Table 9.6 Strengths, weaknesses, and applications of engineering build-up methodology (Courtesy of NASA, 2015)

Strengths	Weaknesses	Applications
Intuitive	Costly; significant effort (time and money) required to create a build-up estimate Susceptible to errors of omission/double counting	• Production estimating • Negotiations • Mature projects • Resource allocation
Defensible	Not readily responsive to "what if" requirements	
Credibility provided by visibility into the BOE for each cost element	New estimates must be "built up" for each alternative scenario	
Severable; the entire estimate is not compromised by the miscalculation of an individual cost element	Cannot by itself provide "statistical" confidence level	
Provides excellent insight into major cost contributors (e.g., high-dollar items).	Does not provide good insight into cost drivers (i.e., parameters that, when increased, cause significant increases in cost)	
Reusable; easily transferable for use and insight into individual project budgets and performer schedules	Relationships/links among cost elements must be "programmed" by the analyst	

9.5 Summary

The NASA (2015) reference is an outstanding source for the details of these cost estimating techniques. That reference provides guidance on developing these cost estimating models:

- Before and after running the model, it is important to check and recheck data entry and formulas to ensure accuracy and to document each input and formula for the detailed basis of estimate.
- Another important step to remember is to conduct a cross-check estimate using an alternative methodology on your point estimate. This is important to ensure a "sanity check" on the original estimate and to show an alternative estimate view of the data.
- In addition, keeping the estimate up to date helps to defend the estimate and reduce updated estimate turnaround time and gives the decision maker a clearer picture for "what if" drills to support major investment and budget decisions.

QUESTION

9-1. Qualitatively develop a simple CER for a home computer system.

PROBLEMS

9-1. Consider the following data collected for software development as a function of thousands of lines of code (KLOC).

KLOC	Staff-Months
7	80
13	106
56	581
100	682
115	854
220	939

a. Using Excel, develop a CER using regression analysis for software development in staff-months.
b. Develop a plot of your regression model and the results as predicted by the semidetached mode using Basic COCOMO. Can you develop a CER using Basic COCOMO?
c. Use Excel Solver to develop a non-linear regression model of the form

$$\text{Staff-Months} = a + b / \text{KLOC} + c \ln(\text{KLOC})$$

Start using a = 1, b = 5, c = 10 as an initial guess for the Solver add-in. Minimize using the sum of the normalized difference squared or

$$\text{Difference} = \left(\begin{array}{l} (\text{Calculated Staff-Months} - \text{Actual Staff-Months}) / \\ \text{Actual Staff-Months} \end{array} \right)^2$$

d. Also fit a quadratic equation or

$$\text{Staff-Months} = a + b * \text{KLOC} + c * \text{KLOC}^2$$

Minimize based on the normalized difference squared.

9-2. A review of several sources of information for lines of code (LOCs) per year is shown in the table below. Use Excel to develop a plot of the various data sources. Then, develop a linear regression of the data and make a prediction for a 750,000 LOC project. Also, what is R^2 for your prediction? Based on your R^2 value, do you have confidence in this prediction?

Project Size Average	COCOMO Average
100,000 LOC	3,200
250,000 LOC	2,600
500,000 LOC	2,000
1,000,000 LOC	1,600

9-3. Your company has used the Basic COCOMO model in the semidetached mode with an effort adjustment factor of 1.75 to estimate software for your line of data acquisition systems. You suspect there is a better PCE and want to develop a more accurate model. Data for all the projects is shown below. Can you recommend a better model?

KLOC	Effort (Professional Staff Months)
12.5	96.5
23	160.8
87	624.5
63	528.2
47	313.3
50	335.8
10	79.1
7	55.7

9-4. Your company manufactures specialty ammunition for the US Army. You are bidding on a new contract for similar ammunition. You review historical results and find out that you have manufactured this type of ammunition four times with the following results:

Lot 1 = 256,000 units @ 853 units/hour
Lot 2 = 332,000 units @ 938 units/hour
Lot 3 = 361,760 units @ 952 units/hour
Lot 4 = 207,000 units @ 690 units/hour

You need to develop a CER using this data. For a lot size of 500,000 units, should you be concerned about how lot size affects the production rate?

9-5. As the owner of a small security firm, you use CERs to develop most of your cost estimates. You are bidding on a contract that is the largest in your company's history and will involve about 20 sensors. Below are your historical records for installing these types of sensors. Using Excel, fit a linear regression model for your CER. Based on R^2 and plot of the data, is a linear model appropriate?

Number of Sensors	Integration and Installation Person–Hours/Sensor
2	12.4
5	16.6
5	14.6
9	27.7
9	28.3
11	42.9
15	77.6

9-6. Consider the following costs of airplanes for the US Air Force and Navy[1]:

Bombers	Nickname	Year	Cost ($M)
B-52	Stratofortress	1962	9.28
B-2	Spirit	1997	737
B-1	Lancer	1998	400

Fighters	Nickname	Year	Cost ($M)
F-14	Tomcat	1973	55
F-15	Eagle	1975	46
A-6E	Intruder	1975	39
A-10	Warthog	1978	11
F-16	Falcon	1983	23
F-117	Nighthawk	1985	110
AV-8B	Harrier	1985	34
F/A 18	Hornet	1985	44
F-22	Raptor	2009	139
F-35	JSF	2018	94.6

a. Develop CERs for both the bombers and fighters. Use historical consumer price index data of January 1913 = 9.8 and January 2018 = 247.8 to determine the inflation rate. Convert all costs to 2018 dollars.
b. The Air Force recently awarded a new bomber contract titled the B-21 Raider at an estimated 2010 cost of $550 million per airplane. Does this agree with your CER?
c. How confident are you in your CERs? Are you using the right grouping to develop the equations?

References

Department of Defense, fall 1995, "Parametric Cost Estimating Handbook," Joint Government/Industry Initiative, accessed 1 August 2018 at www.acqnotes.com/Attachments/Parametric%20Cost%20Estimating%20Guide.doc.

Federal Aviation Administration, n.d., "FAA Procurement Pricing Handbook, Chapter 18. The Learning Curve," accessed August 18, 2018 at https://fast.faa.gov/PPG_Pricing_Handbook.cfm.

Griffis, Fletcher H., and Farr, John V., 1999, *Construction Planning for Engineers*, McGraw-Hill, New York City, NY.

[1] This data was gleaned from a wide variety of sources, including Wikipedia. Unfortunately, different references can provide conflicting numbers. *Popular Mechanics* has a good article on the cost of airplanes at https://www.popularmechanics.com/military/weapons/a21776/f-35-cheaper/.

Martin, James R., n.d., "The Learning Curve or Experience Curve," accessed December 2008 at http://maaw.info/LearningCurveSummary.htm.
NASA, February 2015, "Cost Estimating Handbook," Version 4.0, accessed February 2015 at https://www.nasa.gov/pdf/263676main_2008-NASA-Cost-Handbook-FINAL_v6.pdf.
NetMBA, n.d., "Strategic Management: The Experience Curve," accessed December 2008 at http://www.netmba.com/strategy/experience-curve/.
Wright, T. P., 1936"Factors Affecting the Cost of Airplanes," Presented at the Aircraft Operations Session, 4th Annual Meeting.

section three

Cost management

chapter ten

Costing and managing off-the-shelf systems

10.1 Introduction

Expanding the use of pre-existing hardware and software items for new product development offers owners and developers opportunities for reduced development cycle times, faster cycle times for new technology, potentially lower life cycle cost (LCC), greater reliability and availability, and support from a more diverse industrial base. The use of pre-existing items or those designed for other applications (we will call these off-the-shelf systems) permeates all aspects of modern new product development. "Off-the-shelf" and "open source" are the mantras for developing complex software-centric systems in the global economy of the 21st century.

Using pre-existing commercially available items in complex systems is not new. Most systems being developed today use some pre-existing or developed-elsewhere elements, such as hardware, software, or operating systems. What is new is the availability of more complex systems, not just components that because of modern industry-defined standards are readily adapted to complex systems, systems of systems, and enterprises. The extent to which individual systems use pre-existing items will vary. Some developers will integrate a few pre-existing items, especially for largely custom-built applications. Other developers serve mainly as lead system integrators that build systems that are integrated from multiple pre-existing items purchased from different vendors. There is no single set of rules that covers this broad range of possibilities.

The following definitions are provided:

- A *commercial off-the-shelf (COTS)* component is an item bought from a third-party supplier and integrated into a larger system. Some examples are software subroutines, subsystems, components, libraries, databases, and embedded systems. It might be an application, an application generator, a graphical user interface, or a physical system or component. COTS components are sold, leased, or licensed for a fee, which can include vendor support in fixing defects. What characterizes COTS components is that they already exist and can be used to build a new system by means of integration without the

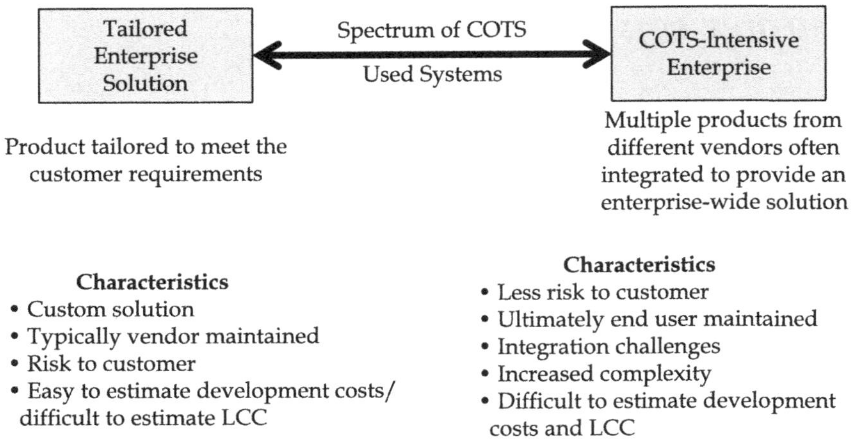

Figure 10.1 Difference between a unique systems solution and COTS-intensive product

need for major modification. The COTS component can supply and support the nearly identical component for numerous customers. The term "COTS" often includes GOTS (government off-the-shelf), MOTS (military off-the-shelf), and non-developmental items. Figure 10.1 illustrates the difference between a COTS tailored enterprise solution and COTS-intensive solution.

The following definitions are used when discussing COTS:

- A *subcontractor, contractor or developer* is often used to describe an organization that is under contract to the owner and is expected to receive a delivered system as specified in a contract (Department of Defense, 2000).
- *Glue code* and *middleware* are synonymous. As the name implies, glue code holds everything together. Glue code and middleware are required to convert data and other information from the format in which the system maintains data to the format required by the COTS component.
- A *GOTS* product is typically developed by a government agency. It is sometimes developed by an external entity but with funding and specification from a government agency. Because agencies can directly control all aspects of GOTS products, these are generally preferred for government purposes because they are recyclable.
- *Open source* describes a broad, general type of software license that makes source code available to the general public with relaxed or non-existent copyright restrictions. The principles as stated say absolutely nothing about trademark or patent use and require absolutely

no cooperation to ensure that any common audit or release regime applies to any derived works. It is an explicit feature of open source that it may put no restrictions on the use, redistribution, organization, nor user whatsoever (Wikipedia, "Open Source").

- *Outsourcing*, also commonly referred to as global sourcing or offshoring by industry professionals, is made up of two words: "out" and "sourcing." Sourcing refers to a number of procurement practices aimed at finding, evaluating, and engaging suppliers of goods and services. Sourcing can be fulfilled by various departments coordinating within the same organization or by involving external parties. Thus, outsourcing refers to the act of transferring work, responsibilities, and decision rights to another entity. Outsourcing is normally conducted with an external party (i.e., external to the unit conducting the sourcing), hence the word "out." A group within an organization can outsource some work to another group or alternatively an organization can outsource work to another organization or person who is external to it (Power et al., 2006).
- "A *MOTS* (either modified or modifiable off-the-shelf, or military off-the-shelf, depending on the context) product is typically a COTS product whose source code can be modified. The product may be customized by the purchaser, by the vendor, or by another party to meet the requirements of the customer. In the military context, MOTS refers to an off-the-shelf product that is developed or customized by a commercial vendor to respond to specific military requirements. Because a MOTS product is adapted for a specific purpose, it can be purchased and used immediately. However, since MOTS software specifications are written by external sources, government agencies are sometimes leery of these products, because they fear that future changes to the product will not be in their control" (Enterprise Linux, 2010).
- *Re-engineering* is the process of examining and altering an existing system to reconstitute it in a new form. This may include reverse engineering (analyzing a system and producing a representation at a higher level of abstraction, such as design from code), forward engineering (using software products derived from an existing system together with new requirements to produce a new system), and translation (transforming source code from one language to another or from one version of a language to another).
- *Software reuse* is the principle of using pre-existing software in multiple applications, for example by deriving application-specific object classes from general-purpose object classes, or using existing class libraries or frameworks for common application functions. Reuse is one of the goals of object technology.
- *Spiral development or model* is a software development process combining elements of both design and prototyping-in-stages in an

effort to combine advantages of top-down and bottom-up concepts. Also known as the spiral life cycle model (or spiral development), it is a systems development method used in IT. This model of development combines the features of the prototyping model and the waterfall model. The spiral model is intended for large, expensive, and complicated projects (Wikipedia, "Spiral Model").
- *Technology refreshment* is the periodic replacement of COTS components (e.g., processors, displays, computer operating systems, commercially available software) within larger systems to assure continued supportability of that system through an indefinite service life.

Whether COTS, MOTS, GOTS, open source, or similar, all are envisioned as cost-effective means of developing a new system. Unfortunately, costing these products means interfacing with existing systems. Little research has been conducted in costing these products, and we do not have a true understanding of the cost savings.

Because of the rapid pace of technology change, the traditional trial-and-error maturing process of technology-intensive systems is often bypassed for the sake of better and more economical performance. Spiral development has become a common development approach because of the rapid pace of technology advancement. Rapid technology insertion using off-the-shelf products poses some unique problems from both a systems integration and a field operational perspective. As shown in Figure 10.2, the increased performance, provided sooner, is often at the expense of mature technology. Integration of these immature COTS systems is becoming a key business driver for many vendors in lieu of equipment development. In this regard, the role of a systems integrator will continue to evolve from providing platform-specific products, systems, and system elements to providing a clearly defined functionality or a solution based on some COTS, MOTS, GOTS, open source, and other components.

Due to shrinking budgets and the continuing exponential growth of technology products in the commercial arena, owners and developers are increasingly turning to COTS sources for components and subsystems as the basis for advanced but affordable systems. Government investments in military and space applications were the primary source of research and development (R&D) funding that spurred development of computing technologies across the board. Since the early 1990s, government investments in technology have been significantly curtailed. By contrast, commercial investments in these technologies have continued to dramatically increase. Today, commercial spending on R&D far outpaces that of the government, and practically all advancements in electronics, computing, and other information-processing areas result from investments from commercial sources. In order to build the most capable and affordable systems, it is now necessary for the large system

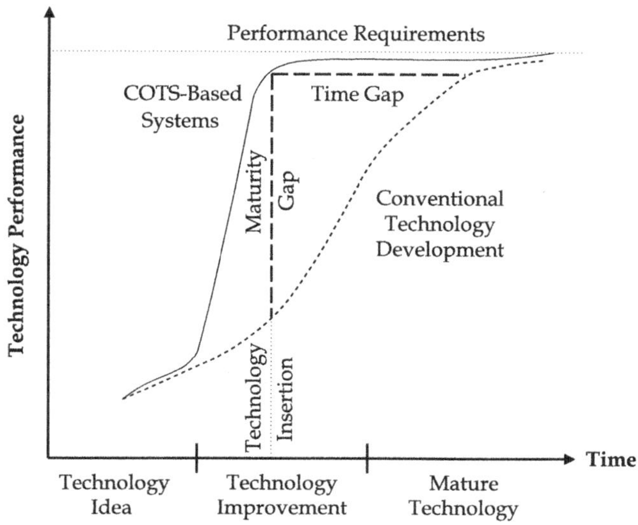

Figure 10.2 A comparison of rapid technology insertion and conventional methods of developing equipment (modified from Schulz et al., 2000)

buyers (especially government) to establish processes, rules, philosophies, and so on to exploit commercial sources of technology. In many cases, this requires that the buyers and developers adopt new methods and procedures that optimize the use and long-term support of systems built primarily from commercial products and technologies.

The use of COTS presents new challenges for other than government buyers to include the designers of large network-centric systems (airplanes, automobiles, etc.). The rapid pace of technology and product evolution combined with strong demand for products has resulted in extremely short product lifetimes and constantly improving performance with decreasing costs. This short product lifetime combined with the difficulty of obtaining product design data can result in system supportability issues shortly after production.

Products based mainly on COTS will inevitably encounter problems or issues associated with supportability. This lack of support comes in a variety of forms throughout the life of the system, including

- Short production runs,
- Difficulty procuring equivalent replacement parts,
- Lack of documentation and knowledge management,
- Lack of hardware, repairs, parts, and facilities for fielded products,
- Lack of vendor support for fixes or updates for earlier software versions, and
- Difficulty in standards migration.

Anyone who has owned a personal computer or mobile phone for several years has dealt with many of these issues.

All these support issues result in rapid obsolescence that must be dealt with throughout the product life cycle. Technology refreshment is a process for periodic replacement of selected system elements before support and obsolescence issues can make operation and maintenance costs prohibitive. This process must coexist with the normal process of periodic functional upgrades required to address evolving requirements. A well-planned technology management program requires balancing functional requirements and obsolescence with cost-effective upgrades.

10.2 Commercial off-the-shelf systems

COTS offers the promise of rapid development and shared risk development costs with other customers. Yet the promise of COTS has not been realized because of the technical risk and the unrealistic expectations of lower cost, higher reliability, immature sourcing strategies, increased project complexity, and, most importantly, integration challenges and costs.

There are several types of COTS products that can be considered for reuse. Some, like the hardware examples below, are relatively easy to integrate, while others, such as software, are challenging:

- Infrastructure: libraries, networking components;
- Application support: database server, application server, Internet server;
- Customizable application/software;
- Services: weather reports, maps, geographic information system, traffic, and so on;
- Embedded systems: device controllers; and
- Hardware: displays, keyboards (modified from Lieberman, 2006).

Since the mid-1990s, most buyers of large systems have made a strong push for COTS-based products. It is hoped that this will reduce costs, make systems flexible and scalable, make it possible to incorporate new technology, and reduce development time. "Shifting to a paradigm in which systems are built primarily of components that are available commercially offers the opportunity to lower costs by sharing them with other users" (Software Engineering Institute, 2007). Table 10.1 lists the various levels of COTS products. Table 10.2 lists some of the disadvantages and advantages of using COTs.

The biggest advantages with implementing COTS can be realized when combining the use of products with an open systems approach. "Open systems emphasize: (1) the use of interface standards and (2) the use of implementations that conform to those standards" (Software Engineering Institute, 2007). This approach maximizes the ability

Table 10.1 Levels of COTS usage (modified from the Defense Science Board, 2009)

Level	Definition of "Commercial Systems"
1	Buy it from a manufacturer—domestic or foreign—and use it as is
2	Buy it from a manufacturer and make minor modifications (e.g., paint it green)
3	Buy if from a manufacturer and make significant modifications (e.g., adding armor, radios)
4	Have a manufacturer make significant modifications before buying it
5	Have a manufacturer gut an existing product and replace most of the parts
6	Have a manufacturer modify a commercial prototype product and replace most of the parts
7	Have a manufacturer assemble a collection of components and independently qualified in different systems
8	The product does not yet exist but requires commercial development and utilizes commercial plans or processes

Table 10.2 Advantages and disadvantages of COTS and custom components and software (Castro, 2008; Boehm and Abts, 1996)

Advantages	Disadvantages
COTS	
• Immediately available • Avoids expensive development • Avoids expensive maintenance • Predictable license fees and performance • Rich functionality • Broadly used, mature technologies • Frequent upgrades often anticipate organization's needs • Dedicated support organization • Hardware/software independence • Tracks technology needs	• Licensing, intellectual property procurement delays • Upfront licensing fees • Recurring maintenance fees • Reliability often unknown or inadequate; scale difficult to change • Too-rich functionality, compromises usability and performance • Constraints on functionality and efficiency • No control over upgrades and maintenance • Dependence on vendor • Integration not always possible; incompatibilities among vendors • Synchronizing multiple-vendor upgrades
Custom	
• Complete freedom • Smaller, often simpler • Often better performance • Control of development and enhancements • Control of reliability trade-offs	• Development expensive and unpredictable • Available data unpredictable • Maintenance expensive • Portability often expensive • Drains expert resources

to make COTS interchangeable in the architecture, thus allowing the program to reap the benefits of competition and to take advantage of technology upgrades. COTS is especially appropriate for horizontal reuse (math routines, user interfaces, graphics, databases, etc.).

Note that COTS does not equal open systems, and many COTS products do not conform to standards. Defining and managing interfaces is the key. Also, an open systems approach requires work early in the development life cycle to identify and select standards, make certain that the standards work for the applications, and find compliant products. COTS, open source, and the like, along with smart buyers, are all needed to minimize LCC.

One consequence of the rapid-fire technology evolution seen in technology-centric industries is the pace of product introduction and consequent short lifetimes. These short product lifetimes are driven directly by the continuing cost reductions resulting from technology evolution, which erodes the cost competitiveness of products very quickly after their introduction. Indeed, the product life of many computer-related products is less than a year. The product development cycle in some cases can be as short as a few months.

One unavoidable consequence of fast-paced product development is short product support lifetimes. Most commercial vendors are unwilling to commit resources to support their products for extended periods after sale. In most cases, commercial and industrial users replace equipment on a regular basis to achieve the cost savings offered by newer products. This makes determining the downstream operation and maintenance costs difficult.

When designing COTS-intensive systems, the primary objective is to achieve a system design that is as independent of the underlying implementation as possible, thus reducing development costs. This goal starts with the development of system requirements specifications, where care should be taken to avoid specifying implementation-level detail as part of a functional requirement. Requirements documents, whether at the system, subsystem, or component level, should describe the system in terms of functional requirements and avoid anticipating design choices within the functional description. As much design leeway as possible should be delegated to the implementers and then properly documented in design descriptions. The specification should allow for, and even anticipate, a wide range of implementations. Carefully structured requirements should allow for future technology refreshment execution with minimal costs.

Designing systems around widely used commercial standards is a practice called open systems design. Selection of a proper set of commercial standards is a critical aspect of the system design and can be a major determiner of the system through LCC. Open system design principles allow for interchangeability of components from multiple vendors and foster a highly competitive market environment that ensures the availability of low-cost products and ready supply. Open systems can dramatically reduce the costs of product improvement programs, development, and downstream operations and support (O&S) costs.

Compared to custom designs, COTS components come with limited documentation. Because of standards and interchangeability of parts from multiple vendors, there is no commercial demand for full design documents for COTS components other than that required for proper interfaces and user data. Therefore, the vendor is the only source of bug fixes, repairs, and spare parts. The role of the vendor is critical beyond the parts selection and acquisition phase of development. When developing a LCC model, increased COTS often translates to increased documentation costs.

The following suggestions can be helpful for evaluating commercial items (Department of Defense, 2000):

- To develop the skills needed,
  - Employ outside experts to support COTS evaluation activities.
  - Train the stakeholders (both developers and buyers) on how to evaluate commercial items.
  - Repeat this training as personnel or the nature of the commercial items being evaluated change.
  - Select a developer who has past experience in integrating commercial items.
- To conduct evaluations,
  - Decide in advance what information you want to gain from the evaluation of a commercial item.
  - Select evaluation techniques based on the type of information required and the importance of the selection to the program.
  - Unless it is impractical, evaluate potential commercial items in a system test bed.
  - Consider both the capabilities of the commercial item and the business practices of the vendor.
  - Take into account the business motivations of the vendors.
  - Understand the vendor's strategy, and talk to other buyers.
  - Understand where you stand in relation to the vendor's other customers.
  - Budget for repeated evaluations throughout the program's life cycle.

10.2.1 Hardware-centric COTS

Tables 10.3 and 10.4 lists some general criteria for selecting vendors to ensure support and minimize LCC. Selecting the right COTS components can dramatically reduce development and O&S costs. The wrong COTS components can lead to unplanned integration and fielding costs.

Table 10.3 Assessment and evaluation methodology for hardware COTS insertion (Verma, 2002)

Assessment and Evaluation System Elements—Vendors, Technologies, and Products							
Supplier Evaluation		Product Evaluation				Technology Evaluation	
Supplier Capability	*Supplier Experience*	*Quality*	*Usability*	*Logistics Support*	*Cost*	*Standards*	*Maturity*
1. Financial strength –Credit line –Research budget –% of revenues to R&D 2. Facilities –Manufacturing facility –Test and integration facility –Computing/ modeling and simulation capability –Design and development capability –Surge capacity	1. History of providing products for application 2. History of providing product support for applications 3. Supplier production maturity 4. Owner/ prime experience with suppliers 5. References 6. Preferred provider list	1. Product availability –Performance history –Field data-based reliability results 2. Built-in test –Coverage –Probability of success –Frequency –Report failures –Log failures –Fault group size 3. Product warranty	Adaptability: 1. Extent of ruggedizing necessary 2. Environment testing conducted—temperature, humidity, shock & vibration, pressure 3. Availability of support tools 4. Availability of software support tools—drivers, compilers, debuggers 5. Availability of interface options 6. Interchangeability	1. Data rights –Training data availability –Technical document availability 2. Supply chain compatibility 3. Lead times –Delivery –Procurement 4. Formal change management process 5. Customer support –Response time	1. Item cost 2. Lot size pricing –Lot size variation/ range –Delivery timing –Modifications to lot sizes, delivery times, and the item itself 3. Support agreement costs 4. Warranty costs 5. Training cost 6. Training data cost	1. Market size of products based on associated standards 2. Any proprietary technology used in product 3. Any proprietary standards used in product 4. Any proprietary interfaces	1. When was the technology introduced? 2. What is the anticipated supported life of the technology? 3. Is the technology scalable?

3. Staffing quality –% degreed engineers –% with advanced degrees 4. Organizational structure –Formal quality and reliability function –Product support function –Help desk –Certification –ISO –SEI 5. Market share 6. Key strategic industrial partners		4. Product safety analysis –Process –History –Certification	7. Hazardous materials 8. Flexibility 9. Potential for growth 10. Product change frequency 11. Product maturity –Obsolescence	–Customer support center (level/staffing/times) –Training support –Installation support 6. Product shelf life –On-board spares perspective –Production inventory perspective	7. Technical documentation cost 8. Development cost 9. Disposal cost 10. Recurring environmental costs	5. Interchange-ability in terms of technologies 6. Number of suppliers supporting the technology

Table 10.4 Software integration activities for COTS-based development and custom development over time (modified from Lombardo, 2008)

COTS Assessment/Selection Criteria							
Vendor Evaluation		Product Evaluation				Technology Evaluation	
Supplier Capability	*Supplier Experience*	*Quality*	*Usability*	*Logistics Support*	*Cost*	*Standards*	*Maturity*
• Financial stability: –Financial credit assessment –Stock and mutual fund assessment • Market share • Availability of software support skills	• Other customer experience in several of the vendor's products • History of support periods for products	• Quality assurance practices • Product functionality • Quality of documentation	• Interchangeability: What is the obsolescence impact? – Number of vendors who have competing products upward/downward compatibility • Ease of installation	• Rate of changes of the product • Does the vendor allow for product customization? • Product availability • Support agreement terms: –Technical support for integration –Response time –Updates to bugs –Support availability of previous releases	• Cost of the software product –Subscription or upgrades on an individual basis • Cost for modifications • Support agreement costs • Number of upgrades expected	• How much of the software contains proprietary technology? • What commercial standards are used?	• When was the technology the software uses introduced (1.0 or 1.X)? • What is the anticipated supported life of the software? • Is the product in its mainstream use or at its tail end?

There are many suggested techniques but no clear solutions for integrating COTS components. Some techniques, such as simulation and prototyping, are typical of larger programs where integration of multiple components is conducted. In these cases, the integrator really is not concerned about the source of the component but only with getting it to function in the system.

10.2.2 *Software-centric COTS*

The costs associated with software development have come to dominate most system development costs. In COTS-based software development, there are many hidden costs associated with the various activities of the component's life cycle, including licensing, royalties, training, pre-integration assessment and evaluation, post-integration certification of compliance with mission-critical requirements, and costs associated with incompatibilities with other software or hardware components. The initial integration costs of COTS products typically occur during the assessment of candidate COTS components and consists mainly of tailoring or the development and testing of glue code.

The most significant variables that influence the LCC of COTS-based systems are the following (in order of impact):

- COTS component integration that needs to be synchronized within a release,
- Technology refresh and renewal cycle times,
- Maintenance workload glue code for glue code updates,
- Maintenance workload to reconfigure updated COTS packages,
- Tracking of new COTS packages release, also called market watch,
- Product evaluation workload during maintenance phase,
- Maintenance workload to update databases,
- Maintenance to migrate to new systems, and
- License costs.

COTS buyers tend to ignore the longer-term maintenance costs associated with the software and only look at the upfront costs, such as the ones associated with functionality, training, and purchase cost. But the true costs of COTS software become apparent during the maintenance phase of the system's life cycle. Maintenance costs can easily equal or exceed the costs of custom-developed applications, and dealing with new releases that incorporate bug fixes and repairs can become a challenge in an already implemented system. During maintenance, COTS products undergo a technology refresh and renewal cycle. Maintainers decide whether to upgrade the COTS products or retain old versions. If they choose to retain the old version of the COTS, they will eventually reach the point where the vendor no longer provides support for that version. If the customer decides to upgrade, they must synchronize the associated update with their release cycle and with product updates other vendors are making in the case of a system comprised of multiple COTS components. In the case of an upgrade, customers must also consider upgrading the glue code.

COTS component upgrade or refresh can represent a major life cycle issue when upgrading a mission-critical operational system without disrupting its normal operations. The greater the number of COTS components in the system, the greater the number of version releases, each one potentially coming out at a different time. The issue becomes when to upgrade the system. If the COTS product is not upgraded after a given period, the vendor will stop maintaining the product version. If the customer decides to freeze the current configuration for a certain number of years and then replace the entire system, they face the risk of continuing to operate with an unsupported component or paying the vendor a premium to continue providing support for that instance of the product.

The major sources of added costs in maintaining systems containing COTS software components compared with custom-developed components are as follows:

- Licensing: Maintenance license fees can increase the overall costs of maintaining a COTS-based system, especially if the vendor increases the price per license over time. In custom-developed systems, there are no licensing fees; in COTS-based systems, the risk of project cancellation increases if the price per license becomes so high that the customer cannot afford it and decides that it is cheaper to start over using custom components instead of COTS components.
- Evaluation of new releases: A major source of cost is associated with COTS component volatility (i.e., the frequency with which vendors release new versions of the product and the significance of the changes in those versions: minor upgrades versus new releases). Compared with custom-developed software, COTS components are controlled by the vendor. The timing and content of releases is at the discretion of the vendor. The customer has to invest a lot of effort in the evaluation and understanding of the implications of an upgrade. Sometimes the evaluation process requires a test bed to replicate the deployed system configurations of software and hardware. This requires a large analysis effort. This need for a test bed and all the analysis involved in order to make the decision to upgrade or not is unique to systems with COTS components.

- Defects: These tend to be more problematic for COTS-intensive systems than for custom code. After documenting and confirming the existence of a defect, the next step is finding the source within the system. In COTS-based systems, it is more difficult to pinpoint the source of the problem. It can be difficult to determine if the defect is coming from the COTS component or from other custom-developed software. Also, if the defect is in the COTS component, there is a chance the vendor will not be able to re-create it because they might have a different hardware/software configuration. All of this can take time and effort, which translates into additional costs. In custom-developed systems, debugging can follow the path through the code without running into component boundaries.
- Vendor support: Vendor support is often used in maintenance to fix defects quickly, provide assistance with the latest product upgrades, or make adjustments to the COTS component in the presence of other product upgrades. The support may range from a 24/7 call service to dedicated on-site staffing. If a defect is found to be caused by the COTS component, it is the vendor who must fix the problem. If a new release of the COTS component has new features or interfaces, the vendor's support may be required to integrate the component into the current system. This support may include some tailoring by the vendor to get the component to work with the existing system architecture. Finally, if a vendor has gone out of business, support may be unavailable.
- Software upgrades: After a new version of a COTS component has been evaluated, the installation of the component into the system may have a ripple effect. Due to new functionalities in the component, the system may require changes to custom code or glue code between components or tailoring of other COTS components. In custom-developed code maintenance, only the fixes and enhancements that are needed are implemented, which minimizes the ripple effects of upgrades.
- Hardware upgrades: Sometimes the upgrade of software components requires new hardware as well, which increases the costs of maintaining the COTS component since the customer has to be up to date with the required hardware in order to obtain the most benefit from the new functionalities of the COTS product. In custom maintenance upgrades, hardware performance is considered part of the upgrade activity. Since only the required features are implemented, there is minimal impact to hardware performance.
- Disabling new features: When upgrading to the COTS component, there may be new features that need to be disabled for security or performance reasons. The added cost is in the form of additional tailoring of the COTS product. This may require finding out how to disable

new features or develop a custom code to hide or disable the new features. Disabling new features is not characteristic of custom systems.

- Early maintenance: Because COTS components continue to evolve in the marketplace, it is possible that upgrades may begin before the system is deployed, particularly if the development spans several years. If the components are not upgraded, it is possible that much of the system may have reached the end of its life before the system is delivered.
- Market watch: Customers may incur additional maintenance costs if they decide to establish a market watch as a risk-mitigation activity in the case the vendor goes out of business. Market watch is a continuous task in which COTS software capabilities and quality is evaluated during the maintenance phase. Establishing a market watch allows the customer to make informed decisions as to the purchase of the COTS component source code or a different component to plan for the event of having an unsupported COTS component when the vendor goes out of business. This activity is not necessary when using custom-developed software.
- Continuous funding: Systems that use COTS components require a more stable funding base. When budgets get tight in a project, funding for maintenance is usually affected. With a custom system, enhancement can be delayed until funding is obtained. The consequences of delaying the funding in a COTS-based system is that licenses may expire, bug fixes and upgrades become unavailable, or vendors go out of business with no resources to exercise a market watch.

10.2.3 Integration costs

The number of COTS components in a system has a large impact on development and O&S costs. For example, the maintenance costs for a single COTS component go down over time as the experience gained by the system maintainers increases, thus improving productivity. The increase in productivity can outpace the increased effort required to maintain the system as the COTS product matures and evolves in divergent directions. However, there is a breakeven point with the number of installed COTS software components (n), where the costs increase disproportionately to the number of COTS products, regardless of the increased productivity. For example, the COTS licensing issue is more complex when dealing with more COTS components. A COTS-intensive system presents multiple licensing strategies, different renewal periods, and different license cost structures. Upgrades can be difficult when dealing with multiple COTS components. The number of possible interactions between components increases exponentially as the number of components increases. When trying to identify defects, the complex interactions of many components makes the task more difficult.

The only estimating tool in the open literature is the COCOTS model (an abbreviation of the longer phrase "Constructive COTS model"). COCOTS is actually an amalgam of four related submodels, each addressing individually what we have identified as the four primary sources of COTS software integration costs. These are costs associated with the effort needed to perform (1) candidate COTS component assessment, (2) COTS component tailoring, (3) the development and testing of any integration or glue code needed to plug a COTS component into a larger system, and (4) increased system-level programming due to volatility in incorporated COTS components (University of Southern California, 2001).

10.3 GOTS

"Government off-the-shelf (GOTS) is a term for software and hardware products that are typically developed by the technical staff of the government agency for which it is created. It is sometimes developed by an external entity, but with funding and specification from the agency. Because agencies can directly control all aspects of GOTS products, these are generally preferred for government purposes" (Wikipedia, "Government Off-the-Shelf").

Many government organizations have developed libraries of software to promote reuse. Unfortunately, these GOTS items are often not used or are lost because of evolving standards (e.g., changes required by the Americans with Disabilities Act), stovepipe development, and poor knowledge management.

10.4 *Software reuse*

The perceived benefit of software reuse was one of the paybacks that the Department of Defense hoped to realize in their effort to implement the Ada programming language as a standard in the 1980s and 1990s. The other goals were reliability, maintainability, and standardization. However, for many reasons, these goals were never realized, and the mandate was abandoned in 1997. This is just one example of how regulation and processes can decrease productivity and innovation and ultimately lead to higher LCC.

As envisioned by the government, software reuse was to be effective across a broad community, achieving what could be defined as systematic software reuse. "Systematic software reuse is a promising means to reduce development cycle time and cost, improve software quality, and leverage existing effort by constructing and applying multi-use assets like architectures, patterns, components, and frameworks" (Schmidt, 2006). The challenges to achieving systematic software reuse fall into two distinct areas: organizational processes and individual programming practices.

The software manufacturing Reuse Reference model provides a good understanding of the technical and non-technical elements or activities that are necessary to establish software reuse practices in an organization. "Technical activities consist of: (1) technologies (tools) that support reuse includes: case-tools and software standards e.g. CORBA, DCOM, and COTS; and (2) software development with reuse processes (technical procedures) include: domain modeling, product-line approach, common architecture, quality control, and best development practices. Non-technical activities (non-technical procedures) include: management of reuse program, market place, financing and marketing forecast, and training" (Rine and Nada, n.d.).

Part of the organization's reuse program must include establishing a library of code that can be categorized and easily retrieved for reuse. This is just one of the upfront administrative costs that the organization must consider for the future return on investment for these types of knowledge management investments.

One area where software reuse has taken hold is in the development of middleware, in particular, middleware that isolates operating system specifics from the application. "Based on the case studies and survey results, we also confirm that the leading indicators of software reuse capability are: product-line approach, architecture which standardizes interfaces and data formats, common software architecture across the product-line, design for manufacturing approach, domain engineering, management which understands reuse issues, software reuse advocate(s) in senior management, state-of-the-art tools and methods, precedence of reusing high level software artifacts such as requirements and design versus just code reuse, and trace end-user requirements to the components that support them" (Rine and Nada, n.d.).

For users of systems that incorporate reuse software, very little will change with respect to LCC for logistical support of the system. Some of the areas that could be affected, positively or negatively, are initial development costs because of increased documentation, scheduled maintenance and overhaul, and lease costs. Repair and scheduled maintenance and overhaul cost impacts depend on who will provide this support. If it is the reuse software vendor or vendors, there could be additional cost and concern about lack of support. Lease costs, interpreted here as license costs for the reuse software, could place a burden on the logistics chain to ensure that they are paid over the years to guarantee continued legal use of a product. Depending on how contracts are written when reuse is executed, there could be a typical government program office that "worries" these issues for the life of the system, or they could become the responsibility of the implementing organization. It is typical that logistics costs are split in this manner.

As noted above, tools that support reuse and standards are a necessity. The costs associated with these tools include procurement, integration, training and administration, and updates, and these must be included in an LCC model. These costs could vary significantly depending on what tools are currently in place and whether the existing tools meet the reuse needs or whether the development environment needs to be completely retooled.

Education of the entire software development life cycle team on this new approach is key. This is an upfront and ongoing cost as the organization adopts reuse techniques and maintains these techniques as personnel or processes change.

Supporting the corporate assets to maintain the software reuse library is a cost that is not specific to a program. This could be a challenge if the corporation does not have a model for "taxing" programs. However, larger corporations could likely adapt the model that they use for the overhead costs for programs that typically include computing resources. The challenge is the cost-benefit trade-off that must be made early in the decision process of whether software reuse is a good alternative or not.

10.4.1 Cost reductions achieved with software reuse

Certainly, the goal of reuse is to reduce development time and costs. An organization that successfully meets the challenge of implementing a culture of software reuse will achieve this goal, but it will not be immediate nor pervasive. Some products will still require standard software development practices. In spite of the fact that software reuse has been a goal of many programs for approximately 20 years, the lack of metrics and documented case studies that can be translated between companies and programs means that there are no hard numbers on the percentage of savings that can be achieved through reuse.

Because of the rapid pace of technology change, the traditional trial-and-error maturing process of technology-intensive systems is often bypassed for the sake of better and more economical performance for many architectures. Rapid technology insertion using off-the-shelf products poses some unique problems from both a systems integration and field operational perspective. As shown in Figure 10.2, the increased performance, provided sooner, is often at the expense of mature technology. Integration of these immature COTS systems is becoming a key business driver for many vendors in lieu of equipment development. In this regard, the role of a systems integrator (IBM, Lockheed Martin, Boeing, etc., market their services as systems integrators) will continue to evolve from providing platform-specific products, systems, and system elements to providing a clearly defined functionality or a solution based on some COTS, MOTS, GOTS, open source, or similar components.

Some literature exists concerning costing software reuse. At best, rules of thumb exist. For example, reusing software takes about 20% of the effort of new development, while rewriting software for reuse takes an extra

50% effort. Some believe that this number is substantially higher. The cost is probably application specific and is greatly affected by the number of external function points. For systems requiring substantial integration, reusing software might cost more than developing new code, whereas one could expect a tenfold return on investment on writing new code in some instances (right use of the code, documented properly, little integration, etc.).

10.5 Open source

Open source doesn't just mean access to the source code. Unfortunately, people often thing that Open source software means that the software is free for anyone to inspect, modify, and submit enhancements. The distribution terms of open-source software must comply with the following criteria (Coar, 2006):

- Free redistribution: The license shall not restrict any party from selling or giving away the software as a component of an aggregate software distribution containing programs from several different sources. The license shall not require a royalty or other fee for such sale.
- Source code: The program must include source code and must allow distribution in source code as well as compiled form. Where some form of a product is not distributed with source code, there must be a well-publicized means of obtaining the source code for no more than a reasonable reproduction cost, preferably by downloading via the Internet without charge. The source code must be in the preferred form in which a programmer would modify the program. Deliberately obfuscated source code is not allowed. Intermediate forms such as the output of a preprocessor or translator are not allowed.
- Derived works: The license must allow modifications and derived works and must allow them to be distributed under the same terms as the license of the original software.
- Integrity of the author's source code: The license may restrict source code from being distributed in modified form only if the license allows the distribution of "patch files" with the source code for the purpose of modifying the program at build time. The license must explicitly permit distribution of software built from modified source code. The license may require derived works to carry a different name or version number from the original software.
- No discrimination against persons or groups: The license must not discriminate against any person or group of persons.
- No discrimination against fields of endeavor: The license must not restrict anyone from making use of the program in a specific field of endeavor. For example, it may not restrict the program from being used in a business or for genetic research.

- Distribution of license: The rights attached to the program must apply to all to whom the program is redistributed without the need for execution of an additional license by those parties.
- License must not be specific to a product: The rights attached to the program must not depend on the program being part of a particular software distribution. If the program is extracted from that distribution and used or distributed within the terms of the program's license, all parties to whom the program is redistributed should have the same rights as those that are granted in conjunction with the original software distribution.
- License must not restrict other software: The license must not place restrictions on other software that is distributed along with the licensed software. For example, the license must not insist that all other programs distributed on the same medium must be open-source software.
- License must be technology neutral: No provision of the license may be predicated on any individual technology or style of interface.

The open-source model touts unrestricted access to the source code to execute or modify at will and support from an ad hoc collection of developers and other users. However, to be accepted, this model must produce software that meets or exceeds the reliability and performance of its proprietary competitors.

The obvious attraction of open source is that it is perceived as being free, but nothing could be further from the truth. The lack of license fees must be balanced with the need for training, support, and maintenance. The biggest advantage of no license fees is the zero cost to scale, meaning that fees do not increase proportional to the number of systems running the software. This can result in huge savings for large companies.

Support is an area that is hard to categorize and cost. Proprietary product support can be costly and can increase as a vendor supports a company's particular version. Open-source support is essentially a network of people on the Internet who can provide timely, but often conflicting, responses.

10.6 Summary

If you lump GOTS, MOTS, and software reuse broadly as off-the-shelf product, there are many challenges in recycling preexisting systems. Although Lieberman (2006) cited the challenges for COTS shown in Table 10.5, they are applicable to all items developed by others.

The challenge of vendor modifications can only be dealt with through communication and planning with the vendor. A suggestion is to make the vendor part of the development team instead of merely a supplier.

Table 10.5 Challenges of integrating COTS products

Challenge	Description
• Sensitivity to vendor modifications and release schedules	• Vendors control the functional roadmap and release schedule for their products; consequently, customers must be prepared to deal with new releases, patches, and functional changes
• Vendor lock-in (entanglement)	• Unless they're extremely careful in developing their system architectures, customers can become tightly dependent on a particular vendor's product or component (e.g., billing engines)
• Deployment complexity for installation and configuration	• Installation of COTS components can involve significant deployment and configuration issues that must be considered early in the product selection process

From a cost standpoint, the cost of upgrades and new releases must be factored into the LCC of the system.

Vendor entanglement can be addressed through the use of an open architecture and open standards. This is easy to say but is very challenging in practice. Once a product is incorporated into a system, there is a tendency to "customize" it for the application. Once a product is customized, typically by the vendor for COTS/GOTS, the system is locked in to that vendor for future modifications to the product. From a costing standpoint, entanglement prevents the system from taking advantage of the benefits of competition with respect to COTS or from being able to move to potentially less costly open-source software.

Finally, deployment complexity can be addressed through good testing, documentation, and training. The users should receive a "system," and the development process should shield them from the intricacies of any reuse products in the system. Costs incurred may include rewriting documentation for reuse components to incorporate this information into system manuals.

To reduce the LCC of COTS-based development, it is important not only to look at the functional capabilities of the COTS component but also to understand the market direction to determine when it is time to upgrade or change vendors in order to maintain a strong position in the market. Also, understanding the long-term costs of using COTS components during the initial product selection allows budgeting for the LCC of the project since no unexpected fees will be discovered later during the operational phase of the product. All of these show the importance of developing a cost estimation tool for the full LCC of COTS-based development.

Many of the risks associated with COTS integration can be attributed to poor initial program planning and tracking. Underestimating the technology risks associated with COTS-based software development can lead to longer schedules, delays, higher development costs, and higher maintenance costs. Applying risk-mitigation strategies while dealing with COTS-based software development can minimize the risks and LCC of using COTS components.

There are no "silver bullets" when dealing with COTS, COTS insertion, GOTS, MOTS, and the like. While there can be significant benefits, they can only be attained by understanding and addressing the challenges of buying COTS versus in-house development. It must also be emphasized that the risks associated with system development do not disappear and usually increase as the amount of COTS systems increases. In summary, no matter how much of a system is comprised of COTS, the system must still be integrated, tested, sustained, and retired.

QUESTIONS

10-1. Consider the technology cycle times shown in the figure below.

a. Discuss the challenges of producing a system with leading-edge technology while conducting systems integration and meaningful independent testing.

b. Military systems often take 15 years or longer from conception to fielding. Discuss the use of modularity as a key performance parameter for assessing platform designs.

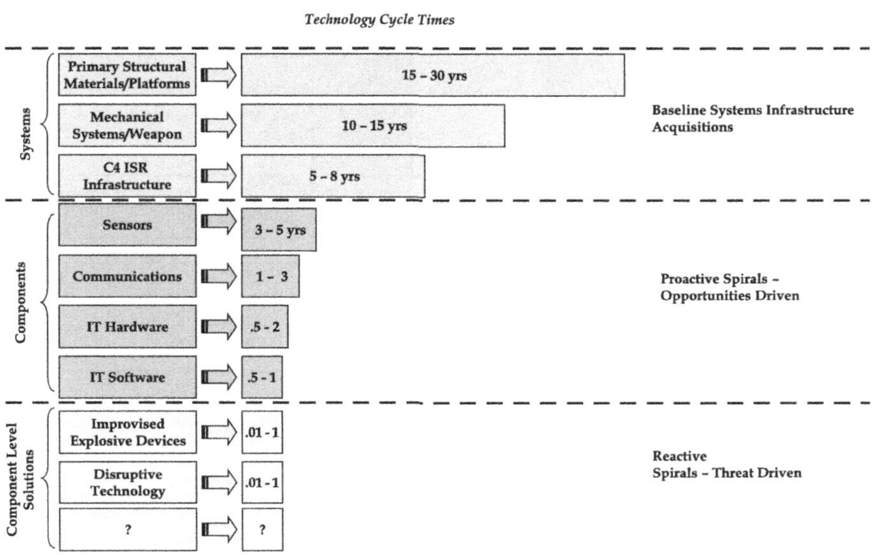

10-2. The figure below (Julian et al., 2011) shows a weighted scoring system for evaluating COTS products. Total ownership cost accounts for 12% (.4 * .3) of the value used for scoring. Would you adjust any of these values? Are LCC categories properly accounted for?

Weighting Scheme

Component 1 Name:	Resistor A
Component 2 Name:	Resistor B
Component 3 Name:	Resistor C

Objective	Evaluation Factor	Weight	Evaluation Element	Weight	Evaluation Measure	Weight
Evaluate and Select COTS Component	Weight Sum:	100%				
	Manufacturer	40%	Weight Sum:	100%		
			Manufacturer Capability	50%	Weight Sum:	100%
					Product Control	25%
					Financial Strength	12%
					Facilities	12%
					Equipment	8%
					Staffing	8%
					Organizational	15%
					Quality Management System	20%
			Manufacturer Experience	50%	Weight Sum:	100%
					Historical Experience/Performance	40%
					SNL's Historical Experience	60%
	Product	40%	Weight Sum:	100%		
			Quality	35%	Weight Sum:	100%
					Reliability	75%
					Maturity	25%
			Technical Specifications	30%	Weight Sum:	100%
					Ability to meet Application Specs	33%
					Ability to meet Manufacturer Specs	33%
					Ability to meet Packaging Specs	33%
			Logistics & Support	5%	Weight Sum:	100%
					Availability	60%
					Customer Support	15%
					Flexibility	25%
			Total Cost of Ownership	30%	Weight Sum:	100%
					Hardware related costs	80%
					Software related costs	0%
					Technology Refresh & Insertion costs	20%
	Technology	20%	Weight Sum:	100%		
			Technology Evaluation	100%	Weight Sum:	100%
					Maturity	70%
					Standards	30%

10-3. If you currently work in industry or have some expereince, answer and/or discuss the following questions:
   a. Does your organization have a software reuse process?
   b. Does your organization have a formal software library?
   c. How do you handle similar yet new development?
   d. How do you cost the integration of existing code?
   e. Do you have any experience in the integration of existing code?

10-4. As 3D printers become more economical and produce more than plastic parts, additive manufacturing will challenge how we buy intellectual property. Spare parts have been a traditional source of major revenue for most developers. Owners have had to re-engineer many of the components of the legacy systems in order to take advantage of additive manufacturing. Should intellectual property be a major cost consideration?

PROBLEMS

10-1. A small software company that develops an integrated business analysis program sees an opportunity to sell some of their software library components for reuse in other commercial products, such as e-commerce websites and banking systems. They identify three products to target for a pilot effort: an investment portfolio management tool, a mortgage payment reduction calculator, and a retirement planning tool. If the pilot is successful, they plan to launch a larger software reuse business. Use the following assumptions:
   - Business intelligence shows that each of these three products could have sales of 10 per month for the first year. After that, we cannot predict sales with any certainty.
   - The price of all three products will be the same.
   - Initial development of the software will require two developers/programmers for a period of 5 weeks for each reuse component. Remember, there are three products.
   - The developers/programmers work 40 hours per week and earn $100 per billable hour.
   - Training the developers/programmers in reuse practices will require 1 week of training per person at a cost of $2,000 plus salary. This week is not considered to be part of the 15 weeks of development time.

- Establishment of a reuse database to catalog the reuse components will require startup costs for a server ($900) and database software ($800). A database administrator will establish and manage this system.
- Since the database is initially small, the administrator will spend 80 hours over the 15-week period to maintain configuration management software, backups, and so on before product launch at the 15-week mark.
- The database administrator earns $90 per hour.
- The database administrator will require 1 week of training at a cost of $3,000.

Given that the products will not be ready for market until 15 weeks after project initiation, and ignoring the time value of money, how much must the company charge for the products in order to break even after the initial 360 sales? Is this a reasonable cost for this product?

10-2. A company needs to implement an Internet server. The first alternative is the Windows IIS and the second is a Linux server with Apache. The company has always used Microsoft products, so the system administrators are trained on these products exclusively.

Microsoft Implementation

- Product cost: Windows Web Server 2015, $469.00 annual license (Internet-facing server only)
- Training: $4,795 for an intensive 7-day course
  - Cost of personnel time: $100 per hour
- Support: MS Software Assurance, updates and support, 25% of license cost annually

Red Hat Linux/Apache Implementation

- License costs: Red Hat Enterprise Linux, standard subscription, $799 annual license
- Training: Red Hat Linux Systems Administration, 4 days, $2,498 Apache and Secure Web Server Administration
  - Cost of personnel time, $100 per hour
- Support: updates and 12 × 5 Internet support included

Three administrators are initially trained for each system. Turnover will result in one additional administrator being trained each year. A course day is 8 hours. Assume all costs will be unaffected by inflation. Compare the two alternatives over 3 years of each implementation. Would your answer change if the company Internet server was a critical asset for the company to conduct its business? Why or why not?

Bibliography

Software Engineering Institute, 1998, "Quotations from Chairman David: A Little Red Book of Truths to Enlighten and Guide on the Long March Toward the COTS Revolution," Prepared for SEI Joint Program Office.

References

Boehm, B., and Abts, C., January 1996, "COTS Integration: Plug and Pray?" *Computer*.

Castro, Lymari, October 2008, "The Risks and Costs of COTS-Based Software Development," EM 620 Project.

Coar, Ken, 7 July 2006, "Open Source Initiative," accessed 20 May 2008 at http://www.opensource.org/docs/osd.

Defense Science Board, February 2009, "Buying Commercial: Gaining the Cost/Schedule Benefits for Defense Systems," Task Force on Integrating Commercial Systems into the DoD, Effectively and Efficiently.

Department of Defense, 26 June 2000, "Commercial Item Acquisition: Considerations and Lessons Learned" accessed 1 August 2018 at https://www.acq.osd.mil/dpap/docs/cotsreport.pdf.

Enterprise Linux, April 2010, "Definition: COTS, MOTS, GOTS, and NOTS," accessed 10 July 2018 at http://searchenterpriselinux.techtarget.com/sDefinition/0,,sid39_gci789218,00.html.

Julian, Cotye S., Lucy, Tana B., and Farr, John V., June 2011, "Commercial-Off-the-Shelf Component Evaluation and Selection Process," *Engineering Management Journal*, Volume 23, Number 2, pp. 63–71.

Lieberman, Benjamin A., 22 August 2006, "COTS Product Integration: An Architectural Approach," IBM, Inc.

Lombardo, Anthony, December 2008, "Evaluating COTS Software," SDOE 620 Project, Stevens Institute of Technology.

Power, M., Desouza, K., and Bonifazi, C., 2006, *The Outsourcing Handbook: How to Implement a Successful Outsourcing Process*, Kogan Page, Philadelphia, PA.

Rine, David C., and Nada, Nader, n.d., "Software Reuse Manufacturing Reference Model: Development and Validation," George Mason University, accessed 1 August 2018 at ftp://ftp.gmu.edu/drine/E-BusMgt-Italy/Software%20Reuse%20Reference%20Model%20(RRM).htm.

Schmidt, Douglas C., 28 September 2006, "Why Software Reuse Has Failed and How to Make It Work for You," accessed 1 August 2018 at http://www.cs.wustl.edu/~schmidt/reuse-lessons.html.

Schulz, Armin P., Clausing, Don P., Fricke, Ernst, and Negele, Herbert, 2000, "Development and Integration of Winning Technologies as Key to Competitive Advantage," *Systems Engineering*, Volume 3, Number 4, pages 180–211.

Software Engineering Institute, 11 January 2007, "COTS and Open Systems—An Overview," accessed 25 April 2008 at http://www.sei.cmu.edu/str/descriptions/cots.html#ndi.

University of Southern California, 1 August 2001, "COCOTS Model Description," accessed 2 June 2008 at http://csse.usc.edu/csse/research/COCOTS/modeldesc.html.

Verma, Dinesh, March 2002, "Crusader Program: Technology Refreshment and Obsolescence Avoidance Strategy," Research Proposal Presented to United Defense, L.P.

Wikipedia, n.d., "Government Off-the-Shelf," accessed 24 January 2008 at http://en.wikipedia.org/wiki/Government_off-the-shelf.

Wikipedia, n.d., "Open Source," accessed 24 January 2008 at http://en.wikipedia.org/wiki/Open_source.

Wikipedia, n.d., "Spiral Model," accessed 8 December 2009 at http://en.wikipedia.org/wiki/Spiral_model.

chapter eleven

Project management's role in life cycle costing

11.1 Introduction

Formal project management (PM) probably predates the construction of the Egyptian pyramids. With the introduction of the Gantt chart in the 1910s, Program Evaluation and Review Technique or PERT by the US Navy in the 1950s, and modern software such as Microsoft Project and Primavera, our ability to plan and track complex projects has grown dramatically. With over 730,000 individuals certified as project management professionals by the Project Management Institute (PMI), most universities offering graduate degrees in PM, and over 38,000 results for a search of PM books on Amazon, few will argue that formal PM has become a critical skill set for anyone in the modern workplace. Thus, in lieu of a chapter summarizing PM, we focused this material on tracking and quantifying costs using formal methods.

A project is a temporary endeavor undertaken to create a unique product, service, or result (PMI, 2013). Formal project or program management is essential in planning, controlling, and communications that, when properly executed, support time and cost estimates needed to manage any project or program. As shown in Figure 11.1, all types of cost estimating techniques are needed throughout the life cycle to develop detailed assessments and then properly track the project. Figure 11.2 demonstrates how requirements map to work breakdown structure (WBS) elements, which ultimately feed a master schedule. This WBS can be used at any point in the life cycle to capture how, what, and when the activities in a project should be accomplished, from which costs are developed.

Much of the literature concerning scheduling and PM concepts focus on the mechanics of PM (networks, critical path method, etc.) and not on how to translate requirements to elements of a WBS and estimate activity duration. Those concepts may also make much more of an impact if they are implemented early in the planning and design stages of a project. Most of the control of the project cost and duration takes place in its early planning stages. In Figure 1.7, you can see that most of the influence of the cost and scheduling of a project occurs in its early planning. As planning

INPUTS	TOOLS AND TECHNIQUES	OUTPUTS
• WBS • Resource requirements • Resource rates • Activity duration estimates • Estimating publications • Historical information • Risks • Architectures • Requirements	• Analogous estimating • Parametric modeling • Bottom-up estimating • Simulation-based costing • Other cost estimating methods • LCC estimate • Integrated product teams	• Cost estimates • Supporting detail • Cost management plan • Integrated master plan

Figure 11.1 Role of PM in cost estimating (modified from Defense Acquisition University, 2003)

proceeds, decisions are made that determine the characteristics of the project, such as the size of the facility, the technology to be used in the facility, and the quality of the facility. As the project moves to design, further decisions nail down cost and schedule. The contractor is left to influence only the cost and scheduling aspects of the project. Managing labor and equipment productivity, smart buying, and the use of good PM practices can accomplish this. Effective PM concepts must be initiated early in the planning of a project and continue throughout its life cycle.

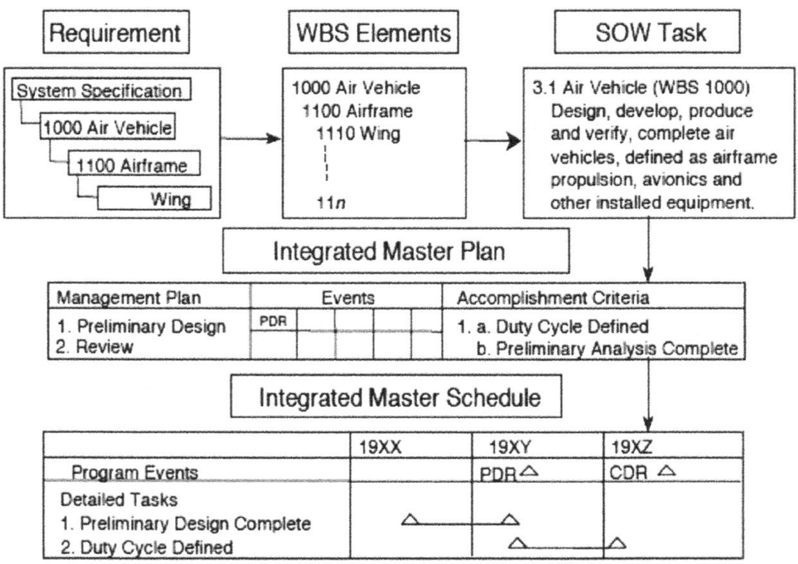

Figure 11.2 Role of WBS in developing an integrated master schedule (Defense Acquisition University, 2001)

Project management has evolved rapidly since the growth of PMI and the project management professional certification. PMI espouses PM processes known as initiating, planning, executing, monitoring and controlling, and closing. There is more than one way to manage a project to successful completion, and one could argue that PM is more of an art than a structured process. These processes are required to track, review, and regulate the progress and performance of the project, to identify any areas in which changes to the plan are required, and to initiate the applicable changes. The initiation process of a new project formally begins by obtaining authorization to start the project (Project Management Institute, 2013).

11.2 Basics of networks

Simply put, time is money, and the duration of any project has a direct bearing on its cost. Because the estimated duration of any project is determined by a schedule, and activities are coordinated by that schedule, project planning is an important element of any engineering or technology project.

Much of the current thinking equates scheduling to a network. A *network* is a diagram of activities joined in interconnected links that reflect relationships between complex, interrelated tasks. It is used to determine the overall duration of a project, to learn about the project, to perform "what if" analyses, and to analyze and settle issues. There are claims that a network can integrate cost and schedule, but, as desirable as this is to cost and schedule management, it is rarely done in practice—even with the advent of such tools as Microsoft Project.

Basically, there are two types of networks. The first is the "activity-on-arrow" network. In this type, the arrows represent activities, while the nodes represent events. This is predominantly taught in most management science and PM courses. The second is the "activity-on-node" or precedence network. In this type of network, the nodes represent activities, while the arrows represent only interrelationships. Modern project planning software uses activity-on-node networks. Both network types have been used in the critical path method (CPM) and in program evaluation and review technique (PERT). However, much of the software is moving toward only supporting the precedence-type networks.

You must remember that a network is not a schedule. The network is used to develop scheduling data. Once the data are developed, you may produce a schedule in the form of a table or bar chart. Networks are the cornerstones of any type of program or project planning. In the engineering profession, the PERT/CPM network is the primary tool for planning (choosing which method or procedure should be used), scheduling (time of the procedures), and controlling (monitoring the procedures). The terms

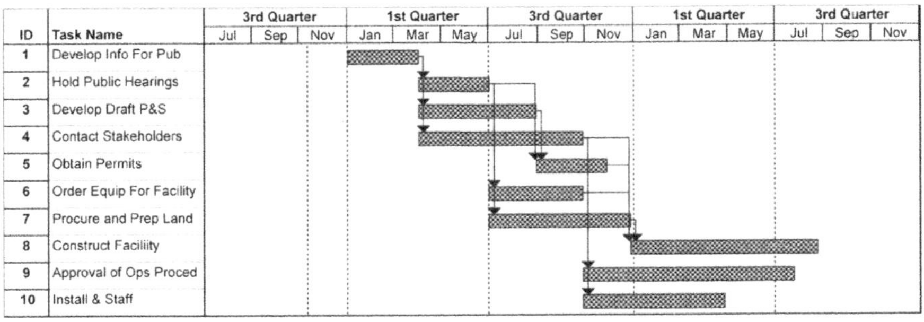

Figure 11.3 Gantt chart of a hazardous waste incineration facility

"PERT" and "CPM" are used interchangeably and together because it has become common terminology for networks in engineering and management when referring to the whole networking technology. We shall use the term "CPM" unless stochastic time duration is introduced.

A project has been defined as a collection of activities interrelated logically or sequentially having a finite scope. Projects related to civil engineering planning are typically non-repetitive and not conducted at the same site. A CPM network consists of the logical relationships between the activities that make up a project. There are two types of CPM networks. The first is the arrow diagramming method (ADM) also called the "activity-on-arrow" method. The second is the precedence diagramming method (PDM). The PDM is increasing in popularity primarily due to some of the newer features of CPM software packages. Primavera and Microsoft Project use only PDM networks. However, most texts use ADM networks to teach the basics of networks.

The bar chart is probably the most often used PM tool. Numerous forms of the bar chart exist for conveying information. The bar (or Gantt) chart can be very effective for conveying a simple schedule. Figure 11.3 is a Gantt chart for the hazardous waste problem. This chart was produced using Microsoft Project.

11.3 Work breakdown structure

Most projects will be broken down into sub-elements. This must be done, at least implicitly, on all but the smallest projects, to schedule the activities. The WBS is used to assign responsibilities, for management control on major complex projects, and to develop activities for project scheduling. It is also used in the planning stage to identify tasks and subtasks. For smaller projects, the WBS can be used for cost accounting and the assignment of responsibilities. Example 11.1 shows a generic WBS. There are no hard-and-fast rules as to the level of detail that goes into developing a WBS.

EXAMPLE 11.1

Figure 11.4 presents a generic WBS for a software development program. Note that NASA (2015) took a life cycle approach to developing the WBS.

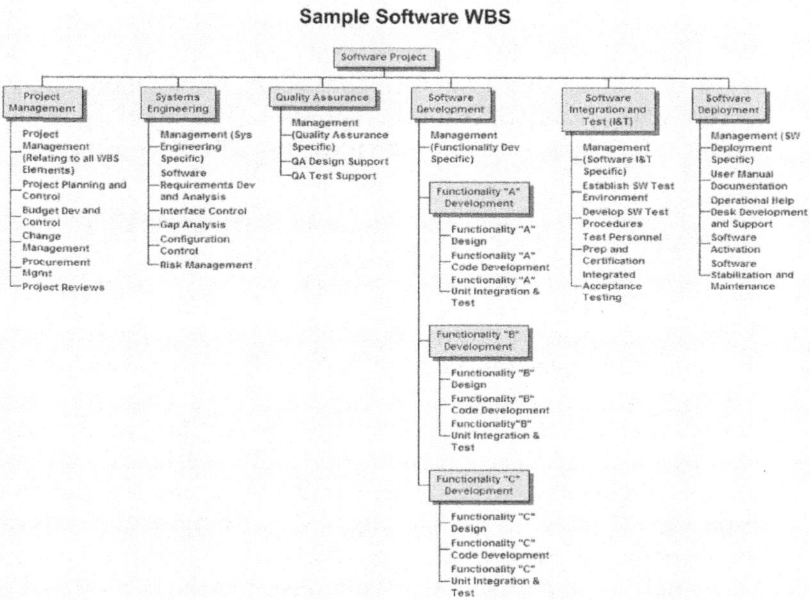

Figure 11.4 Generic software development WBS (NASA, 2015)

11.4 *Progress measurement*

There are numerous methods available to measure work in progress of an activity. This is critical both at the activity or task level and at the project level. Some depend on judgment; some may be calculated analytically. We discuss seven methods used to assess performance or measure work. You must be cognizant of the potential for abuse when at least seven methods exist for measuring performance. The methods are as follows:

1. Units completed: This assumes that the activity is basically linearly scheduled. If an activity consists of the installation 800 linear feet of fiber optic cable and 400 linear feet has been installed, then the progress is 400/800 = 0.5 or 50% complete.

2. Incremental milestones: This method depends on heuristics based on experience. This works well for nonlinear tasks that are treated as lump sum items.
3. Start/finish: Some activities may have actual durations shorter than the reporting period. In that case, it will be 100% complete at the reporting period but accomplished in a shorter duration.
4. Project manager opinion: Painting, landscaping, equipment installation, and similar activities are candidates for this approach. Yes/no such as installed or not installed might be appropriate.
5. Cost ratio: PM, quality assurance, contract administration, and project controls are candidates for this approach. This is not a good measurement for physical activities that require actual work.

$$\text{Percent Complete} = \frac{\text{Actual Cost or Work-Hours to Date}}{\text{Forecast at Completion}} \tag{11.1}$$

6. Weighted or equivalent units: Material weights or some similar unit can be used to measure progress. This is illustrated using the data in Table 11.1.
7. Earned value: This is the preferred method of progress measurement by most agencies and contractors. This method uses person-hours or dollars as a method to weigh the value of one activity with respect to the others. It has nothing to do with the actual consumption of person-hours or dollars on a project. To perform an earned value analysis on a project, first you must find the percentage complete for each activity. This may be done using methods 1 through 6. That percentage complete is then multiplied by the budget in person-hours, person-days, or dollars for that activity. That gives the earned value for that activity, sometimes referred to as the budget

Table 11.1 Weights and quantities for a small steel building (Griffis and Farr, 1999)

Wt.	Subtask	u/m	Quan Total	Equiv stl ton	Quantity to-date	Earned Tons
0.02	Fdn bolts	each	200	10.4	200.0	10.4
0.02	Shim	%	100	10.4	100.0	10.4
0.05	Shakeout	%	100	26.0	100.0	26.0
0.06	Columns	each	84	31.2	74.0	27.5
0.10	Beams	each	859	52.0	0.0	0.0
0.11	Cross-braces	each	837	57.2	0.0	0.0
0.20	Girts & sag rods	bay	38	104.0	0.0	0.0
0.09	Plumb & align	%	100	46.8	5.0	2.3
0.30	Connections	each	2977	156.0	74.0	3.9
0.05	Punchlist	%	100	26.0	0.0	0.0
1.00	STEEL	TON		520.0		80.5

cost of work performed for that activity. The earned value for the project is given by:

$$\text{Percent Complete} = \frac{\text{Earned Work-Hours or Dollars All Accounts}}{\text{Budgeted Work-Hours or Dollars All Accounts}} \quad (11.2)$$

11.5 *Calculating earned value*

These terms are gaining acceptance in earned value progress assessment of networks:

- Budgeted cost of work scheduled (BCWS) or planned value (PV[1]),
- Budgeted cost of work performed (BCWP) or earned value (EV), and
- Actual cost of work performed (ACWP) or actual cost (AC).

Table 11.2 contains additional terms that are often used in earned value analysis.

Note that these calculations can be performed at the task or total project level. Using these numbers as absolute for a project performance by calculating SPI and CPI for the overall budget can be misleading. You must look

Table 11.2 Summary of earned value terms and definitions (modified from Carlos Consulting, n.d.)

Symbol	Description	Formula	Explanation
PV	Planned Value		(*Also known as BCWS—Budgeted Cost of Work Scheduled*) The estimated value of the work planned to be done thus far
AC	Actual Costs		(*Also known as ACWP—Actual Cost of Work Performed*) The total costs incurred to date, based on timesheets, invoices, other expenses, etc.
EV	Earned Value	Sum of All (Task Budget * % Completed)	(*Also known as BCWP—Budgeted Cost of Work Planned*) The estimated value of (intended) work completed thus far, as it relates to the expected deliverables

(*continued*)

[1] The PMI uses a two-letter notation whereas the four-letter notation is common in most PM textbooks.

Table 11.2 *(continued)*

Symbol	Description	Formula	Explanation
CPI	Cost Performance Index	EV/AC	Represents the amount of work being completed on a project for every unit of cost spent
SPI	Schedule Performance Index	EV/PV	Represents how close actual work is to being completed compared to the schedule
BAC	Budget at Completion		The original budget for the total job and is determined at the start of the job but might be updated as the project progresses
EAC	Estimate At Completion	Four Options; See Below:	The amount "currently expected" for the total project cost
EAC (#1)		BAC/CPI	Used if no variances from the BAC have occurred
EAC (#2)		AC + ETC	Used when original estimate is fundamentally flawed
EAC (#3)		AC + BAC – EV	Used when the current variances are atypical
EAC (#4)		AC + [(BAC – EV) /CPI]	Used when current variances are typical
ETC	Estimate to Complete	EAC – AC	From this point on, how much MORE do you expect the cost will be to finish the job
VAC	Variance at Completion	BAC – EAC	How much over or under budget will the project come in at—a negative VAC indicates that the project is forecasted to not be completed within the approved budget
CV	Cost Variance (in dollars)	EV – AC	Negative is over budget, positive is under budget
SV	Schedule Variance (in dollars)	EV – PV	Negative is behind schedule, positive is ahead of schedule, measure of work slippage
PE	Planned Earned	EV/BAC	Project % complete (as it relates to time and deliverables)
PS	Percent Spent	AC/BAC	Project % spent (as it relates to cost)
CSI	Cost Schedule Index	CPI * SPI	The overall efficiency rating—the further the CSI is from 1.0, the more difficulty the project will have in recovering

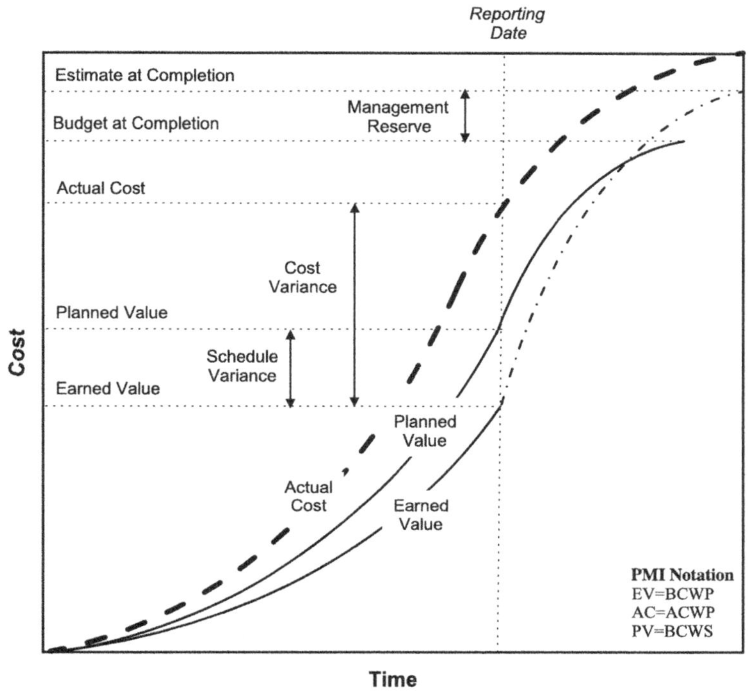

Figure 11.5 Performance measurement parameters

at those tasks on the critical path and separate out non-critical tasks. This can lead to optimistic projections of cost and schedule.

Figures 11.5 and 11.6 show numerical values of those indicators that can be interpreted differently. Figure 11.5 shows how PV (BCWS), EV (BCWP), and AC (ACWP) can be used to reflect various degrees of completion over the life of a project. A time plot of these performance indicators can be useful to identify when projects are starting to fall behind schedule or

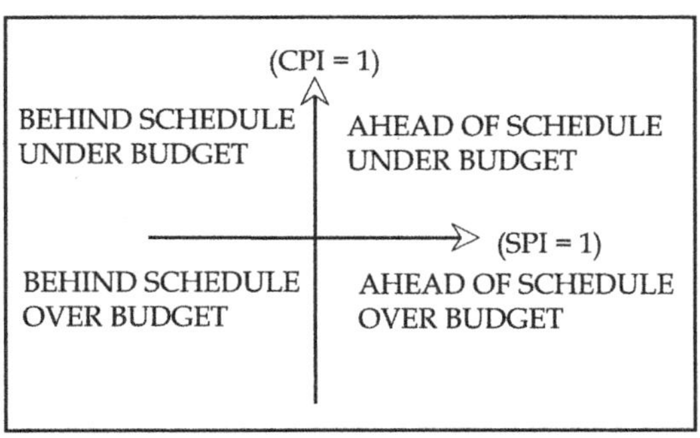

Figure 11.6 Meanings of CPI and SPI

cost overruns are occurring, for example. Example 11.2 demonstrates the basics of earned value assessment.

The techniques for measuring progress and/or completion can dramatically affect earned value analysis. Unfortunately, there is no right or wrong way to measure progress. Under Section 11.4, methods 1, 2, 3, 4, and 6 can certainly be used to assess activity progress. In general, any measurement technique should be objective and quantifiable. Opinions or subjective assessments should be minimized.

EXAMPLE 11.2

As a sub for a major facility upgrade, the prime contractor has asked for the status of a telecommunications revitalization project. You decided to calculate CV and CPI as performance indicators. Your cost accounting section and project manager have provided you with the following information:

Description	Total Cost Forecast	Percent Complete	Cost to Date Actual	Cost to Date Forecast
Building Permits	$ 2,000.00	100.00%	$ 1,253.75	$ 2,000.00
Temp Networks	$ 25,000.00	100.00%	$ 26,497.83	$ 25,000.00
Order Cable Trays	$ 8,000.00	100.00%	$ 7,907.27	$ 8,000.00
Order Routers, Cabling	$ 10,000.00	100.00%	$ 9,017.32	$ 10,000.00
Remove Ceilings	$ 18,000.00	100.00%	$ 11,427.49	$ 18,000.00
Install Cable Trays	$ 28,000.00	100.00%	$ 19,743.19	$ 28,000.00
New Servers	$ 20,000.00	70.00%	$ 11,271.25	$ 14,000.00
Cable TV	$ 10,000.00	10.00%	$ 793.21	$ 1,000.00
Backbone & Routers	$ 20,000.00	5.00%	$ 327.19	$ 1,000.00
LANs	$ 17,500.00	0.00%	$ -	$ -
Connect & Test	$ 15,000.00	0.00%	$ -	$ -
New Ceilings	$ 20,000.00	0.00%	$ -	$ -
Total	$ 193,500.00		$ 88,238.50	$ 107,000.00

Solution

You need two pieces of information to obtain the desired performance indicator: EV and AC (actual cost of work performed). Thus,

$$CV = EV - AC = BCWP - ACWP = \$107{,}000 - \$88{,}238.50 = \$18{,}761.50$$

$$CPI = EV/AC = BCWP/ACWP = \$107{,}000/\$88{,}238.50 = 1.21$$

From Figure 11.6, a CPI>1 indicates that the job is under budget. Although this analysis does not provide an absolute answer as to what is wrong with the project, it provides an indication that management should look into the performance.

Figure 11.6 shows how values of CPI and SPI can be related to schedule and budget.

CPI and SPI can be calculated for a single activity, a group of activities, or a whole project or program. Using SPI may not be a good schedule performance indicator.

11.6 Monte Carlo simulation of networks

Stochastic simulation is a technique by which a computer is used to simulate the actions of a system. Systems can generally be depicted as mathematical or logical models about which specific assumptions are made. If the model is simple enough, it may be solved explicitly. However, most complex, real-world problems cannot be solved explicitly. Thus, simulation is used to evaluate a model numerically. Data are then gathered to estimate the true characteristics of the model. A network must be treated as a system of activities.

The name "Monte Carlo simulation" originated during World War II, when it was used to solve problems related to the development of the atomic bomb. It will be used in this section as a way to study a network or a system. Monte Carlo simulation allows much more flexibility than PERT. You can assign almost any uncertain attribute of a project as a random variable, then simulate the actions of nature, collect data from the model, and predict the characteristics of the system.

The theory behind Monte Carlo simulation is simple. Each activity in the network is assumed to be an independent random variable that behaves according to some known distribution. In simulating a network, the duration of each activity is assumed to follow a probability distribution rather than being a single point estimate. The simulation process will randomly select the duration to add for each activity using random numbers selected from a uniform distribution. A risk profile is the final product. The simple problem in Example 5.7 demonstrates risk versus cost and schedule profiles. For complex systems, these profiles can be used to make informed decisions. Many of the problems contained in Chapter 5 are Monte Carlo simulations of networks.

11.7 Summary

Resources are labor, equipment, material, subcontractors, money, workspace, and anything else needed to perform a project. Resources determine the duration and the cost of a project. They should be viewed as the independent variables of PM. The production rate of driving resources determines the duration of an activity. Similarly, the costs associated with driving resources are fixed regardless of the duration of an activity. If a specified amount of driving resources will complete an activity in a specific period, doubling that resource, from a purely theoretical point of view, should halve the duration, but the cost should remain the same.

Non-driving resources do not determine the duration of an activity. The cost of a non-driving resource may or may not vary with time. For example, the purchase price of installed material should be the same regardless of the installation duration, while the cost of a nighttime security guard will increase as the duration of the activity requiring a guard increases.

Resource allocation and resource leveling are best accomplished using a computer. Any but the most trivial networks are difficult if not impossible to level by hand. You should note, however, that field supervisors are usually adept at establishing crew sizes and managing crews efficiently regardless of any early network gyrations. Therefore, resource leveling is most useful in preliminary planning. Prior to implementation, an experienced project manager should evaluate any computer solution to resource allocation or leveling.

QUESTIONS

11-1. What is the project planning document that should precede the project schedule, and why is it important? How is it useful to a PM?

11-2. As president of You Inc., you are excited about the construction of your new corporate campus. You ask for an update on the project to ensure you are on schedule for the grand opening. The contractor assures you that they are making great progress. Based on the information presented below, should you worry? If yes, why? Assume that most of these tasks are sequential.

Activity	Sched % Complete	Actual Percent Complete	Budget	Cost to Date	ACWP	BCWS	BCWP	CPI	SPI	CV($)	SV
Swing Space	-	-	-	-	-	-	-	-	-	-	-
Selective Demolition	100%	100%	$ 281,100	$ 271,071	$ 271,071	$ 281,100	$ 281,100	1.04	1.00	$ 10,029	$ -
Sitework	100%	100%	$ 1,061,366	$ 1,078,751	$ 1,078,751	$ 1,061,366	$ 1,061,366	0.98	1.00	$ (17,385)	$ -
Building Excavation & Backfill	100%	100%	$ 353,788	$ 359,584	$ 359,584	$ 353,788	$ 353,788	0.98	1.00	$ (5,796)	$ -
Foundation Work	100%	100%	$ 1,002,879	$ 865,962	$ 865,962	$ 1,002,879	$ 1,002,879	1.16	1.00	$ 136,917	$ -
Strucutral Steel	93%	75%	$ 667,300	$ 340,711	$ 340,711	$ 620,589	$ 500,475	1.47	0.81	$ 159,764	$ (120,114)
Concrete Slabs	100%	100%	$ 250,719	$ 216,491	$ 216,491	$ 250,719	$ 250,719	1.16	1.00	$ 34,228	$ -
Masonry	100%	50%	$ 2,016,195	$ 874,093	$ 874,093	$ 2,016,195	$ 1,008,098	1.15	0.50	$ 134,005	$ (1,008,098)
Roofing	100%	85%	$ 766,926	$ 656,447	$ 656,447	$ 766,926	$ 651,887	0.99	0.85	$ (4,560)	$ (115,039)
Mechancial, Electrical, Plumbing	73%	35%	$ 4,012,394	$ 999,156	$ 999,156	$ 2,929,048	$ 1,404,338	1.41	0.48	$ 405,182	$ (1,524,710)
Interior Framing	100%	95%	$ 126,842	$ 115,849	$ 115,849	$ 126,842	$ 120,500	1.04	0.95	$ 4,651	$ (6,342)
Drywall	64%	0%	$ 56,544	$ -	$ -	$ 36,188	$ -	-	0.00	$ -	$ (36,188)
Painting	30%	20%	$ 48,602	$ 9,647	$ 9,647	$ 14,581	$ 9,720	1.01	0.67	$ 73	$ (4,860)
CurtainWall	50%	60%	$ 971,440	$ 701,845	$ 701,845	$ 485,720	$ 582,864	0.83	1.20	$ (118,981)	$ 97,144
Flooring	0%	0%	$ 154,677	$ -	$ -	$ -	$ -	-	-	$ -	$ -
Doors & Hardware	0%	20%	$ 60,748	$ 9,954	$ 9,954	$ -	$ 12,150	1.22	-	$ 2,196	$ 12,150
Punchlist	-	-	-	-	-	-	-	-	-	-	-
Closeout	-	-	-	-	-	-	-	-	-	-	-
Total			**$ 11,831,520**	**$ 6,499,561**	**$ 6,499,561**	**$ 9,945,940**	**$ 7,239,883**	1.11	0.73	**$ 740,322**	**$ (2,706,057)**

PROBLEMS

11-1. A local municipality has issued a request for proposals to conduct a feasibility study to design, implement, and sustain a recycling program in a small town. As PM, you need to develop a bar chart of some type to assign responsibility and maintain a project schedule. You develop these activities, precedence relationships, durations, and responsible individuals.

Activity	Description	Immediate Predecessors	Duration (weeks)	Responsible Organization
A	Review RFP	None	1	Legal and PM
B	Analysis of Community Waste Stream	A	2	Enviro Dept
C	Design of Collection Routes	B	2	Engr Dept
D	Investigate Regulations	A	2	Legal
E	Preliminary Design of Waste Sorting Facility	C	2	Engr Dept
F	Integrate Design	C, D, E	2	PM
G	Review by Legal and Ownership	F	1	All

Develop a Gantt chart to help you manage this project.

11-2. Assume that operations on a work package were expected to cost $1,500 to complete. They were originally scheduled to have been finished today. At this point, however, we have actually expended $1,350, and we estimate that we have completed two-thirds of the work. What are the cost and schedule variances? Calculate CPI, SPI, and ETC and EAC. Interpret the results. (What does it all mean?)

11-3. Find the schedule and cost variances for a project that has an actual cost at month 16 of $540,000, a scheduled cost of $523,000, and an earned value of $535,000. What do the results tell you?

11-4. A sales project at month 5 had an actual cost of $34,000, a planned cost of $42,000, and a value completed of $39,000. Calculate the CPI and SPI, and explain the message they give the program manager.

11-5. A construction project at day 70 has actual costs of $78,000 and a scheduled cost of $84,000. The work package manager estimates value completed of $81,000. Calculate SV and CV, CPI, and SPI. What does this tell you?

11-6. A project was planned to cost $12,000, but the actual cost to date is $10,000 so far, and the value completed is only 70%. Calculate the variances. Should the customer be happy?

11-7. A project to build a new taxiway at Culpepper Airport is 5 days behind at day 65. It had a planned cost of $735,000 for this point in time, but the actual cost is only $550,000. Estimate the variances. What do they say about the health of the project? Reanalyze if the actual cost to date had been $750,000.

11-8. You plan to manufacture printed circuit boards in China using a contractor manufacturer. They have a proven history of delivering their products on time. Because of the cheap labor, you can manufacture the chips using through-hole technology much cheaper than using surface-mount technology and also cut the delivery time from 90 to 60 calendar days. The software will be based on your current operating system. However, some major upgrades will be needed. Your primary software contractor in India will accomplish the changes. By focusing on the major upgrades, you can also reduce the delivery time for the software upgrades to 60 calendar days. Your own in-house staff will be responsible for integration and quality control. Since the software and the circuit boards are contract items, you must carefully monitor the integration phase. After consulting with your team, you develop the following schedule based on a 1 January start date. Assume that all working days are calendar days.

Activity	Description	Immediate Predecessor	Duration (cal days)
1	Software Development	-	60
2	Printed Circuit Board Design and Manufacturing	-	60
3	Develop Performance and Validation Plan	-	45
4	Subsystem Testing	1,2,3	15
5	System Testing	4	15
6	Product Documentation	3	45
7	Validate System Performance	6	15
8	Conduct Reliability Testing	5	30
9	New Product Announcement	-	0
10	Marketing Campaign	9	135
11	Delivery of Voice System 5	8,10	0

Using a Gantt chart developed in Microsoft Project or one of the many free open-source tools, can you meet the 1 July deadline?

11-9. Consider this diagram:

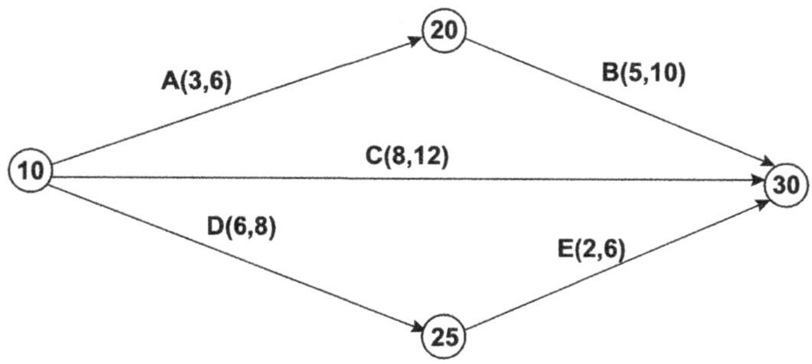

Each activity in the network has a probabilistic duration defined by a uniform distribution (the numbers in parentheses denote the minimum and maximum values for the distribution) as:

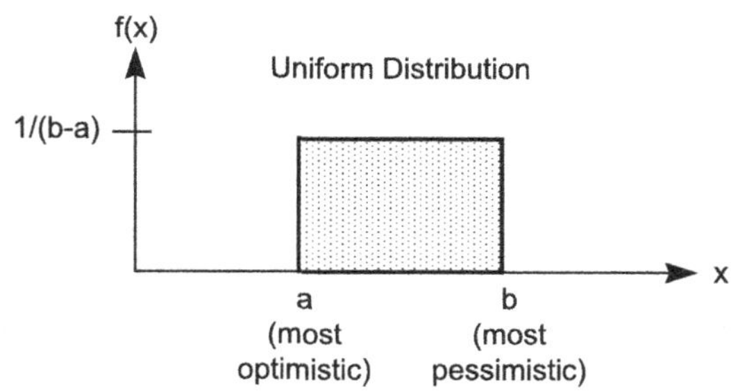

where a and b are shown on the network. Using simulation, determine the percentage of time each path in the critical path and the expected project completion time.

11-10. In November 2016, Ace Builders (your employer) was awarded a contract for the renovation of Buccaneer Stadium to house the New York Neapolitans. The construction must start on 14 January 2017 and must be completed within 15 months. Liquidated damages (penalty clause) of $250,000 per day of delay beyond 31 March 2017 is written into the contract. On the other hand, Ace will receive an early completion bonus of $100,000 per day for each workday the project is completed before 31 March 2017. In the table below is an update of the project as of 10 December 2017. Since this is a highly visible project, you have to brief not only city leadership but also company executives.

Tasks	% Complete	Bid Price	Actual Expenses	Duration (weeks)	Actual
Staging area, mobilization, demolition	100	$ 6,500,000	$ 7,500,000	8	8.5
Excavation—field	100	$ 75,000	$ 75,000	8	8
Subsurface drainage	100	$ 370,000	$ 270,000	14	10
Fill material for field	100	$ 50,000	$ 50,000	4	2
Installation of turf	0	$ 100,000	$0	6	0
Excavation—façade, roof supports, boxes	100	$ 1,000,000	$ 1,000,000	4	5
Pouring concrete footings	100	$ 18,500,000	$ 18,500,000	6	6
Pouring box, roof, and façade supports	100	$ 59,500,000	$ 53,500,000	10	11
Erecting pre-cast concrete façade and boxes	75	$ 75,000,000	$ 57,250,000	16	13
Finishing of boxes and new façade	0	$ 30,000,000	$0	5	0
Painting	10	$ 5,000,000	$ 500,000	4	0.5
Dressing rooms, offices, concession	10	$ 12,000,000	$ 1,200,000	8	1
Prefabricate the retractable roof	90	$ 10,000,000	$ 9,500,000	4	4
Erecting the roof	10	$ 5,250,000	$ 550,000	4	0.5
Scoreboard	75	$ 10,000,000	$ 7,250,000	5	3
New seats	10	$ 32,000,000	$ 3,200,000	3	0.5
Lights and other facilities	5	$ 40,000,000	$ 2,000,000	7	0.1
Wiring, HVAC	15	$ 20,000,000	$ 3,000,000	4	1
Exterior infrastructure improvements	5	$ 19,655,000	$ 982,750	6	0.1
Punch list and demobilization	0	$ 5,000,000	$0	3	0
Total		$350,000,000	$166,327,750		

Based on EV, is this project behind schedule and over cost?

11-11. NASA requires high-performance spaceflight computing capabilities for multiple mission applications associated with both robotic and human space exploration.[2] Traditionally, spacecraft onboard computing systems are single-processor systems based on existing commercial or military computers that are radiation hardened. The systems are implemented and operated at the maximum required mission performance point, where the term "performance point" includes throughput, fault tolerance, and power levels. As NASA considers advanced missions that require both an increase in throughput and wider variations in these operating points, the development of a new processor is needed. This quantum computing–based processor, called "the Quiplet," is needed to provide orders of magnitude improvement in performance and performance-to-power ratio as well as the ability to dynamically set the power-throughput-fault tolerance operating point. The spaceflight computing project will design and deliver radiation-hardened Quiplet devices, associated system software, the software development environment, and the evaluation hardware and software to test the full functionality of the Quiplets.

Work to date by the prime contractor includes a preliminary design phase culminating in a preliminary design review and a detailed design phase culminating in a critical design review; the company has begun the fabrication phase. As the prime contractor, KVF Inc. has subcontracted the preliminary concept support to KF Computing (KFC). KFC, though known to provide superior technical products, has a history of cost overruns and late deliveries. As project manager on the Quiplet program, ensuring the success of this project is key to the future of KVF. This contract is the perfect vehicle to propel KVF to leadership in the military and space applications of quantum computing.

The contract was awarded to KVF, which then subcontracted to KFC in early January with a project start date of 1 February. As a firm-fixed-price contract, there is no room for cost overruns or schedule delays. NASA requires EV updates every quarter. Below is the integrated schedule and quarterly EV report. In preparation for the EV review, you need to be able to answer the following questions:

a. What does this data tell you?
b. Is the project good, bad, healthy, troubled?
c. How would you know?

[2] This is a hypothetical case study developed solely for teaching purposes. The scenario, contractors, and numbers are fictitious.

d. Where would you concentrate your energy?
e. Where would you concentrate KVF's energy?
f. What questions would you ask KFC in front of NASA? One on one without NASA?
g. If senior leadership of KVF doesn't understand EVM, how could you convey tracking projects at the task and project level?
h. What others measures could you use for cost and schedule performance?

			28-Feb-17	28-Feb-17
WBS Description	Total	Actual Expenses (AC)	% Complete	% Schedule
Program Management	$1,677,335	$134,187	8	8
Project Systems Engineering	$1,149,576	$91,966	8	8
Quality Assurance	$527,537	$42,203	8	8
SW Configuration Management	$919,528	$73,562	8	8
Data/Artifacts Management	$667,327	$53,386	8	8
Request for Comments Review	$575,838	$525,000	100	80
Preliminary Concept Engineering Support	$1,767,769	$110,112	5	5
Software Process Support	$143,195	$0	0	0
Engineering Review Board/Configuration Control Board Administrative/Support	$930,107	$0	0	0
System A - System Engineering	$423,109	$0	0	0
System A - Software Engineering	$746,358	$0	0	0
System A - Material	$2,735,870	$0	0	0
System B - System Engineering	$567,344	$0	0	0
System B - Software Engineering	$1,262,545	$0	0	0
System B - Material	$2,469,387	$0	0	0
Integration and Testing Engineering	$1,613,928	$0	0	0
COTS Procurment	$401,959	$0	0	0
COTS HW / SW	$795,201	$0	0	0
Integration and Test	$1,283,328	$0	0	0
Site Survey	$49,556	$0	0	0
Installation Planning & Contract Data Requirements List	$590,767	$0	0	0
Factory Support to Installation	$202,431	$0	0	0
Documentation Support	$329,492	$0	0	0
Training Planning & Contract Data Requirements List	$590,767	$0	0	0
On-Site Installation (Labor)	$853,930	$0	0	0
Project Hand Off	$1,435,825	$0	0	0
Total	**$24,710,009**	**$1,030,416**		

			31-Mar-17	31-Mar-17
WBS Description	Total	Actual Expenses	% Complete	% Schedule
Program Management	$1,677,335	$279,556	17	17
Project Systems Engineering	$1,149,576	$191,596	17	17
Quality Assurance	$527,537	$87,923	17	17
SW Configuration Management	$919,528	$153,255	17	17
Data/Artifacts Management	$667,327	$111,221	17	17
Request for Comments Review	$575,838	$575,838	100	100
Preliminary Concept Engineering Support	$1,767,769	$530,331	20	90
Software Process Support	$143,195	$23,866	20	0
Engineering Review Board/Configuration Control Board Administrative/Support	$930,107	$0	0	0
System A - System Engineering	$423,109	$35,259	10	0
System A - Software Engineering	$746,358	$62,197	10	0
System A - Material	$2,735,870	$227,989	10	0
System B - System Engineering	$567,344	$47,279	8	0
System B - Software Engineering	$1,262,545	$105,212	8	0
System B - Material	$2,469,387	$205,782	8	0
Integration and Testing Engineering	$1,613,928	$134,494	5	0
COTS Procurment	$401,959	$33,497	25	0
COTS HW / SW	$795,201	$66,267	25	0
Integration and Test	$1,283,328	$106,944	5	0
Site Survey	$49,556	$5,000	20	0
Installation Planning & Contract Data Requirements List	$590,767	$0	0	0
Factory Support to Installation	$202,431	$0	0	0
Documentation Support	$329,492	$0	0	0
Training Planning & Contract Data Requirements List	$590,767	$0	0	0
On-Site Installation (Labor)	$853,930	$0	0	0
Project Hand Off	$1,435,825	$0	0	0
Total	**$24,710,009**	$2,983,504		

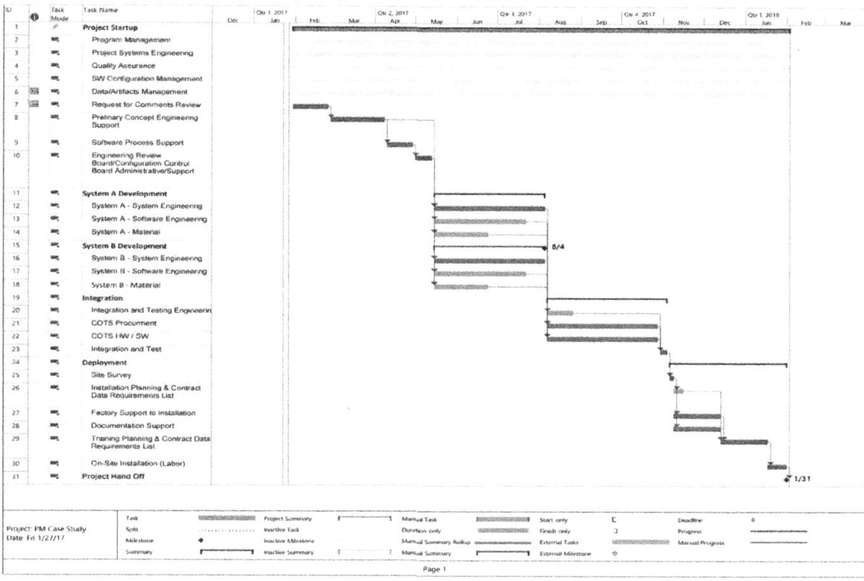

References

Carlos Consulting Group, n.d., "Earned Value Management," accessed 30 January 2017 at http://www.carlosconsulting.com/downloads/EVM_Software_Project.pdf.

Defense Acquisition University, October 2001, "Scheduling Guide for Program Managers," Defense Systems Management College Press, Fort Belvoir, VA.

Defense Acquisition University, June 2003, "U.S. Department of Defense Extension: A Guide to the Project Management Body of Knowledge," PMBOK Guide, First Edition, Version 1.0.

Griffis, Fletcher H., and Farr, John V., 1999, *Construction Planning for Engineers*, McGraw-Hill, New York City, New York.

NASA, January 2010, "Work Breakdown Structure Handbook," accessed 1 August 2018 at https://ntrs.nasa.gov/archive/nasa/casi.ntrs.nasa.gov/20110012671.pdf.

Project Management Institute, 2013, *A Guide to the Project Management Body of Knowledge (PMBOK Guide)*, 5th Edition, Project Management Institute, Newtown Square, PA.

chapter twelve

Use of cost metrics and ratios

12.1 *Introduction*

Cost ratios, indices, and metrics are commonly used in the financial sector to convey the health of a company from both a strategic and operational perspective. The ratios capture both the performance and activity of a company. Engineering projects often use simple metrics as a way to understand the worth or return for a project. Some common metrics are shown in Figure 12.1 (many of these were taken from Marr, 2012). These ratios can play an important role in portfolio analysis, marketing a project, conveying non-economic value, and so on. However, they can also be misleading, especially for public-sector projects.

We are only interested in the project-level metrics shown in this figure. Company financial metrics and ratios are beyond the scope of this text.

12.2 *Benefit-cost ratio*

The use of cost-benefit analysis (CBA) as a simple ratio was a result of the Federal Navigation Act of 1936—even though the concept had been around since the early 1800s (see Melese et al., 2015, for a discussion of the history of cost-benefit analysis). This act required that the US Army Corps of Engineers only construct projects for the improvement of the nation's waterway system when the total benefits of a project exceeded the costs of that project.

In its simplest form, benefit-cost ratio (BCR) is a ratio that is used to define the value of a project divided by the money that will be spent in completing the project at the same point in time. However, accounting for life cycle costs (LCCs) requires the choice of a life cycle, income and expenses, and a minimum acceptable rate of return (MARR). Since some savings may be interpreted as a negative cost to be deducted from the denominator or as a positive benefit to be added to the numerator of the ratio, the BCR is not an absolute numerical measure. As shown in Example 12.1, if NPV of benefits minus costs exceeds 0 for a given MARR,

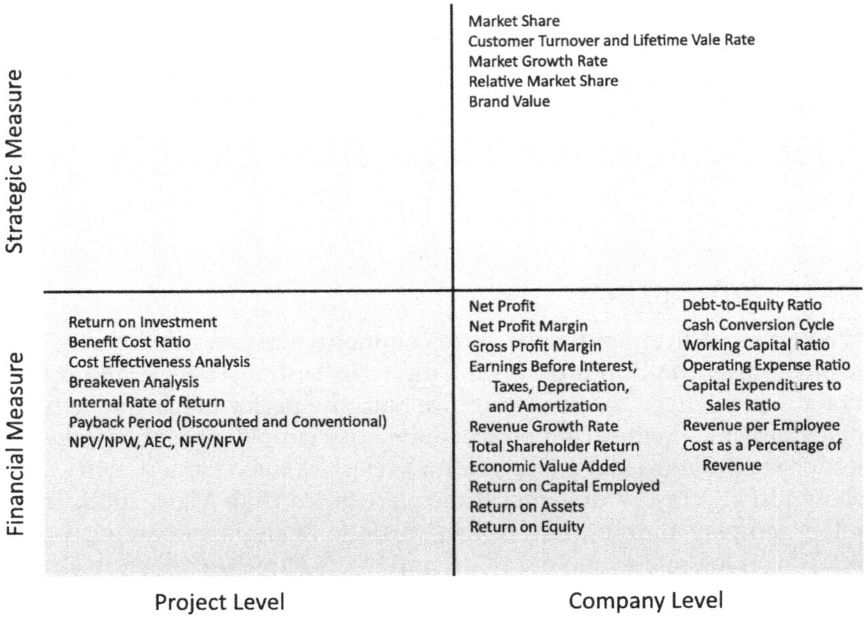

Figure 12.1 Company- and project-level metrics and ratios

then the project is profitable. The BCR can also be calculated using a simple ratio

$$BCR_x = \frac{BPV_x}{CPV_x} \tag{12.1}$$

EXAMPLE 12.1

Consider an investment in environmental remediation for manufacturing a hazardous product. The waste stream is just barely meeting state and EPA standards. The state uses a MARR of 2% for all projects. The benefits and costs are shown in the table below:

Year	0	1	2	3	4	5
Benefits	0	$200,000	$300,000	$400,000	$400,000	$400,000
Costs	$1,000,000	$100,000	$110,000	$121,000	$133,100	$146,400
B-C	-$1,000,000	$100,000	$190,000	$279,000	$266,900	$253,600
Discount Factor	1.02^0	1.02^1	1.02^2	1.02^3	1.02^4	1.02^5
Discounted Annual Cash Flows	-$1,000,000	$98,039	$182,622	$262,908	$246,574	$229,693

The NPV for this project is $19,836. Thus, this project is a go at a discount rate of 2%.

where BPV_x is the present value of the benefits of the project and CPV_x is the present value of the costs of the project. The BCR should be used carefully when comparing competing projects because it does not capture the magnitude of the benefits and costs. It is simply a ratio.

12.3 Return on investment

Return on investment (ROI) is probably the most common performance measure because of its simplicity. It is typically defined as

$$ROI = \frac{\text{Gain from Investment} - \text{Cost of Investment}}{\text{Cost of Investment}} \tag{12.2}$$

The formula was developed by a DuPont explosives salesperson in 1912. Since ROI is usually expressed as a percentage, it is a popular and easy way to convey a metric for investments. A high ROI means the gains from the investment compare favorably to the investment cost. However, for many types of projects, especially infrastructure, ROI is a poor way to evaluate projects as expressed above. To incorporate total ownership costs (TOC) considerations, we should use a modified ROI calculation:

$$ROI = \frac{\left(NPV_{Salvage} + NPV_{Income/Cost\ Avoidance}\right) - \left(NPV_{Design} + NPV_{Construction\ Cost} + NPV_{O\&M\ Costs} + NPV_{Retirement\ Cost}\right)}{NPV_{Design} + NPV_{Construction\ Cost} + NPV_{O\&M\ Costs} + NPV_{Retirement\ Cost}} \tag{12.3}$$

ROI can vary from –1 to a large number. A ROI of –1 means that the project simply does not provide a quantifiable financial gain from the investment. That does not mean it is not a viable project but simply that there is no cost avoidance or intrinsic financial value to the project. Example 12.2 demonstrates how to calculate both TOC and ROI for a simple problem.

12.4 Cost-benefit analysis

In today's resource-constrained and regulated environment, with easy access to information, all professionals must exercise wise stewardship of every dollar. For every proposed idea or program, portfolio decisions must be made and justified based on trade-offs. Therefore, it is important to provide an accurate and complete picture of both the cost estimates and the benefits to be derived.

EXAMPLE 12.2

Some wastewater treatment plants use ultraviolet (UV) light combined with granulated activated carbon (GAC) to reduce the concentrations of hazardous materials in wastewater. These systems typically involve inserting UV and GAC treatments into the treatment process, similar to the process shown in the figure below.

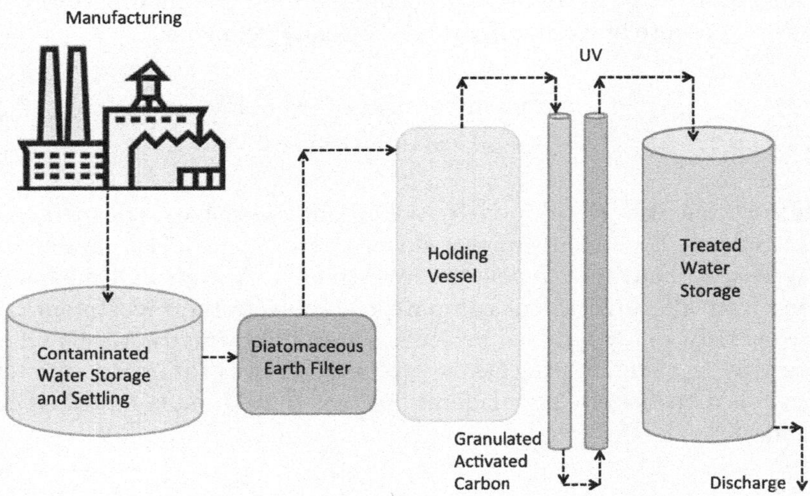

A proposed plant upgrade involves the following financials:

Design: $50,000—Design costs could be significant because of the use of new technology (UV treatment) for this type of application.

Construction: $100,000—Actual construction costs should be easy to identify.

Income/cost avoidance: This project reduces treatment cost and provides a cost avoidance of $10,000 per year in fines.

Operations and maintenance: Operations: $9,000 per year—UV treatment requires significant electricity, GAC must be replenished, UV bulbs replaced, and so on.

Maintenance: $5,000 per year—Water storage and settling must be cleaned out.

Retirement/salvage: $100,000—Given the environmental risks associated with manufacturing, cleanup is often a significant hidden cost.

The town board has asked that you provide the ROI for this upgrade.

Solution

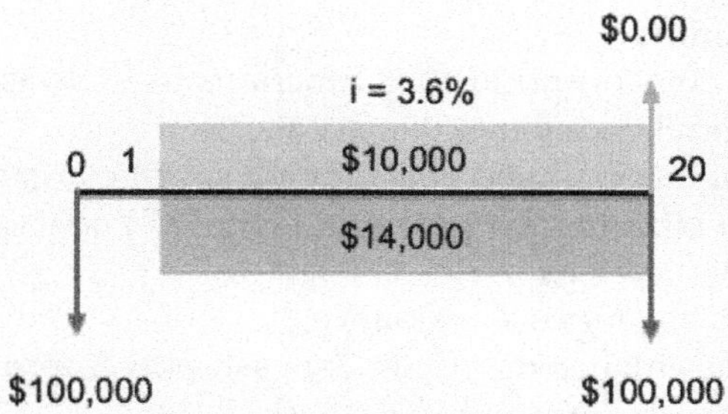

To determine the NPV for the modifications to the wastewater treatment plant, the design and construction cost, operating and maintenance cost, and the retirement cost all have to be determined. For the wastewater treatment plant, the design and construction cost is $100,000, the annual income and cost avoidance is $10,000, the annual operating and maintenance cost is $14,000, and the retirement cost is $100,000. Before calculating the EAC, NPV must be established. The calculation we use for NPV is the design and construction cost of $100,000 plus the annual maintenance cost of $14,000, discounted over 20 years at a rate of 3.6%. We treat the operating cost as an annuity value since it is the same every year for 20 years. Finally, we add the retirement cost of the plant at a cost of $100,000, discounted over 20 years at 3.6%.

Calculation

$$\begin{aligned}\text{NPV} &= -100{,}000 + (10{,}000*14.0847) + (-14{,}000*14.0847) \\ &\quad - 100{,}000(0.4930)\\ \text{NPV} &= -\$205{,}639\\ \text{EAC} &= -\$205{,}639*0.0710\\ \text{EAC} &= -\$14{,}600.40\end{aligned}$$

$$\text{ROI} = \frac{(10{,}000*14.0847) - (100{,}000) - (14{,}000*14.0847) - (100{,}000*.4930)}{(100{,}000) + (14{,}000*14.0847) + (100{,}000*.4930)}$$

$$\text{ROI} = -0.5934$$

Once the NPV is found using this calculation, we use the A/P function to find the EAC for the project.

A CBA provides decision makers with facts, data, and analysis required to make an informed decision. Specifically, a CBA (US Army, 2012)

- Is a decision support tool that documents the predicted effect of actions under consideration to solve a problem or take advantage of an opportunity,
- Is a structured proposal that functions as a decision package for organizational decision makers,
- Defines a solution aimed at achieving specific organizational objectives by quantifying the potential financial impacts and other business benefits, such as
  - Savings and/or cost avoidance,
  - Revenue enhancements and/or cash-flow improvements,
  - Performance improvements, and
  - Reduction or elimination of a capability gap.
- Considers all benefits including non-financial or difficult-to-quantify benefits of a specific course of action or alternative,
- Analyzes needs and problems, their proposed alternative solutions, and risks to lead the analyst to a recommended choice before a significant amount of funds are invested by the stakeholders,
- Must be tailored to fit the problem, because finding the optimal solution is the focus of the CBA,
- Supports the decision-making process but will not make a final decision; that will be the responsibility of the decision maker or leadership, and
- Is not a substitute for sound judgment, management, or control.

These eight steps are shown graphically in Figure 12.2.

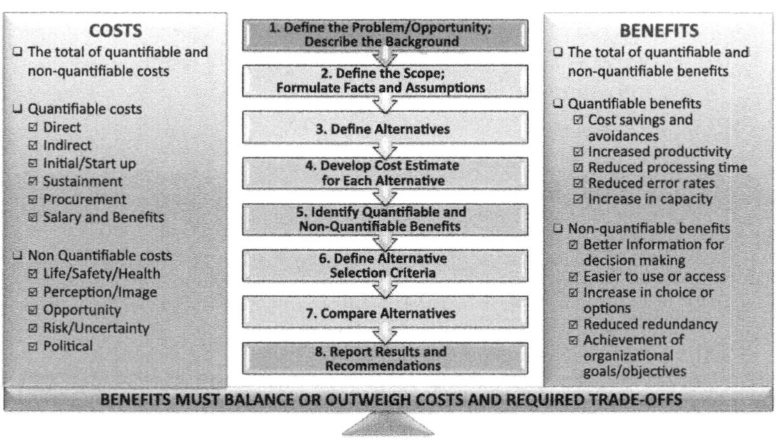

Figure 12.2 The eight-step CBA process (US Army, 2012)

Identifying quantifiable (both economic and non-economic) and difficult-to-quantify benefits (Step 5) is by far the biggest challenge for conducting CBA analysis. Basic engineering economics and decision analysis tools exist for conducting the other steps in the CBA process, and these are well understood.

12.5 *Multi-objective decision analysis approach for quantifying benefits*

Economics or positive ROI are often used as the primary consideration for strategic and project investments. However, an important component of good decisions is to capture the total value of the proposition. This not only provides a means to capture the total value of competing alternatives but also allows for articulating and measuring the value proposition. Techniques such as multi-objective decision analysis (MODA) can be used to quantify value. Note that cost is often not included in a value model and should be used to compare the value results for various alternatives. MODA ranks alternatives to assist in the selection of a preferred alternative. Specifically, it is useful in enhancing decision making for allocation of resources and solidifying support for a particular portfolio of projects. This methodology is well suited to portfolio prioritization and optimization. Using MODA can help to identify an appropriate mix of projects to maximize overall value and conduct tradeoff analysis.

The MODA process begins with the development of a value hierarchy. The three core functions and subfunctions are further broken down into objectives. The objectives identified can be further broken down into supporting evaluation criteria (value measures) in the value hierarchy model.

MODA is dependent on stakeholder analysis and requires global and local weights in the value hierarchy model. Appropriate evaluation criteria and local weights will be determined based on current information.

Multi-objective decision or value analysis (Kirkwood, 1997) uses an overall value function that combines the multiple evaluation measures into a single measure of the overall value of each evaluation alternative or portfolio of projects. Thus, different mixes of projects in a portfolio may be compared to determine the appropriate mix for maximizing value. Multi-objective value analysis is useful for structuring the judgments used in assessing the value of projects that comprise a portfolio in an organization with multiple and conflicting objectives. Multi-objective value analysis methods are based on structured objectives, evaluation measures, value functions, and weights.

A multiple criteria value function based on weights and scores is used to rank alternatives. An additive value function is used for this research since it is common (Keeney, 1992). The additive multi-criteria function $V(a_i)$ can be expressed as

$$\text{Project Value} = \sum_{n=1}^{N} \omega_i V(x_i) \tag{12.4}$$

where

ω_i = the global weight of the i_{th} evaluation measure
x_i = the score of a project on evaluation measure i
$V(x_i)$ = the value function that returns a value for each evaluation measure score
N = the number of evaluation measures

where

$$\sum_{i=1}^{N} \omega_i = 1 \tag{12.5}$$

When weights have been determined for the current situation, the model can be used to find the right mix of projects to maximize value and support a combination of core outcomes within a fixed budget portfolio. The mix of projects with the highest overall score adds the most value. Projects can then be viewed as a function of cost or some other variable to make logical and defensible decisions.

When using MODA, a structured approach must be taken to develop the weights, objectives, and functions. Often this involves defining objectives and functions based on the experience of the authors, literature, focus groups, surveys, and input from subject-matter experts. Ideally, stakeholders should be involved at all levels. In general, there is often very little disagreement on the objectives and functions and how to quantify the functions. However, stakeholder interests are reflected when assigning the weights. For example, one group of stakeholders might place a higher value on maximizing ROI. Another group of stakeholders, such as the facility contractor, would place a greater weight on maximizing strategic investment. Stakeholder buy-in is critical with all parties agreeing to the framework. Sensitivity analysis can play a key role here to show how varying the weights over different ranges can have little or major impact on the objective function. Example 12.3 demonstrates the use of MODA for a simple problem.

EXAMPLE 12.3

Consider the value hierarchy shown in Figures 12.3 and 12.4. This value hierarchy from the overall objective down to the evaluation measures for a data set from the National Reconnaissance Office (NRO) of research and development projects (see Farr and Parnell, 2000). Figures 12.3 and 12.4 contain one of the value measures. Figure 12.5 contains two of the scoring functions used for

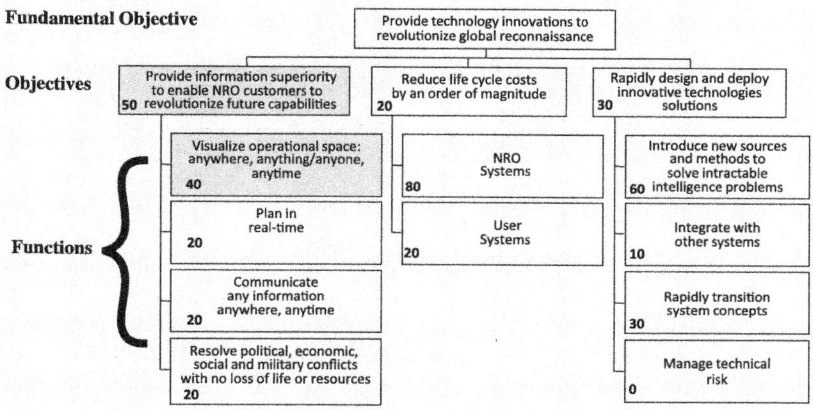

Figure 12.3 Value analysis high-level model for the NRO data set

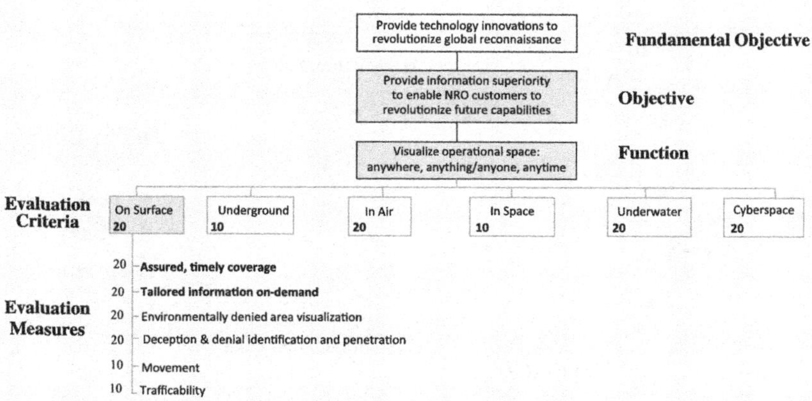

Figure 12.4 Evaluation criteria and evaluation measures for a function for the NRO data set

(continued)

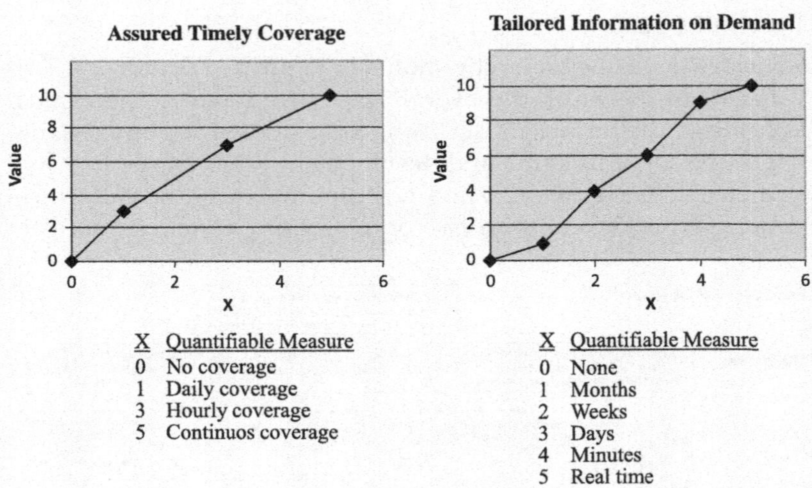

Figure 12.5 Scoring functions for two evaluation measures

evaluating project value. Figure 12.6 is a plot of the cumulative value when the projects are sorted by budget. The lowest-budget projects are put in the portfolio before the higher-budget projects.

This is an example of how a CBA study can be used to look across a portfolio of projects.

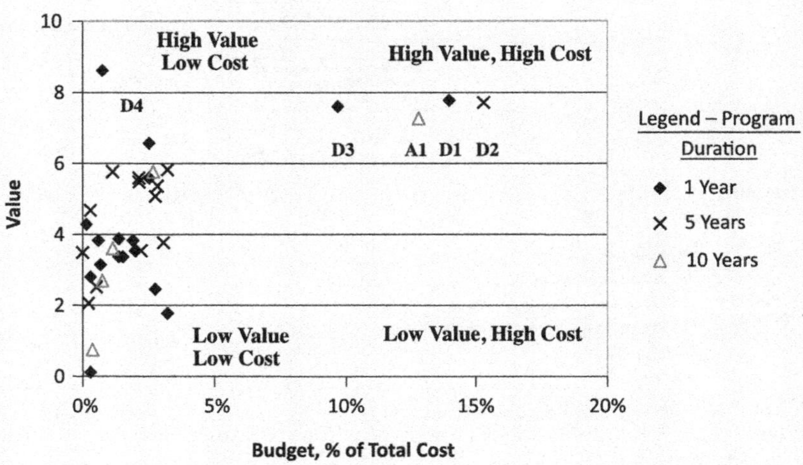

Figure 12.6 Plot of value versus percentage of the total portfolio cost

12.6 Breakeven analysis

A basic consideration cited in any financial decision is when the project will break even. In other words, at what point in the life of the project will the initial investment be recouped? In general, we prefer for a project to break even earlier in the life of the project versus later. Many factors contribute to this preference, but perhaps the most important is risk. Risk increases over time because of unforeseen events and because the values we use for our calculations are more difficult to accurately predict further in the future. In engineering economics, the two most common forms of breakeven analysis are breakeven quantity analysis and payback period analysis.

12.6.1 Breakeven quantity

For certain projects, it can be valuable for decision makers to understand the sales quantity required in order to recoup the initial investment for the project. This quantity could then be compared with demand projections. While these calculations can become very complex when the volatility of input prices and interest rates (just to name a few) are included in the analysis, below is a very simple example of this calculation. In order to determine the breakeven quantity, we will use the following equation:

$$\text{Breakeven Quantity} = \text{Initial Investment}/\text{Marginal Profit} \tag{12.6}$$

For example, a project requires an initial investment of $1 million, and each unit sold produces a marginal profit of $100. Therefore, the breakeven quantity is $1,000,000/$100 = 10,000. This concept is illustrated graphically in Figure 12.7.

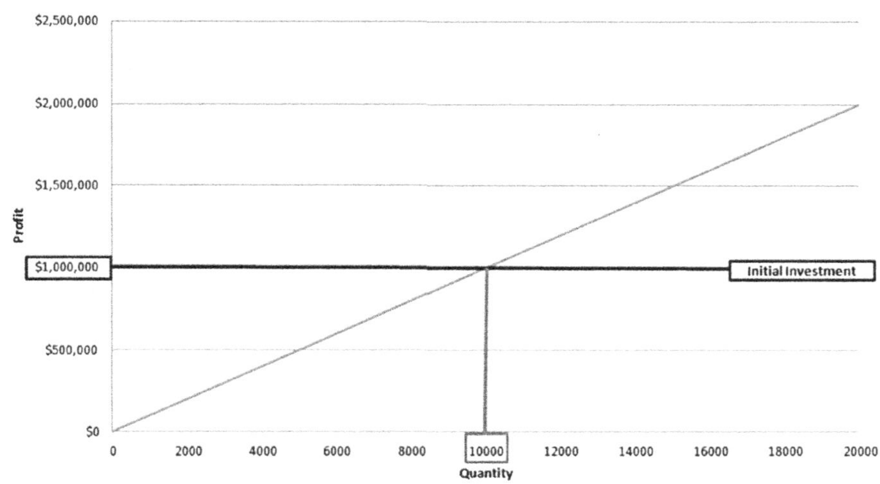

Figure 12.7 Illustration of breakeven analysis

As previously mentioned, this problem makes numerous simplifying assumptions, but it demonstrates the breakeven concept. Of note, more involved breakeven analysis would incorporate the time value of money (TVM) by accounting for the time period when cash flows resulting from sales actually occur.

12.6.2 *Conventional payback period*

Conventional payback period (CPP) is the simplest of the project selection methods and can easily be conducted on the back of a napkin. This is the greatest strength of this method. It is simple to understand and to calculate.

A key assumption when calculating both conventional and discounted payback period is that end of time period cash flows is actually occurring at a continuous rate throughout the period. This can be confusing because it is counter to the assumptions we make when evaluating cash flow series (i.e., that all cash flows occurring during a time period take place at the end of the time period). For example, if a firm received \$365 in revenue during year 1, we assume that the revenue was received at a constant rate of \$1 per day. The following variables will be used when determining the payback period for a project.

The equation for CPP equation is

$$\text{Payback Period} = \left(\frac{-\left(\text{Cumulative CF at the start of the Payback Period Interval}\right)}{\text{Net CF during the Payback Period Interval}} \right) + \left(\text{Initial Time Period of the Payback Period Interval}\right) \tag{12.7}$$

Example 12.4 illustrates this concept in a manner that is much simpler to understand than Equation 12.7. Consider a project with the following yearly cash flows (cash flow column). As you would expect, the project requires an initial investment but then earns a return on that investment over the life of the project.

A key weakness of the CPP method is that it violates the TVM by summing cash flows across time periods without adjusting value. For example, in this method, a \$35,000 cash flow in year 3 has the same value as a \$35,000 cash flow in year 6. This weakness is addressed in the next method discussed, the discounted payback period (DPP) method. Example 12.4 demonstrates the CPP calculation.

12.6.3 *Discounted payback period*

Calculating the discounted payback period is a straightforward process that only requires one additional step beyond the CPP. As the name implies, all the cash flows must be discounted when conducting

EXAMPLE 12.4

Determine the CPP for a project that has the following cash flows:

End of Year	Cash Flow
0	–$70,000
1	$20,000
2	$19,000
3	$18,000
4	$17,000
5	$16,000
6	$15,000

$$-70k + 20k + 19k + 18k = -13k \text{ at 3 years}$$

$$-13k + 17k = 4k \Rightarrow \text{Payback period} = 3 + 13/17 = 3.76 \text{ years}$$

this analysis. Therefore, we adjust our variables and equation to reflect discounting. This is shown in Example 12.5.

Comparing the DPP to the CPP, we see that the DPP takes longer because of the TVM. This is typical given the traditional cash flow

EXAMPLE 12.5

Determine the DPP for the cash flows in Example 12.4.

If the firm evaluates projects at an effective annual interest rate of 10%, we determine the DPP as shown below.

$$-70k + 20k/1.1 = -51.8k \text{ at } t = 1;$$

$$-51.8k + 19k/1.1\text{^}2 = -43.7k \text{ at } t = 2;$$

$$-43.7k(1.1) + 18k = -30.07k \text{ at } t = 3;$$

$$-30.07k(1.1) + 17k = -16.08k \text{ at } t = 4;$$

$$-16.08k(1.1) + 16k = -1.685k \text{ at } t = 5;$$

$$-1.685k(1.1) + 15k = 13.15k \text{ at } t = 6 \Rightarrow$$

$$\text{Payback} = 5 + 1.685/15 = 5.11 \text{ Years}$$

(continued)

The best way to make this calculation is to use a table:

End of Year	Cash Flow	Discounted Amount	Sum
0	–$70,000	–$70,000	($70,000)
1	$20,000	$18,181	($51,818)
2	$19,000	$15,702	($36,115)
3	$18,000	$13,523	($22,592)
4	$17,000	$11,611	($10,980)
5	$16,000	$ 9,934	($1,046)
6	$15,000	$ 8,467	$7,421

The discounted amounts are calculated as –$70,000/1.1^0+ $20,000/1.1^+$19,000.....$15,000/1.1^6

Thus, the discounted payback period is 5 + $1,046/$8,467 = 5.12 Years

structure of a project. The negative cash flows occur at the beginning of the project and are thus unchanged when discounted, unlike the positive cash flows that occur later and see their value reduced due to discounting.

12.6.4 *Breakeven charts*

A breakeven chart is when fixed costs, variable costs, and revenue are plotted as a function of sales or capacity. The chart is generated with some simplifying assumptions, such as (see Walsh, 2012):

- Fixed costs are totally fixed for all levels of production,
- The variable costs do not change as a function of production levels, and
- The revenue is the same for all levels of sales/production.

Example 12.6 illustrates a simple breakeven chart problem.

12.7 *Internal rate of return*

The internal rate of return (IRR) is defined as the interest rate that sets the NPV of the stream of future cash flows equal to the initial investment. The IRR must be used with caution because it fails to capture the magnitude of the investment. If a project consists of a single cost at the beginning and generates a stream of net benefits afterward, a unique value of IRR indicates the return over cost per period from funds that remain invested in the project. But, the IRR does not consider the external reinvestment opportunities related to the timing and intensity of the outlays

EXAMPLE 12.6

Consider the costs of operating a water treatment plant. The fixed cost to operate the facility is \$5 million annually. The variable cost to treat the water is \$2/1,000 gallons. As you are developing the budget for next year, you want to develop a simple graphic for revenue and costs to show to the city planner. Currently, you charge \$4/1,000 gallons to your customers, and last year the facility processed about 3 million gallons. You decide to use a simple breakeven chart to illustrate the cost and revenue structure for the facility.

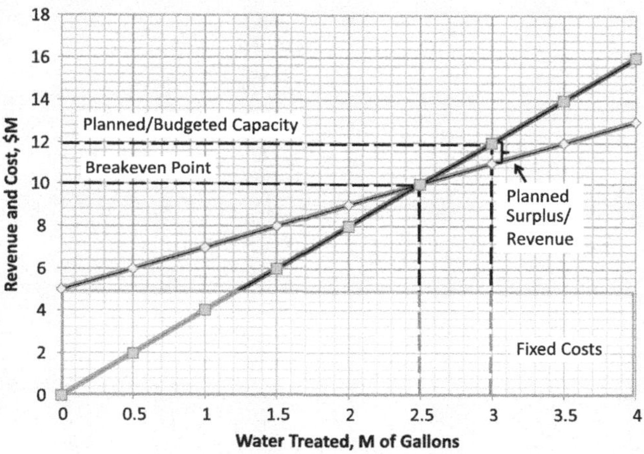

and returns at the intermediate points over the life cycle. There may exist multiple values of IRR for complex cash flows. (The IRR for complex cash flows must be calculated via software such as Excel.)

There are three methods used to solve for IRR by hand. The first step is to determine if the cash flow series you are analyzing is simple or complex. Simple or complex depends on the number of sign changes in a cash flow series. More than one sign change means that the cash flow series is complex. Table 12.1 provides an illustration.

Table 12.1 Simple versus complex cash flows

Time Period							
0	1	2	3	4	5	# Changes	Type
-	+	+	+	+	+	1	Simple
-	+	-	+	+	-	4	Complex
-	-	+	+	0	+	1	Simple
-	+	+	-	0	+	3	Complex

12.7.1 The compounding equation

When a cash flow series only contains one initial investment and one future payment (one negative cash flow and one positive cash flow), the compounding equation can be applied to determine the IRR as shown in Example 12.7.

12.7.2 The quadratic equation

When a cash flow series contains one initial investment and two future payments (one negative cash flow and two positive cash flows), the quadratic formula can be used.

n	Cash Flow
0	–$2,000
1	$1,100
2	$1,400

In order to use the quadratic formula, construct the formula for the net future worth (NFW) for the cash flow series.

$$\text{NFW} = -2000(1+i)^2 + 1100(1+i)^1 + 1400$$

Set NFW equal to zero

$$-2000(1+i)^2 + 1100(1+i)^1 + 1400 = 0$$

The NFW is in the same format as the quadratic formula ($ax^2 + bx + c = 0$) and thus, we can apply

$$x = \frac{-b \pm \sqrt{b^2 - 4(a)(c)}}{2(a)} \tag{12.8}$$

EXAMPLE 12.7

Determine the rate of return for the following cash flows:

n	Cash Flow
0	–$1,000
1	
2	$1,350

Using $F = P(1 + i)^n$ or $\$1,350 = \$1,000(1 + i)^2$

i = IRR = 16.19%

From the NFW equation: a = –2,000, b = 1,100, c = 1,400

$$1+i=\frac{-1100\pm\sqrt{1100^2-4(-2000)(1400)}}{2(-2000)}$$

$$i=15.57\%$$

i = IRR = 15.57%

This method has limited utility because it can only be applied to three cash flows.

12.7.3 *Trial-and-error method*

The final method for calculating IRR by hand is the trial-and-error method. Unlike the previous two methods, which are used for very special cases, the trial-and-error method can be used for any simple cash flow, as shown in Example 12.8.

EXAMPLE 12.8

Consider the following cash flows:

n	Cash Flow
0	–$2,000
1	$1,100
2	$1,400

Construct an equation for the NPV of the cash flow series.

$$\text{NPV}=1{,}026(1+i)^{-3}+1{,}400(1+i)^{-2}+1{,}100(1+i)^{-1}+(-2{,}000)$$

Now we begin the iterative process of guessing a value for the interest rate and then checking the resulting NPV value. We know that the IRR equals the interest at which the NPV equals zero. Our first guess should be the MARR because we anticipate that the IRR will be in the range of the MARR, and just knowing whether the IRR is greater or less than the MARR is valuable. Our MARR in this example is 20%. Therefore,

$$1{,}026(1+.2)^{-3}+1{,}400(1+.2)^{-2}+1{,}100(1+.2)^{-1}+(-2{,}000)=\$482.63$$

(continued)

A positive NPV means that the interest rate we guessed is too low. Next, we guess 40%.

$$1{,}026(1+.4)^{-3}+1{,}400(1+.4)^{-2}+1{,}100(1+.4)^{-1}+(-2{,}000)=-\$126.09$$

A negative NPV means that the interest rate we guessed is too high. We also know that 20%>IRR<40%. Next, we guess 35%

$$1{,}026(1+.35)^{-3}+1{,}400(1+.35)^{-2}+1{,}100(1+.35)^{-1}+(-2{,}000)=0$$

Therefore, IRR = 35%. The Goal Seek command in Excel allows you to simplify this process and obtain an accurate value.

12.7.4 *IRR in Excel*

The =IRR function in Excel is an extremely powerful tool for calculating IRR, particularly given the tedious nature of the trial-and-error method. In order for the =IRR function to return a value, there must be at least one negative value in the string of cash flows it is asked to evaluate. Also, the function provides an entry for a guess for the IRR. The MARR should be entered as the guess. However, most often the guess is not required because the default value for the guess in Excel is 10%. That guess is normally close enough to the actual IRR for Excel to return the correct value. If Excel is unable to determine the value, "#NUM" will be displayed. Example 12.9 shows how the IRR tool can be used in Excel.

Note that when comparing alternative projects using IRR, it does not capture the magnitude of the cash flows.

EXAMPLE 12.9

Given the tax flows below, determine the IRR.

	A	B
1	n	Cash Flow
2	0	($2,000.00)
3	1	$1,100.00
4	2	$1,400.00
5	3	$1,025.00
6	IRR	35%

IRR is determined by using the Excel function = IRR(B2:B5)

12.7.5 Incremental IRR

As noted previously, comparing projects using IRR is problematic because the method does not account for the magnitude of the cash flows. For example, consider the following two cash flow series:

n	Project A	Project B
0	–$5	–$10,000
1	$3	$5,000
2	$3	$5,000
3	$3	$5,000
IRR	36%	23%
NPV	$2.46	$2,434.26

Based solely on IRR, a firm would pick Project A. However, based on NPV using a 10% discount rate, Project B is the clear choice.

The incremental IRR method accounts for the magnitude of the cash flows and thus provides additional information for the project selection process. In order to conduct incremental IRR analysis, use the following steps:

Step 1: Compute the IRRs for all alternatives, and screen out any alternatives with IRR < MARR.

Step 2: Subtract the cash flow of any lower first-cost alternative (A) from any higher first-cost alternative (B) to find the incremental investment.

Step 3: Calculate the IRR of the incremental investment.

Decision rules:

$IRR_{B\text{-}A} > MARR$,	Select B, discard A
$IRR_{B\text{-}A} = MARR$,	Select either A or B
$IRR_{B\text{-}A} < MARR$,	Select A, discard B

Step 4: As long as there is more than one alternative remaining, return to Step 2.

Returning to our example and assuming the MARR for our firm equals 20%:

Step 1: Project A IRR = 36% > MARR, Project B IRR = 23% > MARR

Step 2: Project A is alternative A, Project B is alternative B.

n	Project A	Project B	B-A
0	-$5	-$10,000	-$9,995
1	$3	$5,000	$4,997
2	$3	$5,000	$4,997
3	$3	$5,000	$4,997

Step 3: IRR is calculated using Excel.

n	Project A	Project B	B-A
0	-$5	-$10,000	-$9,995
1	$3	$5,000	$4,997
2	$3	$5,000	$4,997
3	$3	$5,000	$4,997
IRR	36%	23%	23%

Step 4: B-A IRR = 23%, $IRR_{B\text{-}A}$ > MARR, Select B (Project B), discard A.

12.8 Summary

Numerous metrics and ratios are used in the financial world to quantify the health of a project or company. Metrics are used at the project, program, and enterprise level to track, measure, and report. For example, Walsh (2012) lists over 100 ratios for managers to understand financial and operational performance. Unfortunately, the kind of LCC analysis that engineers conduct does not translate to simple ratios and metrics. Detailed LCC models are needed to make procurement and design decisions, not simple ratios and metrics.

QUESTION

12-1. Is there an advantage to using NPW over NFW?

PROBLEMS

12-1. Your manager asked you to evaluate a financial proposition where the firm wants to select one of two possible projects. The cash flows for the projects appear in the table below, and the company has a MARR of 15%. Using NPV, AEW, and IRR, which project would you recommend? Do not calculate an incremental IRR.

Time	Project A	Project B
0	–$100	–$200
1		$100
2	$300	$200

12-2. Given the following net cash flows, use IRR in conjunction with incremental analysis to identify the best project. The MARR is 8%, compounded annually.

Year (n)	Project A	Project B	Project C	Project D
0	–$105,300	–$65,000	–$37,000	–$79,000
1	$0	$16,500	$31,000	$20,000
2	$0	$29,000	$32,270	$31,000
3	$133,020	$31,500		$43,000
IRR	?	8.11%	?	8.14%

a. Manually solve for the IRR for Project A and Project C. Check your answers using the IRR function in Excel.
b. Write down the equation you would use to solve for the IRR for Project B.
c. Based on the IRRs, which projects would you continue to consider?

12-3. Your manager asked you to evaluate a financial proposition where the firm wants to select one of three possible projects. The cash flows for the projects appear in the table below, and the company has a MARR of 10%. Using NPV and IRR, which project would you recommend? Do not calculate an incremental IRR.

Time	Project A	Project B	Project C
0	–100	–150	–300
1	50		100
2	100	190	200

12-4. The town of Highland Falls is evaluating four mutually exclusive beautification projects. Given the following net cash flows, use IRR in conjunction with incremental analysis to identify the best project. The MARR is 7%, compounded annually.

Year (n)	Project A	Project B	Project C	Project D
0	$–65,000	$–37,000	$–83,000	$–79,000
1	$ 16,500	$–16,500	$0	$ 20,000
2	$ 29,000	$ 31,000	$0	$ 31,000
3	$ 30,000	$ 31,000	$102,000	$ 41,000
IRR	7.18%	6.96%		7.15%

a. Solve for the IRR for Project C.
b. Write down the equation you would use to solve for the IRR for Project B. Do not solve.
c. Based on the IRRs, which projects would you continue to consider and why?
d. Use incremental analysis to identify the best project. Explain your decision, and show your decision criteria.
e. What is your recommendation and why?

12-5. Bacon Burritos (BBs) hired you to help them choose between four mutually exclusive projects, each with a 4-year life span. The cash flows for each project are shown in the table below. BB's MARR is 25%, and their burritos are delicious.

Year	Project A	Project B	Project C	Project D
0	–$ 35,000	–$20,000	–$100,000	–$45,000
1	$ 5,000	–$ 2,000	$ 40,000	$20,000
2	$ 30,000	$28,000	–$ 15,000	$32,000
3	$ 15,000	$15,000	$ 40,000	$20,000
4	$ 30,000	$35,000	$ 55,000	$40,000
IRR	35.44%	56.23%	6.58%	44.46%

a. Classify each project cash flow schedule as "simple" or "non-simple."
b. Based on the given IRRs, which project(s) would you continue to consider, and why?
c. Write the equation you would use to solve for the IRR for Project A.
d. Set up the incremental IRR table.
e. If the $IRR_{incr} = 89.27\%$ for Projects D-A and the $IRR_{incr} = 24.26\%$ for Projects D-B, what is the best project?

12-6. You have decided to open a microbrewery selling university-themed beers as one of your product lines. You need to conduct a breakeven analysis of sales and price. You know your fixed costs, but based on your marketing plan you are not sure of pricing and sales. Your company MARR is 5%, and fixed costs include an initial capital investment of $250,000 with fixed annual expenses of $100,000. Use a 5-year planning horizon, and assume that these expenses will not grow with inflation.
a. Draw the cash flow diagram.
b. Populate the following table, and plot the NPW results in a meaningful plot. (Make sure you label your axis, and your plot should contain three curves.)

Cost/Sales	$1.00	$1.25	$1.50
150,000			
200,000			
250,000			

c. Make a pricing recommendation as a function of sales. You would like to make a minimum of $100,000 for your work on this project.

12-7. A corporation is trying to decide whether to buy the patent for a product designed by another company. The decision to buy will mean an investment of $10 million, and the demand for the product is not known. If demand is light, the company expects a return of $1.2 million each year for 3 years. If demand is moderate, the return will be $2.5 million each year for 4 years, and high demand means a return of $7 million each year for 4 years. It is estimated that the probability of a high demand is .3, and the probability of a light demand is .2. The firm's weighted average cost of capital is 12%. Calculate the expected NPW of the patent. On this basis, should the company make the investment? (All figures represent after-tax values.)

12-8. You have been approached by a developer who wants you to invest in a strip mall complex. Based on development costs, rent, and expenses, they present the following cash flows to you:

Time Period	Cash Flows
0	–$7,000,000
1	$2,500,000
2	$1,900,000
3	$1,800,000
4	$1,800,000
5	$1,900,000
6	$1,500,000

Using a cost of funds of 10%, determine the conventional payback period and the discounted payback period. Would you make this investment? Justify your answer.

12-9. Suppose the following cash flow represents the base case for a potential investment project. Conduct a sensitivity analysis.

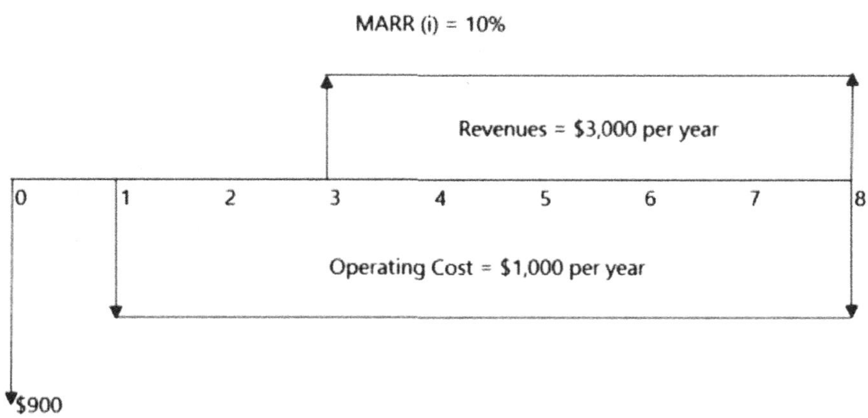

Variable	−30%	−20%	−10%	0	+10%	+20%	+30%
Revenues				$3,000/$4,563			
MARR				10%/$4,563			
Operating Costs				$1,000/$4,563			

a. Complete the above table for the appropriate changes in variable values. Fill in the table with the new revenues, MARR, and operating costs based on the variable changes.
b. Find the net present worth for all the missing blocks in the table below.

Variable	−30%	−20%	−10%	0	+10%	+20%	+30%
Revenues				$3,000/$4,563			
MARR				10%/$4,563			
Operating Costs				$1,000/$4,563			

c. Plot the NPV for all three variables from the table in (b) (revenue, MARR, and operating costs). Be sure to correctly label each line appropriately with a legend.
d. Which input variable shows the most sensitivity to NPV? Why did you choose this variable?

12-10. In order to start your small custom screen printing business, you develop the following costs:

Item	Cost	Cost/Per
Computer	$1,100.00	
Color Printer	$ 499.00	
Design Software	$ 120.00	
T-Shirts	$ 67.00	per order
Ink	$ 50.00	per order
Brushes	$ 34.00	
Silk Screens	$ 235.00	
High-Speed Internet Service	$ 40.00	per month
Paper	$ 15.00	
Printer Cartridges	$ 60.00	
Business License	$ 95.00	per year
Phone	$ 30.00	per month
Advertising Fliers	$ 200.00	
Business Taxes	$ 300.00	per year
Additional Electricity Charges	$ 102.00	per month
Shipping and Handling Costs	$ 17.50	per order

Assume you will only do production runs of 50 shirts, which constitutes an order. For your first year, you would like to make a profit of $25,000. You plan to sell the shirts at $5 each or $250/order, which will include shipping and handling.

a. Identify the fixed and variable costs.
b. Develop a plot of number of orders or production volume as a function of net income. Plot total revenue, total cost, and profit. (Hint: set the revenue equal to the fixed plus variable costs, and vary the number of orders. Profit is simply revenue – total costs.)

12-11. Power Corp's 2018 cost, production, and sales projections for its new generator are as follows:

Manufacturing Costs and Factors		Non-Manufacturing Costs	
Direct Labor	$45 per hour	Marketing	$115,000 per year
Direct Materials	$118 per unit	Administrative	$140,000 per year
Overhead	$155,000 plus $7 per unit	Other Overhead	$10 per unit
2017 Annual Maximum Production Hours = 7,000 Hours			
15 Units Produced per Hour			
Projected Selling price = $200 per unit			

a. What is Power Corp's maximum production level?
b. What is Power Corp's average unit cost at maximum production level?
c. What is Power Corp's breakeven annual production level at the projected sales price?

12-12. Determine the payback period for the project cash flows below, ignoring the time value of money.

Time Period	Cash Flows	Cumulative Cash Flow	Calculations
0	–$70,000		
1	$20,000		
2	$19,000		
3	$18,000		
4	$17,000		
5	$16,000		
6	$15,000		

If the firm evaluates projects at an effective annual interest rate of 10%, determine the discounted payback period of this project.

Time Period	Cash Flows	Present Value	Cumulative Cash Flow
0	–$70,000		
1	$20,000		
2	$19,000		
3	$18,000		
4	$17,000		
5	$16,000		
6	$15,000		

References

Farr, John V., and Parnell, Gregory S., October 2000, "A Comparison of Portfolio Analysis Techniques for Research and Development Program," 21st Annual American Society of Engineering Management Conference, Washington, DC, pp. 293–304.

Keeney, Ralph J., 1992, Value Focused Thinking – A Path to Creative Decision-making, Harvard University Press, Cambridge, MA.

Kirkwood, C., 1997, *Strategic Decision Making: Multiobjective Decision Analysis with Spreadsheets*, Wadsworth Publishing Company, Belmont, CA.

Marr, Bernard, 2012, *Key Performance Indicators*, Pearson Education, London, United Kingdom.

Melese, F., Richter, A., and Solomon, B., 2015, "Military Cost-Benefit Analysis (CBA): Theory & Practice," Naval Postgraduate School, accessed 18 December 2017 at https://my.nps.edu/documents/103424423/107207147/Melese_Richter_Military_Cost-Benefit_chapter_2015.pdf/049d9071-7a52-4322-92af-d4696d425374.

US Army, 22 February 2012, "Cost Benefit Analysis Guide," 3rd Edition, Office of the Deputy Assistant Secretary of the Army (Cost and Economics), Washington, DC.

Walsh, Claran, 2012, *Key Performance Indicators*, 4th Edition, Pearson Education, London, United Kingdom.

appendix

Interest factors

Interest Rate 0.25%

N	(F/P,i,N)	(P/F,i,N)	(F/A,i,N)	(A/F,i,n)	(P/A,i,N)	(A/P,i,N)	(A/G,i,N)	(P/G,i,N)
1	1.0025	0.9975	1.0000	1.0000	0.9975	1.0025	0.0000	0.0000
2	1.0050	0.9950	2.0025	0.4994	1.9925	0.5019	0.4994	0.9950
3	1.0075	0.9925	3.0075	0.3325	2.9851	0.3350	0.9983	2.9801
4	1.0100	0.9901	4.0150	0.2491	3.9751	0.2516	1.4969	5.9503
5	1.0126	0.9876	5.0251	0.1990	4.9627	0.2015	1.9950	9.9007
6	1.0151	0.9851	6.0376	0.1656	5.9478	0.1681	2.4927	14.8263
7	1.0176	0.9827	7.0527	0.1418	6.9305	0.1443	2.9900	20.7223
8	1.0202	0.9802	8.0704	0.1239	7.9107	0.1264	3.4869	27.5839
9	1.0227	0.9778	9.0905	0.1100	8.8885	0.1125	3.9834	35.4061
10	1.0253	0.9753	10.1133	0.0989	9.8639	0.1014	4.4794	44.1842
11	1.0278	0.9729	11.1385	0.0898	10.8368	0.0923	4.9750	53.9133
12	1.0304	0.9705	12.1664	0.0822	11.8073	0.0847	5.4702	64.5886
13	1.0330	0.9681	13.1968	0.0758	12.7753	0.0783	5.9650	76.2053
14	1.0356	0.9656	14.2298	0.0703	13.7410	0.0728	6.4594	88.7587
15	1.0382	0.9632	15.2654	0.0655	14.7042	0.0680	6.9534	102.2441
16	1.0408	0.9608	16.3035	0.0613	15.6650	0.0638	7.4469	116.6567
17	1.0434	0.9584	17.3443	0.0577	16.6235	0.0602	7.9401	131.9917
18	1.0460	0.9561	18.3876	0.0544	17.5795	0.0569	8.4328	148.2446
19	1.0486	0.9537	19.4336	0.0515	18.5332	0.0540	8.9251	165.4106
20	1.0512	0.9513	20.4822	0.0488	19.4845	0.0513	9.4170	183.4851
21	1.0538	0.9489	21.5334	0.0464	20.4334	0.0489	9.9085	202.4634
22	1.0565	0.9466	22.5872	0.0443	21.3800	0.0468	10.3995	222.3410
23	1.0591	0.9442	23.6437	0.0423	22.3241	0.0448	10.8901	243.1131
24	1.0618	0.9418	24.7028	0.0405	23.2660	0.0430	11.3804	264.7753
25	1.0644	0.9395	25.7646	0.0388	24.2055	0.0413	11.8702	287.3230
26	1.0671	0.9371	26.8290	0.0373	25.1426	0.0398	12.3596	310.7516
27	1.0697	0.9348	27.8961	0.0358	26.0774	0.0383	12.8485	335.0566
28	1.0724	0.9325	28.9658	0.0345	27.0099	0.0370	13.3371	360.2334
29	1.0751	0.9301	30.0382	0.0333	27.9400	0.0358	13.8252	386.2776
30	1.0778	0.9278	31.1133	0.0321	28.8679	0.0346	14.3130	413.1847
31	1.0805	0.9255	32.1911	0.0311	29.7934	0.0336	14.8003	440.9502
32	1.0832	0.9232	33.2716	0.0301	30.7166	0.0326	15.2872	469.5696
33	1.0859	0.9209	34.3547	0.0291	31.6375	0.0316	15.7736	499.0386
34	1.0886	0.9186	35.4406	0.0282	32.5561	0.0307	16.2597	529.3528
35	1.0913	0.9163	36.5292	0.0274	33.4724	0.0299	16.7454	560.5076
36	1.0941	0.9140	37.6206	0.0266	34.3865	0.0291	17.2306	592.4988
37	1.0968	0.9118	38.7146	0.0258	35.2982	0.0283	17.7154	625.3219
38	1.0995	0.9095	39.8114	0.0251	36.2077	0.0276	18.1998	658.9727
39	1.1023	0.9072	40.9109	0.0244	37.1149	0.0269	18.6838	693.4468
40	1.1050	0.9050	42.0132	0.0238	38.0199	0.0263	19.1673	728.7399
41	1.1078	0.9027	43.1182	0.0232	38.9226	0.0257	19.6505	764.8476
42	1.1106	0.9004	44.2260	0.0226	39.8230	0.0251	20.1332	801.7658
43	1.1133	0.8982	45.3366	0.0221	40.7212	0.0246	20.6156	839.4900
44	1.1161	0.8960	46.4499	0.0215	41.6172	0.0240	21.0975	878.0162
45	1.1189	0.8937	47.5661	0.0210	42.5109	0.0235	21.5789	917.3400
46	1.1217	0.8915	48.6850	0.0205	43.4024	0.0230	22.0600	957.4572
47	1.1245	0.8893	49.8067	0.0201	44.2916	0.0226	22.5407	998.3637
48	1.1273	0.8871	50.9312	0.0196	45.1787	0.0221	23.0209	1040.0552
49	1.1301	0.8848	52.0585	0.0192	46.0635	0.0217	23.5007	1082.5276
50	1.1330	0.8826	53.1887	0.0188	46.9462	0.0213	23.9802	1125.7767

Interest Rate 0.50%

N	(F/P,i,N)	(P/F,i,N)	(F/A,i,N)	(A/F,i,n)	(P/A,i,N)	(A/P,i,N)	(A/G,i,N)	(P/G,i,N)
1	1.0050	0.9950	1.0000	1.0000	0.9950	1.0050	0.0000	0.0000
2	1.0100	0.9901	2.0050	0.4988	1.9851	0.5038	0.4988	0.9901
3	1.0151	0.9851	3.0150	0.3317	2.9702	0.3367	0.9967	2.9604
4	1.0202	0.9802	4.0301	0.2481	3.9505	0.2531	1.4938	5.9011
5	1.0253	0.9754	5.0503	0.1980	4.9259	0.2030	1.9900	9.8026
6	1.0304	0.9705	6.0755	0.1646	5.8964	0.1696	2.4855	14.6552
7	1.0355	0.9657	7.1059	0.1407	6.8621	0.1457	2.9801	20.4493
8	1.0407	0.9609	8.1414	0.1228	7.8230	0.1278	3.4738	27.1755
9	1.0459	0.9561	9.1821	0.1089	8.7791	0.1139	3.9668	34.8244
10	1.0511	0.9513	10.2280	0.0978	9.7304	0.1028	4.4589	43.3865
11	1.0564	0.9466	11.2792	0.0887	10.6770	0.0937	4.9501	52.8526
12	1.0617	0.9419	12.3356	0.0811	11.6189	0.0861	5.4406	63.2136
13	1.0670	0.9372	13.3972	0.0746	12.5562	0.0796	5.9302	74.4602
14	1.0723	0.9326	14.4642	0.0691	13.4887	0.0741	6.4190	86.5835
15	1.0777	0.9279	15.5365	0.0644	14.4166	0.0694	6.9069	99.5743
16	1.0831	0.9233	16.6142	0.0602	15.3399	0.0652	7.3940	113.4238
17	1.0885	0.9187	17.6973	0.0565	16.2586	0.0615	7.8803	128.1231
18	1.0939	0.9141	18.7858	0.0532	17.1728	0.0582	8.3658	143.6634
19	1.0994	0.9096	19.8797	0.0503	18.0824	0.0553	8.8504	160.0360
20	1.1049	0.9051	20.9791	0.0477	18.9874	0.0527	9.3342	177.2322
21	1.1104	0.9006	22.0840	0.0453	19.8880	0.0503	9.8172	195.2434
22	1.1160	0.8961	23.1944	0.0431	20.7841	0.0481	10.2993	214.0611
23	1.1216	0.8916	24.3104	0.0411	21.6757	0.0461	10.7806	233.6768
24	1.1272	0.8872	25.4320	0.0393	22.5629	0.0443	11.2611	254.0820
25	1.1328	0.8828	26.5591	0.0377	23.4456	0.0427	11.7407	275.2686
26	1.1385	0.8784	27.6919	0.0361	24.3240	0.0411	12.2195	297.2281
27	1.1442	0.8740	28.8304	0.0347	25.1980	0.0397	12.6975	319.9523
28	1.1499	0.8697	29.9745	0.0334	26.0677	0.0384	13.1747	343.4332
29	1.1556	0.8653	31.1244	0.0321	26.9330	0.0371	13.6510	367.6625
30	1.1614	0.8610	32.2800	0.0310	27.7941	0.0360	14.1265	392.6324
31	1.1672	0.8567	33.4414	0.0299	28.6508	0.0349	14.6012	418.3348
32	1.1730	0.8525	34.6086	0.0289	29.5033	0.0339	15.0750	444.7618
33	1.1789	0.8482	35.7817	0.0279	30.3515	0.0329	15.5480	471.9055
34	1.1848	0.8440	36.9606	0.0271	31.1955	0.0321	16.0202	499.7583
35	1.1907	0.8398	38.1454	0.0262	32.0354	0.0312	16.4915	528.3123
36	1.1967	0.8356	39.3361	0.0254	32.8710	0.0304	16.9621	557.5598
37	1.2027	0.8315	40.5328	0.0247	33.7025	0.0297	17.4317	587.4934
38	1.2087	0.8274	41.7354	0.0240	34.5299	0.0290	17.9006	618.1054
39	1.2147	0.8232	42.9441	0.0233	35.3531	0.0283	18.3686	649.3883
40	1.2208	0.8191	44.1588	0.0226	36.1722	0.0276	18.8359	681.3347
41	1.2269	0.8151	45.3796	0.0220	36.9873	0.0270	19.3022	713.9372
42	1.2330	0.8110	46.6065	0.0215	37.7983	0.0265	19.7678	747.1886
43	1.2392	0.8070	47.8396	0.0209	38.6053	0.0259	20.2325	781.0815
44	1.2454	0.8030	49.0788	0.0204	39.4082	0.0254	20.6964	815.6087
45	1.2516	0.7990	50.3242	0.0199	40.2072	0.0249	21.1595	850.7631
46	1.2579	0.7950	51.5758	0.0194	41.0022	0.0244	21.6217	886.5376
47	1.2642	0.7910	52.8337	0.0189	41.7932	0.0239	22.0831	922.9252
48	1.2705	0.7871	54.0978	0.0185	42.5803	0.0235	22.5437	959.9188
49	1.2768	0.7832	55.3683	0.0181	43.3635	0.0231	23.0035	997.5116
50	1.2832	0.7793	56.6452	0.0177	44.1428	0.0227	23.4624	1035.6966

Interest Rate 1.00%

N	(F/P,i,N)	(P/F,i,N)	(F/A,i,N)	(A/F,i,n)	(P/A,i,N)	(A/P,i,N)	(A/G,i,N)	(P/G,i,N)
1	1.0100	0.9901	1.0000	1.0000	0.9901	1.0100	0.0000	0.0000
2	1.0201	0.9803	2.0100	0.4975	1.9704	0.5075	0.4975	0.9803
3	1.0303	0.9706	3.0301	0.3300	2.9410	0.3400	0.9934	2.9215
4	1.0406	0.9610	4.0604	0.2463	3.9020	0.2563	1.4876	5.8044
5	1.0510	0.9515	5.1010	0.1960	4.8534	0.2060	1.9801	9.6103
6	1.0615	0.9420	6.1520	0.1625	5.7955	0.1725	2.4710	14.3205
7	1.0721	0.9327	7.2135	0.1386	6.7282	0.1486	2.9602	19.9168
8	1.0829	0.9235	8.2857	0.1207	7.6517	0.1307	3.4478	26.3812
9	1.0937	0.9143	9.3685	0.1067	8.5660	0.1167	3.9337	33.6959
10	1.1046	0.9053	10.4622	0.0956	9.4713	0.1056	4.4179	41.8435
11	1.1157	0.8963	11.5668	0.0865	10.3676	0.0965	4.9005	50.8067
12	1.1268	0.8874	12.6825	0.0788	11.2551	0.0888	5.3815	60.5687
13	1.1381	0.8787	13.8093	0.0724	12.1337	0.0824	5.8607	71.1126
14	1.1495	0.8700	14.9474	0.0669	13.0037	0.0769	6.3384	82.4221
15	1.1610	0.8613	16.0969	0.0621	13.8651	0.0721	6.8143	94.4810
16	1.1726	0.8528	17.2579	0.0579	14.7179	0.0679	7.2886	107.2734
17	1.1843	0.8444	18.4304	0.0543	15.5623	0.0643	7.7613	120.7834
18	1.1961	0.8360	19.6147	0.0510	16.3983	0.0610	8.2323	134.9957
19	1.2081	0.8277	20.8109	0.0481	17.2260	0.0581	8.7017	149.8950
20	1.2202	0.8195	22.0190	0.0454	18.0456	0.0554	9.1694	165.4664
21	1.2324	0.8114	23.2392	0.0430	18.8570	0.0530	9.6354	181.6950
22	1.2447	0.8034	24.4716	0.0409	19.6604	0.0509	10.0998	198.5663
23	1.2572	0.7954	25.7163	0.0389	20.4558	0.0489	10.5626	216.0660
24	1.2697	0.7876	26.9735	0.0371	21.2434	0.0471	11.0237	234.1800
25	1.2824	0.7798	28.2432	0.0354	22.0232	0.0454	11.4831	252.8945
26	1.2953	0.7720	29.5256	0.0339	22.7952	0.0439	11.9409	272.1957
27	1.3082	0.7644	30.8209	0.0324	23.5596	0.0424	12.3971	292.0702
28	1.3213	0.7568	32.1291	0.0311	24.3164	0.0411	12.8516	312.5047
29	1.3345	0.7493	33.4504	0.0299	25.0658	0.0399	13.3044	333.4863
30	1.3478	0.7419	34.7849	0.0287	25.8077	0.0387	13.7557	355.0021
31	1.3613	0.7346	36.1327	0.0277	26.5423	0.0377	14.2052	377.0394
32	1.3749	0.7273	37.4941	0.0267	27.2696	0.0367	14.6532	399.5858
33	1.3887	0.7201	38.8690	0.0257	27.9897	0.0357	15.0995	422.6291
34	1.4026	0.7130	40.2577	0.0248	28.7027	0.0348	15.5441	446.1572
35	1.4166	0.7059	41.6603	0.0240	29.4086	0.0340	15.9871	470.1583
36	1.4308	0.6989	43.0769	0.0232	30.1075	0.0332	16.4285	494.6207
37	1.4451	0.6920	44.5076	0.0225	30.7995	0.0325	16.8682	519.5329
38	1.4595	0.6852	45.9527	0.0218	31.4847	0.0318	17.3063	544.8835
39	1.4741	0.6784	47.4123	0.0211	32.1630	0.0311	17.7428	570.6616
40	1.4889	0.6717	48.8864	0.0205	32.8347	0.0305	18.1776	596.8561
41	1.5038	0.6650	50.3752	0.0199	33.4997	0.0299	18.6108	623.4562
42	1.5188	0.6584	51.8790	0.0193	34.1581	0.0293	19.0424	650.4514
43	1.5340	0.6519	53.3978	0.0187	34.8100	0.0287	19.4723	677.8312
44	1.5493	0.6454	54.9318	0.0182	35.4555	0.0282	19.9006	705.5853
45	1.5648	0.6391	56.4811	0.0177	36.0945	0.0277	20.3273	733.7037
46	1.5805	0.6327	58.0459	0.0172	36.7272	0.0272	20.7524	762.1765
47	1.5963	0.6265	59.6263	0.0168	37.3537	0.0268	21.1758	790.9938
48	1.6122	0.6203	61.2226	0.0163	37.9740	0.0263	21.5976	820.1460
49	1.6283	0.6141	62.8348	0.0159	38.5881	0.0259	22.0178	849.6237
50	1.6446	0.6080	64.4632	0.0155	39.1961	0.0255	22.4363	879.4176

Interest Rate 2.00%

N	(F/P,i,N)	(P/F,i,N)	(F/A,i,N)	(A/F,i,n)	(P/A,i,N)	(A/P,i,N)	(A/G,i,N)	(P/G,i,N)
1	1.0200	0.9804	1.0000	1.0000	0.9804	1.0200	0.0000	0.0000
2	1.0404	0.9612	2.0200	0.4950	1.9416	0.5150	0.4950	0.9612
3	1.0612	0.9423	3.0604	0.3268	2.8839	0.3468	0.9868	2.8458
4	1.0824	0.9238	4.1216	0.2426	3.8077	0.2626	1.4752	5.6173
5	1.1041	0.9057	5.2040	0.1922	4.7135	0.2122	1.9604	9.2403
6	1.1262	0.8880	6.3081	0.1585	5.6014	0.1785	2.4423	13.6801
7	1.1487	0.8706	7.4343	0.1345	6.4720	0.1545	2.9208	18.9035
8	1.1717	0.8535	8.5830	0.1165	7.3255	0.1365	3.3961	24.8779
9	1.1951	0.8368	9.7546	0.1025	8.1622	0.1225	3.8681	31.5720
10	1.2190	0.8203	10.9497	0.0913	8.9826	0.1113	4.3367	38.9551
11	1.2434	0.8043	12.1687	0.0822	9.7868	0.1022	4.8021	46.9977
12	1.2682	0.7885	13.4121	0.0746	10.5753	0.0946	5.2642	55.6712
13	1.2936	0.7730	14.6803	0.0681	11.3484	0.0881	5.7231	64.9475
14	1.3195	0.7579	15.9739	0.0626	12.1062	0.0826	6.1786	74.7999
15	1.3459	0.7430	17.2934	0.0578	12.8493	0.0778	6.6309	85.2021
16	1.3728	0.7284	18.6393	0.0537	13.5777	0.0737	7.0799	96.1288
17	1.4002	0.7142	20.0121	0.0500	14.2919	0.0700	7.5256	107.5554
18	1.4282	0.7002	21.4123	0.0467	14.9920	0.0667	7.9681	119.4581
19	1.4568	0.6864	22.8406	0.0438	15.6785	0.0638	8.4073	131.8139
20	1.4859	0.6730	24.2974	0.0412	16.3514	0.0612	8.8433	144.6003
21	1.5157	0.6598	25.7833	0.0388	17.0112	0.0588	9.2760	157.7959
22	1.5460	0.6468	27.2990	0.0366	17.6580	0.0566	9.7055	171.3795
23	1.5769	0.6342	28.8450	0.0347	18.2922	0.0547	10.1317	185.3309
24	1.6084	0.6217	30.4219	0.0329	18.9139	0.0529	10.5547	199.6305
25	1.6406	0.6095	32.0303	0.0312	19.5235	0.0512	10.9745	214.2592
26	1.6734	0.5976	33.6709	0.0297	20.1210	0.0497	11.3910	229.1987
27	1.7069	0.5859	35.3443	0.0283	20.7069	0.0483	11.8043	244.4311
28	1.7410	0.5744	37.0512	0.0270	21.2813	0.0470	12.2145	259.9392
29	1.7758	0.5631	38.7922	0.0258	21.8444	0.0458	12.6214	275.7064
30	1.8114	0.5521	40.5681	0.0246	22.3965	0.0446	13.0251	291.7164
31	1.8476	0.5412	42.3794	0.0236	22.9377	0.0436	13.4257	307.9538
32	1.8845	0.5306	44.2270	0.0226	23.4683	0.0426	13.8230	324.4035
33	1.9222	0.5202	46.1116	0.0217	23.9886	0.0417	14.2172	341.0508
34	1.9607	0.5100	48.0338	0.0208	24.4986	0.0408	14.6083	357.8817
35	1.9999	0.5000	49.9945	0.0200	24.9986	0.0400	14.9961	374.8826
36	2.0399	0.4902	51.9944	0.0192	25.4888	0.0392	15.3809	392.0405
37	2.0807	0.4806	54.0343	0.0185	25.9695	0.0385	15.7625	409.3424
38	2.1223	0.4712	56.1149	0.0178	26.4406	0.0378	16.1409	426.7764
39	2.1647	0.4619	58.2372	0.0172	26.9026	0.0372	16.5163	444.3304
40	2.2080	0.4529	60.4020	0.0166	27.3555	0.0366	16.8885	461.9931
41	2.2522	0.4440	62.6100	0.0160	27.7995	0.0360	17.2576	479.7535
42	2.2972	0.4353	64.8622	0.0154	28.2348	0.0354	17.6237	497.6010
43	2.3432	0.4268	67.1595	0.0149	28.6616	0.0349	17.9866	515.5253
44	2.3901	0.4184	69.5027	0.0144	29.0800	0.0344	18.3465	533.5165
45	2.4379	0.4102	71.8927	0.0139	29.4902	0.0339	18.7034	551.5652
46	2.4866	0.4022	74.3306	0.0135	29.8923	0.0335	19.0571	569.6621
47	2.5363	0.3943	76.8172	0.0130	30.2866	0.0330	19.4079	587.7985
48	2.5871	0.3865	79.3535	0.0126	30.6731	0.0326	19.7556	605.9657
49	2.6388	0.3790	81.9406	0.0122	31.0521	0.0322	20.1003	624.1557
50	2.6916	0.3715	84.5794	0.0118	31.4236	0.0318	20.4420	642.3606

Interest Rate	3.00%							
N	(F/P,i,N)	(P/F,i,N)	(F/A,i,N)	(A/F,i,n)	(P/A,i,N)	(A/P,i,N)	(A/G,i,N)	(P/G,i,N)
1	1.0300	0.9709	1.0000	1.0000	0.9709	1.0300	0.0000	0.0000
2	1.0609	0.9426	2.0300	0.4926	1.9135	0.5226	0.4926	0.9426
3	1.0927	0.9151	3.0909	0.3235	2.8286	0.3535	0.9803	2.7729
4	1.1255	0.8885	4.1836	0.2390	3.7171	0.2690	1.4631	5.4383
5	1.1593	0.8626	5.3091	0.1884	4.5797	0.2184	1.9409	8.8888
6	1.1941	0.8375	6.4684	0.1546	5.4172	0.1846	2.4138	13.0762
7	1.2299	0.8131	7.6625	0.1305	6.2303	0.1605	2.8819	17.9547
8	1.2668	0.7894	8.8923	0.1125	7.0197	0.1425	3.3450	23.4806
9	1.3048	0.7664	10.1591	0.0984	7.7861	0.1284	3.8032	29.6119
10	1.3439	0.7441	11.4639	0.0872	8.5302	0.1172	4.2565	36.3088
11	1.3842	0.7224	12.8078	0.0781	9.2526	0.1081	4.7049	43.5330
12	1.4258	0.7014	14.1920	0.0705	9.9540	0.1005	5.1485	51.2482
13	1.4685	0.6810	15.6178	0.0640	10.6350	0.0940	5.5872	59.4196
14	1.5126	0.6611	17.0863	0.0585	11.2961	0.0885	6.0210	68.0141
15	1.5580	0.6419	18.5989	0.0538	11.9379	0.0838	6.4500	77.0002
16	1.6047	0.6232	20.1569	0.0496	12.5611	0.0796	6.8742	86.3477
17	1.6528	0.6050	21.7616	0.0460	13.1661	0.0760	7.2936	96.0280
18	1.7024	0.5874	23.4144	0.0427	13.7535	0.0727	7.7081	106.0137
19	1.7535	0.5703	25.1169	0.0398	14.3238	0.0698	8.1179	116.2788
20	1.8061	0.5537	26.8704	0.0372	14.8775	0.0672	8.5229	126.7987
21	1.8603	0.5375	28.6765	0.0349	15.4150	0.0649	8.9231	137.5496
22	1.9161	0.5219	30.5368	0.0327	15.9369	0.0627	9.3186	148.5094
23	1.9736	0.5067	32.4529	0.0308	16.4436	0.0608	9.7093	159.6566
24	2.0328	0.4919	34.4265	0.0290	16.9355	0.0590	10.0954	170.9711
25	2.0938	0.4776	36.4593	0.0274	17.4131	0.0574	10.4768	182.4336
26	2.1566	0.4637	38.5530	0.0259	17.8768	0.0559	10.8535	194.0260
27	2.2213	0.4502	40.7096	0.0246	18.3270	0.0546	11.2255	205.7309
28	2.2879	0.4371	42.9309	0.0233	18.7641	0.0533	11.5930	217.5320
29	2.3566	0.4243	45.2189	0.0221	19.1885	0.0521	11.9558	229.4137
30	2.4273	0.4120	47.5754	0.0210	19.6004	0.0510	12.3141	241.3613
31	2.5001	0.4000	50.0027	0.0200	20.0004	0.0500	12.6678	253.3609
32	2.5751	0.3883	52.5028	0.0190	20.3888	0.0490	13.0169	265.3993
33	2.6523	0.3770	55.0778	0.0182	20.7658	0.0482	13.3616	277.4642
34	2.7319	0.3660	57.7302	0.0173	21.1318	0.0473	13.7018	289.5437
35	2.8139	0.3554	60.4621	0.0165	21.4872	0.0465	14.0375	301.6267
36	2.8983	0.3450	63.2759	0.0158	21.8323	0.0458	14.3688	313.7028
37	2.9852	0.3350	66.1742	0.0151	22.1672	0.0451	14.6957	325.7622
38	3.0748	0.3252	69.1594	0.0145	22.4925	0.0445	15.0182	337.7956
39	3.1670	0.3158	72.2342	0.0138	22.8082	0.0438	15.3363	349.7942
40	3.2620	0.3066	75.4013	0.0133	23.1148	0.0433	15.6502	361.7499
41	3.3599	0.2976	78.6633	0.0127	23.4124	0.0427	15.9597	373.6551
42	3.4607	0.2890	82.0232	0.0122	23.7014	0.0422	16.2650	385.5024
43	3.5645	0.2805	85.4839	0.0117	23.9819	0.0417	16.5660	397.2852
44	3.6715	0.2724	89.0484	0.0112	24.2543	0.0412	16.8629	408.9972
45	3.7816	0.2644	92.7199	0.0108	24.5187	0.0408	17.1556	420.6325
46	3.8950	0.2567	96.5015	0.0104	24.7754	0.0404	17.4441	432.1856
47	4.0119	0.2493	100.3965	0.0100	25.0247	0.0400	17.7285	443.6515
48	4.1323	0.2420	104.4084	0.0096	25.2667	0.0396	18.0089	455.0255
49	4.2562	0.2350	108.5406	0.0092	25.5017	0.0392	18.2852	466.3031
50	4.3839	0.2281	112.7969	0.0089	25.7298	0.0389	18.5575	477.4803

Interest Rate 4.00%

N	(F/P,i,N)	(P/F,i,N)	(F/A,i,N)	(A/F,i,n)	(P/A,i,N)	(A/P,i,N)	(A/G,i,N)	(P/G,i,N)
1	1.0400	0.9615	1.0000	1.0000	0.9615	1.0400	0.0000	0.0000
2	1.0816	0.9246	2.0400	0.4902	1.8861	0.5302	0.4902	0.9246
3	1.1249	0.8890	3.1216	0.3203	2.7751	0.3603	0.9739	2.7025
4	1.1699	0.8548	4.2465	0.2355	3.6299	0.2755	1.4510	5.2670
5	1.2167	0.8219	5.4163	0.1846	4.4518	0.2246	1.9216	8.5547
6	1.2653	0.7903	6.6330	0.1508	5.2421	0.1908	2.3857	12.5062
7	1.3159	0.7599	7.8983	0.1266	6.0021	0.1666	2.8433	17.0657
8	1.3686	0.7307	9.2142	0.1085	6.7327	0.1485	3.2944	22.1806
9	1.4233	0.7026	10.5828	0.0945	7.4353	0.1345	3.7391	27.8013
10	1.4802	0.6756	12.0061	0.0833	8.1109	0.1233	4.1773	33.8814
11	1.5395	0.6496	13.4864	0.0741	8.7605	0.1141	4.6090	40.3772
12	1.6010	0.6246	15.0258	0.0666	9.3851	0.1066	5.0343	47.2477
13	1.6651	0.6006	16.6268	0.0601	9.9856	0.1001	5.4533	54.4546
14	1.7317	0.5775	18.2919	0.0547	10.5631	0.0947	5.8659	61.9618
15	1.8009	0.5553	20.0236	0.0499	11.1184	0.0899	6.2721	69.7355
16	1.8730	0.5339	21.8245	0.0458	11.6523	0.0858	6.6720	77.7441
17	1.9479	0.5134	23.6975	0.0422	12.1657	0.0822	7.0656	85.9581
18	2.0258	0.4936	25.6454	0.0390	12.6593	0.0790	7.4530	94.3498
19	2.1068	0.4746	27.6712	0.0361	13.1339	0.0761	7.8342	102.8933
20	2.1911	0.4564	29.7781	0.0336	13.5903	0.0736	8.2091	111.5647
21	2.2788	0.4388	31.9692	0.0313	14.0292	0.0713	8.5779	120.3414
22	2.3699	0.4220	34.2480	0.0292	14.4511	0.0692	8.9407	129.2024
23	2.4647	0.4057	36.6179	0.0273	14.8568	0.0673	9.2973	138.1284
24	2.5633	0.3901	39.0826	0.0256	15.2470	0.0656	9.6479	147.1012
25	2.6658	0.3751	41.6459	0.0240	15.6221	0.0640	9.9925	156.1040
26	2.7725	0.3607	44.3117	0.0226	15.9828	0.0626	10.3312	165.1212
27	2.8834	0.3468	47.0842	0.0212	16.3296	0.0612	10.6640	174.1385
28	2.9987	0.3335	49.9676	0.0200	16.6631	0.0600	10.9909	183.1424
29	3.1187	0.3207	52.9663	0.0189	16.9837	0.0589	11.3120	192.1206
30	3.2434	0.3083	56.0849	0.0178	17.2920	0.0578	11.6274	201.0618
31	3.3731	0.2965	59.3283	0.0169	17.5885	0.0569	11.9371	209.9556
32	3.5081	0.2851	62.7015	0.0159	17.8736	0.0559	12.2411	218.7924
33	3.6484	0.2741	66.2095	0.0151	18.1476	0.0551	12.5396	227.5634
34	3.7943	0.2636	69.8579	0.0143	18.4112	0.0543	12.8324	236.2607
35	3.9461	0.2534	73.6522	0.0136	18.6646	0.0536	13.1198	244.8768
36	4.1039	0.2437	77.5983	0.0129	18.9083	0.0529	13.4018	253.4052
37	4.2681	0.2343	81.7022	0.0122	19.1426	0.0522	13.6784	261.8399
38	4.4388	0.2253	85.9703	0.0116	19.3679	0.0516	13.9497	270.1754
39	4.6164	0.2166	90.4091	0.0111	19.5845	0.0511	14.2157	278.4070
40	4.8010	0.2083	95.0255	0.0105	19.7928	0.0505	14.4765	286.5303
41	4.9931	0.2003	99.8265	0.0100	19.9931	0.0500	14.7322	294.5414
42	5.1928	0.1926	104.8196	0.0095	20.1856	0.0495	14.9828	302.4370
43	5.4005	0.1852	110.0124	0.0091	20.3708	0.0491	15.2284	310.2141
44	5.6165	0.1780	115.4129	0.0087	20.5488	0.0487	15.4690	317.8700
45	5.8412	0.1712	121.0294	0.0083	20.7200	0.0483	15.7047	325.4028
46	6.0748	0.1646	126.8706	0.0079	20.8847	0.0479	15.9356	332.8104
47	6.3178	0.1583	132.9454	0.0075	21.0429	0.0475	16.1618	340.0914
48	6.5705	0.1522	139.2632	0.0072	21.1951	0.0472	16.3832	347.2446
49	6.8333	0.1463	145.8337	0.0069	21.3415	0.0469	16.6000	354.2689
50	7.1067	0.1407	152.6671	0.0066	21.4822	0.0466	16.8122	361.1638

Interest Rate 5.00%								
N	(F/P,i,N)	(P/F,i,N)	(F/A,i,N)	(A/F,i,n)	(P/A,i,N)	(A/P,i,N)	(A/G,i,N)	(P/G,i,N)
1	1.0500	0.9524	1.0000	1.0000	0.9524	1.0500	0.0000	0.0000
2	1.1025	0.9070	2.0500	0.4878	1.8594	0.5378	0.4878	0.9070
3	1.1576	0.8638	3.1525	0.3172	2.7232	0.3672	0.9675	2.6347
4	1.2155	0.8227	4.3101	0.2320	3.5460	0.2820	1.4391	5.1028
5	1.2763	0.7835	5.5256	0.1810	4.3295	0.2310	1.9025	8.2369
6	1.3401	0.7462	6.8019	0.1470	5.0757	0.1970	2.3579	11.9680
7	1.4071	0.7107	8.1420	0.1228	5.7864	0.1728	2.8052	16.2321
8	1.4775	0.6768	9.5491	0.1047	6.4632	0.1547	3.2445	20.9700
9	1.5513	0.6446	11.0266	0.0907	7.1078	0.1407	3.6758	26.1268
10	1.6289	0.6139	12.5779	0.0795	7.7217	0.1295	4.0991	31.6520
11	1.7103	0.5847	14.2068	0.0704	8.3064	0.1204	4.5144	37.4988
12	1.7959	0.5568	15.9171	0.0628	8.8633	0.1128	4.9219	43.6241
13	1.8856	0.5303	17.7130	0.0565	9.3936	0.1065	5.3215	49.9879
14	1.9799	0.5051	19.5986	0.0510	9.8986	0.1010	5.7133	56.5538
15	2.0789	0.4810	21.5786	0.0463	10.3797	0.0963	6.0973	63.2880
16	2.1829	0.4581	23.6575	0.0423	10.8378	0.0923	6.4736	70.1597
17	2.2920	0.4363	25.8404	0.0387	11.2741	0.0887	6.8423	77.1405
18	2.4066	0.4155	28.1324	0.0355	11.6896	0.0855	7.2034	84.2043
19	2.5270	0.3957	30.5390	0.0327	12.0853	0.0827	7.5569	91.3275
20	2.6533	0.3769	33.0660	0.0302	12.4622	0.0802	7.9030	98.4884
21	2.7860	0.3589	35.7193	0.0280	12.8212	0.0780	8.2416	105.6673
22	2.9253	0.3418	38.5052	0.0260	13.1630	0.0760	8.5730	112.8461
23	3.0715	0.3256	41.4305	0.0241	13.4886	0.0741	8.8971	120.0087
24	3.2251	0.3101	44.5020	0.0225	13.7986	0.0725	9.2140	127.1402
25	3.3864	0.2953	47.7271	0.0210	14.0939	0.0710	9.5238	134.2275
26	3.5557	0.2812	51.1135	0.0196	14.3752	0.0696	9.8266	141.2585
27	3.7335	0.2678	54.6691	0.0183	14.6430	0.0683	10.1224	148.2226
28	3.9201	0.2551	58.4026	0.0171	14.8981	0.0671	10.4114	155.1101
29	4.1161	0.2429	62.3227	0.0160	15.1411	0.0660	10.6936	161.9126
30	4.3219	0.2314	66.4388	0.0151	15.3725	0.0651	10.9691	168.6226
31	4.5380	0.2204	70.7608	0.0141	15.5928	0.0641	11.2381	175.2333
32	4.7649	0.2099	75.2988	0.0133	15.8027	0.0633	11.5005	181.7392
33	5.0032	0.1999	80.0638	0.0125	16.0025	0.0625	11.7566	188.1351
34	5.2533	0.1904	85.0670	0.0118	16.1929	0.0618	12.0063	194.4168
35	5.5160	0.1813	90.3203	0.0111	16.3742	0.0611	12.2498	200.5807
36	5.7918	0.1727	95.8363	0.0104	16.5469	0.0604	12.4872	206.6237
37	6.0814	0.1644	101.6281	0.0098	16.7113	0.0598	12.7186	212.5434
38	6.3855	0.1566	107.7095	0.0093	16.8679	0.0593	12.9440	218.3378
39	6.7048	0.1491	114.0950	0.0088	17.0170	0.0588	13.1636	224.0054
40	7.0400	0.1420	120.7998	0.0083	17.1591	0.0583	13.3775	229.5452
41	7.3920	0.1353	127.8398	0.0078	17.2944	0.0578	13.5857	234.9564
42	7.7616	0.1288	135.2318	0.0074	17.4232	0.0574	13.7884	240.2389
43	8.1497	0.1227	142.9933	0.0070	17.5459	0.0570	13.9857	245.3925
44	8.5572	0.1169	151.1430	0.0066	17.6628	0.0566	14.1777	250.4175
45	8.9850	0.1113	159.7002	0.0063	17.7741	0.0563	14.3644	255.3145
46	9.4343	0.1060	168.6852	0.0059	17.8801	0.0559	14.5461	260.0844
47	9.9060	0.1009	178.1194	0.0056	17.9810	0.0556	14.7226	264.7281
48	10.4013	0.0961	188.0254	0.0053	18.0772	0.0553	14.8943	269.2467
49	10.9213	0.0916	198.4267	0.0050	18.1687	0.0550	15.0611	273.6418
50	11.4674	0.0872	209.3480	0.0048	18.2559	0.0548	15.2233	277.9148

Interest Rate 6.00%

N	(F/P,i,N)	(P/F,i,N)	(F/A,i,N)	(A/F,i,n)	(P/A,i,N)	(A/P,i,N)	(A/G,i,N)	(P/G,i,N)
1	1.0600	0.9434	1.0000	1.0000	0.9434	1.0600	0.0000	0.0000
2	1.1236	0.8900	2.0600	0.4854	1.8334	0.5454	0.4854	0.8900
3	1.1910	0.8396	3.1836	0.3141	2.6730	0.3741	0.9612	2.5692
4	1.2625	0.7921	4.3746	0.2286	3.4651	0.2886	1.4272	4.9455
5	1.3382	0.7473	5.6371	0.1774	4.2124	0.2374	1.8836	7.9345
6	1.4185	0.7050	6.9753	0.1434	4.9173	0.2034	2.3304	11.4594
7	1.5036	0.6651	8.3938	0.1191	5.5824	0.1791	2.7676	15.4497
8	1.5938	0.6274	9.8975	0.1010	6.2098	0.1610	3.1952	19.8416
9	1.6895	0.5919	11.4913	0.0870	6.8017	0.1470	3.6133	24.5768
10	1.7908	0.5584	13.1808	0.0759	7.3601	0.1359	4.0220	29.6023
11	1.8983	0.5268	14.9716	0.0668	7.8869	0.1268	4.4213	34.8702
12	2.0122	0.4970	16.8699	0.0593	8.3838	0.1193	4.8113	40.3369
13	2.1329	0.4688	18.8821	0.0530	8.8527	0.1130	5.1920	45.9629
14	2.2609	0.4423	21.0151	0.0476	9.2950	0.1076	5.5635	51.7128
15	2.3966	0.4173	23.2760	0.0430	9.7122	0.1030	5.9260	57.5546
16	2.5404	0.3936	25.6725	0.0390	10.1059	0.0990	6.2794	63.4592
17	2.6928	0.3714	28.2129	0.0354	10.4773	0.0954	6.6240	69.4011
18	2.8543	0.3503	30.9057	0.0324	10.8276	0.0924	6.9597	75.3569
19	3.0256	0.3305	33.7600	0.0296	11.1581	0.0896	7.2867	81.3062
20	3.2071	0.3118	36.7856	0.0272	11.4699	0.0872	7.6051	87.2304
21	3.3996	0.2942	39.9927	0.0250	11.7641	0.0850	7.9151	93.1136
22	3.6035	0.2775	43.3923	0.0230	12.0416	0.0830	8.2166	98.9412
23	3.8197	0.2618	46.9958	0.0213	12.3034	0.0813	8.5099	104.7007
24	4.0489	0.2470	50.8156	0.0197	12.5504	0.0797	8.7951	110.3812
25	4.2919	0.2330	54.8645	0.0182	12.7834	0.0782	9.0722	115.9732
26	4.5494	0.2198	59.1564	0.0169	13.0032	0.0769	9.3414	121.4684
27	4.8223	0.2074	63.7058	0.0157	13.2105	0.0757	9.6029	126.8600
28	5.1117	0.1956	68.5281	0.0146	13.4062	0.0746	9.8568	132.1420
29	5.4184	0.1846	73.6398	0.0136	13.5907	0.0736	10.1032	137.3096
30	5.7435	0.1741	79.0582	0.0126	13.7648	0.0726	10.3422	142.3588
31	6.0881	0.1643	84.8017	0.0118	13.9291	0.0718	10.5740	147.2864
32	6.4534	0.1550	90.8898	0.0110	14.0840	0.0710	10.7988	152.0901
33	6.8406	0.1462	97.3432	0.0103	14.2302	0.0703	11.0166	156.7681
34	7.2510	0.1379	104.1838	0.0096	14.3681	0.0696	11.2276	161.3192
35	7.6861	0.1301	111.4348	0.0090	14.4982	0.0690	11.4319	165.7427
36	8.1473	0.1227	119.1209	0.0084	14.6210	0.0684	11.6298	170.0387
37	8.6361	0.1158	127.2681	0.0079	14.7368	0.0679	11.8213	174.2072
38	9.1543	0.1092	135.9042	0.0074	14.8460	0.0674	12.0065	178.2490
39	9.7035	0.1031	145.0585	0.0069	14.9491	0.0669	12.1857	182.1652
40	10.2857	0.0972	154.7620	0.0065	15.0463	0.0665	12.3590	185.9568
41	10.9029	0.0917	165.0477	0.0061	15.1380	0.0661	12.5264	189.6256
42	11.5570	0.0865	175.9505	0.0057	15.2245	0.0657	12.6883	193.1732
43	12.2505	0.0816	187.5076	0.0053	15.3062	0.0653	12.8446	196.6017
44	12.9855	0.0770	199.7580	0.0050	15.3832	0.0650	12.9956	199.9130
45	13.7646	0.0727	212.7435	0.0047	15.4558	0.0647	13.1413	203.1096
46	14.5905	0.0685	226.5081	0.0044	15.5244	0.0644	13.2819	206.1938
47	15.4659	0.0647	241.0986	0.0041	15.5890	0.0641	13.4177	209.1681
48	16.3939	0.0610	256.5645	0.0039	15.6500	0.0639	13.5485	212.0351
49	17.3775	0.0575	272.9584	0.0037	15.7076	0.0637	13.6748	214.7972
50	18.4202	0.0543	290.3359	0.0034	15.7619	0.0634	13.7964	217.4574

Interest Rate	8.00%							
N	(F/P,i,N)	(P/F,i,N)	(F/A,i,N)	(A/F,i,n)	(P/A,i,N)	(A/P,i,N)	(A/G,i,N)	(P/G,i,N)
1	1.0800	0.9259	1.0000	1.0000	0.9259	1.0800	0.0000	0.0000
2	1.1664	0.8573	2.0800	0.4808	1.7833	0.5608	0.4808	0.8573
3	1.2597	0.7938	3.2464	0.3080	2.5771	0.3880	0.9487	2.4450
4	1.3605	0.7350	4.5061	0.2219	3.3121	0.3019	1.4040	4.6501
5	1.4693	0.6806	5.8666	0.1705	3.9927	0.2505	1.8465	7.3724
6	1.5869	0.6302	7.3359	0.1363	4.6229	0.2163	2.2763	10.5233
7	1.7138	0.5835	8.9228	0.1121	5.2064	0.1921	2.6937	14.0242
8	1.8509	0.5403	10.6366	0.0940	5.7466	0.1740	3.0985	17.8061
9	1.9990	0.5002	12.4876	0.0801	6.2469	0.1601	3.4910	21.8081
10	2.1589	0.4632	14.4866	0.0690	6.7101	0.1490	3.8713	25.9768
11	2.3316	0.4289	16.6455	0.0601	7.1390	0.1401	4.2395	30.2657
12	2.5182	0.3971	18.9771	0.0527	7.5361	0.1327	4.5957	34.6339
13	2.7196	0.3677	21.4953	0.0465	7.9038	0.1265	4.9402	39.0463
14	2.9372	0.3405	24.2149	0.0413	8.2442	0.1213	5.2731	43.4723
15	3.1722	0.3152	27.1521	0.0368	8.5595	0.1168	5.5945	47.8857
16	3.4259	0.2919	30.3243	0.0330	8.8514	0.1130	5.9046	52.2640
17	3.7000	0.2703	33.7502	0.0296	9.1216	0.1096	6.2037	56.5883
18	3.9960	0.2502	37.4502	0.0267	9.3719	0.1067	6.4920	60.8426
19	4.3157	0.2317	41.4463	0.0241	9.6036	0.1041	6.7697	65.0134
20	4.6610	0.2145	45.7620	0.0219	9.8181	0.1019	7.0369	69.0898
21	5.0338	0.1987	50.4229	0.0198	10.0168	0.0998	7.2940	73.0629
22	5.4365	0.1839	55.4568	0.0180	10.2007	0.0980	7.5412	76.9257
23	5.8715	0.1703	60.8933	0.0164	10.3711	0.0964	7.7786	80.6726
24	6.3412	0.1577	66.7648	0.0150	10.5288	0.0950	8.0066	84.2997
25	6.8485	0.1460	73.1059	0.0137	10.6748	0.0937	8.2254	87.8041
26	7.3964	0.1352	79.9544	0.0125	10.8100	0.0925	8.4352	91.1842
27	7.9881	0.1252	87.3508	0.0114	10.9352	0.0914	8.6363	94.4390
28	8.6271	0.1159	95.3388	0.0105	11.0511	0.0905	8.8289	97.5687
29	9.3173	0.1073	103.9659	0.0096	11.1584	0.0896	9.0133	100.5738
30	10.0627	0.0994	113.2832	0.0088	11.2578	0.0888	9.1897	103.4558
31	10.8677	0.0920	123.3459	0.0081	11.3498	0.0881	9.3584	106.2163
32	11.7371	0.0852	134.2135	0.0075	11.4350	0.0875	9.5197	108.8575
33	12.6760	0.0789	145.9506	0.0069	11.5139	0.0869	9.6737	111.3819
34	13.6901	0.0730	158.6267	0.0063	11.5869	0.0863	9.8208	113.7924
35	14.7853	0.0676	172.3168	0.0058	11.6546	0.0858	9.9611	116.0920
36	15.9682	0.0626	187.1021	0.0053	11.7172	0.0853	10.0949	118.2839
37	17.2456	0.0580	203.0703	0.0049	11.7752	0.0849	10.2225	120.3713
38	18.6253	0.0537	220.3159	0.0045	11.8289	0.0845	10.3440	122.3579
39	20.1153	0.0497	238.9412	0.0042	11.8786	0.0842	10.4597	124.2470
40	21.7245	0.0460	259.0565	0.0039	11.9246	0.0839	10.5699	126.0422
41	23.4625	0.0426	280.7810	0.0036	11.9672	0.0836	10.6747	127.7470
42	25.3395	0.0395	304.2435	0.0033	12.0067	0.0833	10.7744	129.3651
43	27.3666	0.0365	329.5830	0.0030	12.0432	0.0830	10.8692	130.8998
44	29.5560	0.0338	356.9496	0.0028	12.0771	0.0828	10.9592	132.3547
45	31.9204	0.0313	386.5056	0.0026	12.1084	0.0826	11.0447	133.7331
46	34.4741	0.0290	418.4261	0.0024	12.1374	0.0824	11.1258	135.0384
47	37.2320	0.0269	452.9002	0.0022	12.1643	0.0822	11.2028	136.2739
48	40.2106	0.0249	490.1322	0.0020	12.1891	0.0820	11.2758	137.4428
49	43.4274	0.0230	530.3427	0.0019	12.2122	0.0819	11.3451	138.5480
50	46.9016	0.0213	573.7702	0.0017	12.2335	0.0817	11.4107	139.5928

Interest Rate 10.00%

N	(F/P,i,N)	(P/F,i,N)	(F/A,i,N)	(A/F,i,n)	(P/A,i,N)	(A/P,i,N)	(A/G,i,N)	(P/G,i,N)
1	1.1000	0.9091	1.0000	1.0000	0.9091	1.1000	0.0000	0.0000
2	1.2100	0.8264	2.1000	0.4762	1.7355	0.5762	0.4762	0.8264
3	1.3310	0.7513	3.3100	0.3021	2.4869	0.4021	0.9366	2.3291
4	1.4641	0.6830	4.6410	0.2155	3.1699	0.3155	1.3812	4.3781
5	1.6105	0.6209	6.1051	0.1638	3.7908	0.2638	1.8101	6.8618
6	1.7716	0.5645	7.7156	0.1296	4.3553	0.2296	2.2236	9.6842
7	1.9487	0.5132	9.4872	0.1054	4.8684	0.2054	2.6216	12.7631
8	2.1436	0.4665	11.4359	0.0874	5.3349	0.1874	3.0045	16.0287
9	2.3579	0.4241	13.5795	0.0736	5.7590	0.1736	3.3724	19.4215
10	2.5937	0.3855	15.9374	0.0627	6.1446	0.1627	3.7255	22.8913
11	2.8531	0.3505	18.5312	0.0540	6.4951	0.1540	4.0641	26.3963
12	3.1384	0.3186	21.3843	0.0468	6.8137	0.1468	4.3884	29.9012
13	3.4523	0.2897	24.5227	0.0408	7.1034	0.1408	4.6988	33.3772
14	3.7975	0.2633	27.9750	0.0357	7.3667	0.1357	4.9955	36.8005
15	4.1772	0.2394	31.7725	0.0315	7.6061	0.1315	5.2789	40.1520
16	4.5950	0.2176	35.9497	0.0278	7.8237	0.1278	5.5493	43.4164
17	5.0545	0.1978	40.5447	0.0247	8.0216	0.1247	5.8071	46.5819
18	5.5599	0.1799	45.5992	0.0219	8.2014	0.1219	6.0526	49.6395
19	6.1159	0.1635	51.1591	0.0195	8.3649	0.1195	6.2861	52.5827
20	6.7275	0.1486	57.2750	0.0175	8.5136	0.1175	6.5081	55.4069
21	7.4002	0.1351	64.0025	0.0156	8.6487	0.1156	6.7189	58.1095
22	8.1403	0.1228	71.4027	0.0140	8.7715	0.1140	6.9189	60.6893
23	8.9543	0.1117	79.5430	0.0126	8.8832	0.1126	7.1085	63.1462
24	9.8497	0.1015	88.4973	0.0113	8.9847	0.1113	7.2881	65.4813
25	10.8347	0.0923	98.3471	0.0102	9.0770	0.1102	7.4580	67.6964
26	11.9182	0.0839	109.1818	0.0092	9.1609	0.1092	7.6186	69.7940
27	13.1100	0.0763	121.0999	0.0083	9.2372	0.1083	7.7704	71.7773
28	14.4210	0.0693	134.2099	0.0075	9.3066	0.1075	7.9137	73.6495
29	15.8631	0.0630	148.6309	0.0067	9.3696	0.1067	8.0489	75.4146
30	17.4494	0.0573	164.4940	0.0061	9.4269	0.1061	8.1762	77.0766
31	19.1943	0.0521	181.9434	0.0055	9.4790	0.1055	8.2962	78.6395
32	21.1138	0.0474	201.1378	0.0050	9.5264	0.1050	8.4091	80.1078
33	23.2252	0.0431	222.2515	0.0045	9.5694	0.1045	8.5152	81.4856
34	25.5477	0.0391	245.4767	0.0041	9.6086	0.1041	8.6149	82.7773
35	28.1024	0.0356	271.0244	0.0037	9.6442	0.1037	8.7086	83.9872
36	30.9127	0.0323	299.1268	0.0033	9.6765	0.1033	8.7965	85.1194
37	34.0039	0.0294	330.0395	0.0030	9.7059	0.1030	8.8789	86.1781
38	37.4043	0.0267	364.0434	0.0027	9.7327	0.1027	8.9562	87.1673
39	41.1448	0.0243	401.4478	0.0025	9.7570	0.1025	9.0285	88.0908
40	45.2593	0.0221	442.5926	0.0023	9.7791	0.1023	9.0962	88.9525
41	49.7852	0.0201	487.8518	0.0020	9.7991	0.1020	9.1596	89.7560
42	54.7637	0.0183	537.6370	0.0019	9.8174	0.1019	9.2188	90.5047
43	60.2401	0.0166	592.4007	0.0017	9.8340	0.1017	9.2741	91.2019
44	66.2641	0.0151	652.6408	0.0015	9.8491	0.1015	9.3258	91.8508
45	72.8905	0.0137	718.9048	0.0014	9.8628	0.1014	9.3740	92.4544
46	80.1795	0.0125	791.7953	0.0013	9.8753	0.1013	9.4190	93.0157
47	88.1975	0.0113	871.9749	0.0011	9.8866	0.1011	9.4610	93.5372
48	97.0172	0.0103	960.1723	0.0010	9.8969	0.1010	9.5001	94.0217
49	106.7190	0.0094	1057.1896	0.0009	9.9063	0.1009	9.5365	94.4715
50	117.3909	0.0085	1163.9085	0.0009	9.9148	0.1009	9.5704	94.8889

Interest Rate 15.00%

N	(F/P,i,N)	(P/F,i,N)	(F/A,i,N)	(A/F,i,n)	(P/A,i,N)	(A/P,i,N)	(A/G,i,N)	(P/G,i,N)
1	1.1500	0.8696	1.0000	1.0000	0.8696	1.1500	0.0000	0.0000
2	1.3225	0.7561	2.1500	0.4651	1.6257	0.6151	0.4651	0.7561
3	1.5209	0.6575	3.4725	0.2880	2.2832	0.4380	0.9071	2.0712
4	1.7490	0.5718	4.9934	0.2003	2.8550	0.3503	1.3263	3.7864
5	2.0114	0.4972	6.7424	0.1483	3.3522	0.2983	1.7228	5.7751
6	2.3131	0.4323	8.7537	0.1142	3.7845	0.2642	2.0972	7.9368
7	2.6600	0.3759	11.0668	0.0904	4.1604	0.2404	2.4498	10.1924
8	3.0590	0.3269	13.7268	0.0729	4.4873	0.2229	2.7813	12.4807
9	3.5179	0.2843	16.7858	0.0596	4.7716	0.2096	3.0922	14.7548
10	4.0456	0.2472	20.3037	0.0493	5.0188	0.1993	3.3832	16.9795
11	4.6524	0.2149	24.3493	0.0411	5.2337	0.1911	3.6549	19.1289
12	5.3503	0.1869	29.0017	0.0345	5.4206	0.1845	3.9082	21.1849
13	6.1528	0.1625	34.3519	0.0291	5.5831	0.1791	4.1438	23.1352
14	7.0757	0.1413	40.5047	0.0247	5.7245	0.1747	4.3624	24.9725
15	8.1371	0.1229	47.5804	0.0210	5.8474	0.1710	4.5650	26.6930
16	9.3576	0.1069	55.7175	0.0179	5.9542	0.1679	4.7522	28.2960
17	10.7613	0.0929	65.0751	0.0154	6.0472	0.1654	4.9251	29.7828
18	12.3755	0.0808	75.8364	0.0132	6.1280	0.1632	5.0843	31.1565
19	14.2318	0.0703	88.2118	0.0113	6.1982	0.1613	5.2307	32.4213
20	16.3665	0.0611	102.4436	0.0098	6.2593	0.1598	5.3651	33.5822
21	18.8215	0.0531	118.8101	0.0084	6.3125	0.1584	5.4883	34.6448
22	21.6447	0.0462	137.6316	0.0073	6.3587	0.1573	5.6010	35.6150
23	24.8915	0.0402	159.2764	0.0063	6.3988	0.1563	5.7040	36.4988
24	28.6252	0.0349	184.1678	0.0054	6.4338	0.1554	5.7979	37.3023
25	32.9190	0.0304	212.7930	0.0047	6.4641	0.1547	5.8834	38.0314
26	37.8568	0.0264	245.7120	0.0041	6.4906	0.1541	5.9612	38.6918
27	43.5353	0.0230	283.5688	0.0035	6.5135	0.1535	6.0319	39.2890
28	50.0656	0.0200	327.1041	0.0031	6.5335	0.1531	6.0960	39.8283
29	57.5755	0.0174	377.1697	0.0027	6.5509	0.1527	6.1541	40.3146
30	66.2118	0.0151	434.7451	0.0023	6.5660	0.1523	6.2066	40.7526
31	76.1435	0.0131	500.9569	0.0020	6.5791	0.1520	6.2541	41.1466
32	87.5651	0.0114	577.1005	0.0017	6.5905	0.1517	6.2970	41.5006
33	100.6998	0.0099	664.6655	0.0015	6.6005	0.1515	6.3357	41.8184
34	115.8048	0.0086	765.3654	0.0013	6.6091	0.1513	6.3705	42.1033
35	133.1755	0.0075	881.1702	0.0011	6.6166	0.1511	6.4019	42.3586
36	153.1519	0.0065	1014.3457	0.0010	6.6231	0.1510	6.4301	42.5872
37	176.1246	0.0057	1167.4975	0.0009	6.6288	0.1509	6.4554	42.7916
38	202.5433	0.0049	1343.6222	0.0007	6.6338	0.1507	6.4781	42.9743
39	232.9248	0.0043	1546.1655	0.0006	6.6380	0.1506	6.4985	43.1374
40	267.8635	0.0037	1779.0903	0.0006	6.6418	0.1506	6.5168	43.2830
41	308.0431	0.0032	2046.9539	0.0005	6.6450	0.1505	6.5331	43.4128
42	354.2495	0.0028	2354.9969	0.0004	6.6478	0.1504	6.5478	43.5286
43	407.3870	0.0025	2709.2465	0.0004	6.6503	0.1504	6.5609	43.6317
44	468.4950	0.0021	3116.6334	0.0003	6.6524	0.1503	6.5725	43.7235
45	538.7693	0.0019	3585.1285	0.0003	6.6543	0.1503	6.5830	43.8051
46	619.5847	0.0016	4123.8977	0.0002	6.6559	0.1502	6.5923	43.8778
47	712.5224	0.0014	4743.4824	0.0002	6.6573	0.1502	6.6006	43.9423
48	819.4007	0.0012	5456.0047	0.0002	6.6585	0.1502	6.6080	43.9997
49	942.3108	0.0011	6275.4055	0.0002	6.6596	0.1502	6.6146	44.0506
50	1083.6574	0.0009	7217.7163	0.0001	6.6605	0.1501	6.6205	44.0958

Interest Rate 20.00%

N	(F/P,i,N)	(P/F,i,N)	(F/A,i,N)	(A/F,i,n)	(P/A,i,N)	(A/P,i,N)	(A/G,i,N)	(P/G,i,N)
1	1.2000	0.8333	1.0000	1.0000	0.8333	1.2000	0.0000	0.0000
2	1.4400	0.6944	2.2000	0.4545	1.5278	0.6545	0.4545	0.6944
3	1.7280	0.5787	3.6400	0.2747	2.1065	0.4747	0.8791	1.8519
4	2.0736	0.4823	5.3680	0.1863	2.5887	0.3863	1.2742	3.2986
5	2.4883	0.4019	7.4416	0.1344	2.9906	0.3344	1.6405	4.9061
6	2.9860	0.3349	9.9299	0.1007	3.3255	0.3007	1.9788	6.5806
7	3.5832	0.2791	12.9159	0.0774	3.6046	0.2774	2.2902	8.2551
8	4.2998	0.2326	16.4991	0.0606	3.8372	0.2606	2.5756	9.8831
9	5.1598	0.1938	20.7989	0.0481	4.0310	0.2481	2.8364	11.4335
10	6.1917	0.1615	25.9587	0.0385	4.1925	0.2385	3.0739	12.8871
11	7.4301	0.1346	32.1504	0.0311	4.3271	0.2311	3.2893	14.2330
12	8.9161	0.1122	39.5805	0.0253	4.4392	0.2253	3.4841	15.4667
13	10.6993	0.0935	48.4966	0.0206	4.5327	0.2206	3.6597	16.5883
14	12.8392	0.0779	59.1959	0.0169	4.6106	0.2169	3.8175	17.6008
15	15.4070	0.0649	72.0351	0.0139	4.6755	0.2139	3.9588	18.5095
16	18.4884	0.0541	87.4421	0.0114	4.7296	0.2114	4.0851	19.3208
17	22.1861	0.0451	105.9306	0.0094	4.7746	0.2094	4.1976	20.0419
18	26.6233	0.0376	128.1167	0.0078	4.8122	0.2078	4.2975	20.6805
19	31.9480	0.0313	154.7400	0.0065	4.8435	0.2065	4.3861	21.2439
20	38.3376	0.0261	186.6880	0.0054	4.8696	0.2054	4.4643	21.7395
21	46.0051	0.0217	225.0256	0.0044	4.8913	0.2044	4.5334	22.1742
22	55.2061	0.0181	271.0307	0.0037	4.9094	0.2037	4.5941	22.5546
23	66.2474	0.0151	326.2369	0.0031	4.9245	0.2031	4.6475	22.8867
24	79.4968	0.0126	392.4842	0.0025	4.9371	0.2025	4.6943	23.1760
25	95.3962	0.0105	471.9811	0.0021	4.9476	0.2021	4.7352	23.4276
26	114.4755	0.0087	567.3773	0.0018	4.9563	0.2018	4.7709	23.6460
27	137.3706	0.0073	681.8528	0.0015	4.9636	0.2015	4.8020	23.8353
28	164.8447	0.0061	819.2233	0.0012	4.9697	0.2012	4.8291	23.9991
29	197.8136	0.0051	984.0680	0.0010	4.9747	0.2010	4.8527	24.1406
30	237.3763	0.0042	1181.8816	0.0008	4.9789	0.2008	4.8731	24.2628
31	284.8516	0.0035	1419.2579	0.0007	4.9824	0.2007	4.8908	24.3681
32	341.8219	0.0029	1704.1095	0.0006	4.9854	0.2006	4.9061	24.4588
33	410.1863	0.0024	2045.9314	0.0005	4.9878	0.2005	4.9194	24.5368
34	492.2235	0.0020	2456.1176	0.0004	4.9898	0.2004	4.9308	24.6038
35	590.6682	0.0017	2948.3411	0.0003	4.9915	0.2003	4.9406	24.6614
36	708.8019	0.0014	3539.0094	0.0003	4.9929	0.2003	4.9491	24.7108
37	850.5622	0.0012	4247.8112	0.0002	4.9941	0.2002	4.9564	24.7531
38	1020.6747	0.0010	5098.3735	0.0002	4.9951	0.2002	4.9627	24.7894
39	1224.8096	0.0008	6119.0482	0.0002	4.9959	0.2002	4.9681	24.8204
40	1469.7716	0.0007	7343.8578	0.0001	4.9966	0.2001	4.9728	24.8469
41	1763.7259	0.0006	8813.6294	0.0001	4.9972	0.2001	4.9767	24.8696
42	2116.4711	0.0005	10577.3553	0.0001	4.9976	0.2001	4.9801	24.8890
43	2539.7653	0.0004	12693.8263	0.0001	4.9980	0.2001	4.9831	24.9055
44	3047.7183	0.0003	15233.5916	0.0001	4.9984	0.2001	4.9856	24.9196
45	3657.2620	0.0003	18281.3099	0.0001	4.9986	0.2001	4.9877	24.9316
46	4388.7144	0.0002	21938.5719	0.0000	4.9989	0.2000	4.9895	24.9419
47	5266.4573	0.0002	26327.2863	0.0000	4.9991	0.2000	4.9911	24.9506
48	6319.7487	0.0002	31593.7436	0.0000	4.9992	0.2000	4.9924	24.9581
49	7583.6985	0.0001	37913.4923	0.0000	4.9993	0.2000	4.9935	24.9644
50	9100.4382	0.0001	45497.1908	0.0000	4.9995	0.2000	4.9945	24.9698

Index

Made in the USA
Coppell, TX
08 March 2023

13992844R00208